Dimensions of Ring Theory

Mathematics and Its Applications

Dimensions of Ring Theory

by

Constantin Năstăsescu

University of Bucharest, Romania

and

Freddy van Oystaeyen

Department of Mathematics and Computer Science,
University of Antwerp, Belgium

D. Reidel Publishing Company

A MEMBER OF THE KLUWER ACADEMIC PUBLISHERS GROUP

Dordrecht / Boston / Lancaster / Tokyo

Library of Congress Cataloging in Publication Data

CIP

Năstăsescu, C. (Constantin), 1943–
Dimensions of ring theory.

(Mathematics and its applications)
Bibliography: p.
Includes index.
1. Rings (Algebra) 2. Dimension theory (Algebra) I. Oystaeyen, F. van, 1947– . II. Title. III. Series: Mathematics and its applications (D. Reidel Publishing Company)
QA247.N334 1987 512'.4 87–4537
ISBN-13: 978-94-010-8207-5 e-ISBN-13: 978-94-009-3835-9
DOI: 10.1007/978-94-009-3835-9

Published by D. Reidel Publishing Company,
P.O. Box 17, 3300 AA Dordrecht, Holland.

Sold and distributed in the U.S.A. and Canada
by Kluwer Academic Publishers,
101 Philip Drive, Assinippi Park, Norwell, MA 02061, U.S.A.

In all other countries, sold and distributed
by Kluwer Academic Publishers Group,
P.O. Box 322, 3300 AH Dordrecht, Holland.

Contents

Series Editor's Preface

Approach your problems from the right end and begin with the answers. Then one day, perhaps you will find the final question.

'The Hermit Clad in Crane Feathers' in R. van Gulik's *The Chinese Maze Murders*.

It isn't that they can't see the solution. It is that they can't see the problem.

G.K. Chesterton. *The Scandal of Father Brown* 'The point of a Pin'.

Growing specialization and diversification have brought a host of monographs and textbooks on increasingly specialized topics. However, the "tree" of knowledge of mathematics and related fields does not grow only by putting forth new branches. It also happens, quite often in fact, that branches which were thought to be completely disparate are suddenly seen to be related.

Further, the kind and level of sophistication of mathematics applied in various sciences has changed drastically in recent years: measure theory is used (non-trivially) in regional and theoretical economics; algebraic geometry interacts with physics; the Minkowsky lemma, coding theory and the structure of water meet one another in packing and covering theory; quantum fields, crystal defects and mathematical programming profit from homotopy theory; Lie algebras are relevant to filtering; and prediction and electrical engineering can use Stein spaces. And in addition to this there are such new emerging subdisciplines as "experimental mathematics", "CFD", "completely integrable systems", "chaos, synergetics and large-scale order", which are almost impossible to fit into the existing classification schemes. They draw upon widely different sections of mathematics. This programme, Mathematics and Its Applications, is devoted to new emerging (sub)disciplines and to such (new) interrelations as exempla gratia:

- a central concept which plays an important role in several different mathematical and/or scientific specialized areas;
- new applications of the results and ideas from one area of scientific endeavour into another;
- influences which the results, problems and concepts of one field of enquiry have and have had on the development of another.

The Mathematics and Its Applications programme tries to make available a careful selection of books which fit the philosophy outlined above. With such books, which are stimulating rather than definitive, intriguing rather than encyclopaedic, we hope to contribute something towards better communication among the practitioners in diversified fields.

The idea of 'dimension', or more generally 'cardinal invariants' is fundamental in many parts of mathematics. Also in algebra, in casu ring and module theory. Here, however, the non-super-expert finds a bewildering variety of notions of dimension; Kroll dimension , (co)homological dimension, Gabriel dimension, Goldie rank, Gelfand-Kirillov dimension, ... It is far from easy to sort out what the interrelations and differences between all these notions are, when they are defined, when they are equal, ... Thus it seemed to me that there was reason for a book on the various ideas of dimension in algebra and their interrelations especially because these notions are so useful both in algebra itself and in its manyfold applications. The present authors were willing to take up this considerable challenge, and have produced precisely the kind of book a non-expert needs to find his way through this, at first sight, rather bewildering maze.

The unreasonable effectiveness of mathematics in science ...

Eugene Wigner

Well, if you know of a better 'ole, go to it.

Bruce Bairnsfather

What is now proved was once only imagined.

William Blake

As long as algebra and geometry proceeded along separate paths, their advance was slow and their applications limited.

But when these sciences joined company they drew from each other fresh vitality and thenceforward marched on at a rapid pace towards perfection.

Joseph Louis Lagrange.

Bussum, March 1987

Michiel Hazewinkel

ACKNOWLEDGMENT

Everyone who ever had to say thank you to a large group of friends and colleagues knows how easy this really is ... except if you commit the common error of trying to be original, as most people do. The best by far is to be *natural.* So we recognize that this book is a product of the delicate equilibrium in the mathematical ecosystem created for us by our colleagues. Michiel Hazewinkel planted the seed of this book in the peat of our brain and when it germinated, almost unexpectedly, he provided continuous support as a fertilizer. Serban Raianu rooted a few sections. The development of certain branches benefited from some organic material, e.g. papers etc..., provided by Jan Krempa, Tom Lenagan, André Leroy, Jan Okninski, Edmund Puczylowski a.o. .

Money is the filty lucre (mud in Dutch!) of the earth, so we needed some of that too and we were happy to receive some from the National Foundation for Scientific Research (NFWO).

Our publisher, D. Reidel, helped with the actual production of this fruit of our work. Finally we are glad to acknowledge that, even with all the help we have received, this book would be another kind of organic soup if it were not for Eddy Ruyssers who typed, organized the lay-out, printed and reprinted, generated a direct index, etc..., like the real TEX–wizard he is. Thanks to everybody and even more to those we forgot to mention.

F. Van Oystaeyen
C. Năstăsescu

INTRODUCTION

Could it be that Ring Theory is the part of mathematics where chaos is most present? Researchers in this field will recognize an apparent state of disorder caused on one hand by the existence of a large variety of different kinds of rings evolving from various areas of applications, while on the other hand few methods of general applicability are available. There seems to be a pseudo-philosophical basis for the idea that dimension theory may be a general applicable method in all fields of mathematics. Indeed, the concept of *"dimension"* is fundamental in the logical interpretation of our observations of reality and it must be inherent in the structure of human thought. But how can one then avoid the basic contradiction in trying to solve specific problems by a general method? We have tried to do so by proposing dimension theory, not as a general method, but as a unifying concept allowing for the differentiation of specific dimension-notions prompted by the consideration of different problems. Geometric intuition about dimension of spaces may without much trouble be extended to abstract algebraic objects like vector spaces and algebras over fields, but, when applied to rings and modules, this intuition will fail on many occasions. Therefore one quickly learns to comply with the fact that different problems in Ring Theory require different notions of "dimension". Recent developments have established that this diversification does not lead to: just an enlargement of the available body of abstract nonsense. On the contrary, these dimension theories provide more depth to known techniques as well as new (constructive) methods for obtaining structural results and examples.

Perhaps Euclidean intuition made us present a three dimensional picture of rings here. We concentrated attention upon Krull dimension, Gabriel dimension, global (homological) dimension, allowing for some variants like: classical Krull dimension, weak (flat) dimension, Goldie rank. Of course, there is a fourth dimension. Here too it is a sort of twilight-zone because the Gelfand-Kirillov dimension is not defined for general rings but it restricts attention to algebras over fields. Hoping to decrease the entropy of the theory we have chosen to use a unified approach where possible and consequently the theory of Krull dimension, Goldie dimension and Gabriel dimension has been founded on a lattice-theoretic approach. This necessitated the introduction of some rather formal notions from general lattice-theory but on the other hand it allowed to key-down the importance of category-theory to an absolute minimum.

Chapter 1 and Chapter 3 give a self-contained account of the general dimension theory in modular lattices and ordered sets.

Chapter 2 may be viewed as a crash-course on rings and modules; it can be used as a course on graduate level. The presence of this chapter makes the book completely self-contained except for the sections listed as addenda (i.e. 4.2.8., 4.2.9., 10.3.2., 10.3.3.) where we refer on some occasions to other publications for general knowledge concerning some types of rings e.g. P.I. rings, enveloping algebras, semigroup rings,...

For more detail on each separate chapter we refer to the bibliographical comments at the end of each chapter. The relative dimensions in Chapter 7 have been introduced mainly because relative Krull dimension may be used as an alternative approach to the non-commutative principal ideal theorem, which we treat by other methods in Section 6.6.. At first reading Chapter 7 may be omitted without destroying the continuity of the book. As a final remark concerning the self-containdness of this book let us point out that we have avoided the use of derived functors and properties of $Ext^n(-,-)$ as much as possible in Chapter 9. In fact for most results one only has to define what it means to have $Ext(-,-) = 0$ but for the more specific results at the end of the chapter some knowledge of homological algebra is beneficial. We composed the book with the following ideas in mind. After having introduced the general theory concerning dimension, eventually first on lattices and then specializing to modules and rings, we immediately turn to the structural properties of rings having such dimension (specializing to rings with finite dimension where this is essential). Wherever the possibility arises we compare the different dimensions e.g. Krull dimension and Gabriel dimension, Krull dimension and homological dimension, classical Krull dimension and Gelfand-Kirillov dimension. These dimensions only relate well when one considers suitably restricted classes of rings (e.g. Section 4.2., Chapter 6, Section 9.3. and 9.4., Section 10.3.) and it is exactly here that we achieve contact with contemporary developments in the field.

We have included over two hundred exercises. A large part of these are suitable for students and non-specialists and they have an ability-training function, other exercises stem from recent publications which have not (or not completely) been embodied in this book so they must be considered as "difficult" exercises.

Chapter 1

FINITENESS CONDITIONS FOR LATTICES.

§1.1 Lattices

Let $\leq$ denote a partial ordering on a nonempty set L and let $<$ be defined by: a $<$ b with a,b $\in$ L if and only if a $\leq$ b and a $\neq$ b. If a,b $\in$ L and b $\leq$ a then a is said to *contain* b. If M is a subset of *L*, then an a $\in$ L such that x $\leq$ a, resp. a $\leq$ x, for all x $\in$ M is said to be an *upper*, resp. *lower*, *bound* for the set M. An element a $\in$ L is said to be the *supremum*, resp. *infemum*, of M if a is the least upperbound, resp. the largest lower bound, for M (i.e. a is an upper (resp. lower) bound for M and if a' is another upper (resp. lower) bound for M then we have a $\leq$ a', resp. a' $\leq$ a). A supremum, resp. infemum of M is unique and we will denote it by sup(M) or $\bigvee_{x \in M} x$, resp. inf(M) or $\bigwedge_{x \in M} x$. When using the notation $\bigvee_{x \in M} x$, resp. $\bigwedge_{x \in M} x$, we will also refer to these elements as the *join*, resp. *meet*, of the elements of M. In case M = $\{x_1, ..., x_n\}$, we also write sup(M) = $\bigvee_{i=1}^n x_i = x_1 \vee ... \vee x_n$ and inf(M) = $\bigwedge_{i=1}^n x_i = x_1 \wedge x_2 \wedge ... x_n$.

A *lattice* is a partially ordered set L such that every pair a,b $\in$ L allows a supremum and an infemum for {a,b} $\subset$ L; we write a $\vee$ b = sup{a,b} and a $\wedge$ b = inf{a,b}. A lattice L is said to be *complete* if each subset of L has a supremum and infemum. To a partially ordered set (L,$\leq$) we may associate the *opposite* which is the set L but ordered by the opposite relation $\geq$ i.e. we have a $\geq$ b if and only if b $\leq$ a. If L is a lattice then L° denotes the opposite or the *dual* lattice. For a subset M of L, sup(M) resp inf(M) becomes inf(M), resp. sup(M), in L°. Clearly a partially ordered set L is a lattice, resp. a complete lattice, if and only if L° is a lattice, resp. a complete lattice.

The lattice L is *modular* if for each a,b,c ∈ L such that b ≤ a, we have $a \wedge (b \vee c) = b \vee (a \wedge c)$; again it is clear that L is modular if and only if L° is modular. A nonempty subset L' of L is a *sublattice* if for all a,b ∈ L' we have a ∨ b ∈ L' and a ∧ b ∈ L'.

If M and L are partially ordered sets, a function f:M→L is said to be *isotone*, or *increasing*, resp. *strictly increasing*, if for a ≤ b in L we have that f(a) ≤ f(b), resp. a < b implies f(a) < f(b). A *lattice morphism* $\psi : M \to L$ is a function ψ satisfying : $\psi(a \vee b) = \psi(a) \vee \psi(b)$, $\psi(a \wedge b) = \psi(a) \wedge \psi(b)$ for all a,b ∈ M. Obviously, every lattice morphism is increasing. A bijective lattice morphism is called a *lattice isomorphism* and the inverse function is then also a lattice isomorphism; if there exists a lattice isomorphism ψ: L→ L' then we say that L and L' are *isomorphic* and we write L ≅ L'. Isomorphism defines an equivalence relation on lattices.

We now give some examples which are widely used in algebra plus some very particular ones highlighting the failure of some desirable property introduced above.

1.1.1. Let (L,≤) be a lattice and a ≤ b in L. Write $[a,b] = \{x \in L, a \leq x \leq b\}$. The set [a,b] with the induced partial ordering is a sublattice of L, called the *interval* determined by a and b.

1.1.2. Every totally ordered set is a lattice. Recall that a partially ordered set (L,≤) is totally ordered if for each pair a,b ∈ L either a ≤ b or b ≤ a.

1.1.3. Consider partially ordered sets (L,≤) and (L',≤). (Note that we sometimes use the symbol ≤ for different orderings; this will usually not lead to ambiguity because the set where the relation is defined will be clear from the context). On the Cartesian product L × L' we define a partial order by: (x,y) ≤ (x',y') if and only if x ≤ x' and y ≤ y', where x,x' ∈ L and y,y' ∈ L'. If L and L' are lattices then so is L × L' and if L and L' are complete then L × L' is complete.

1.1.4. For any set X, the set $\mathcal{G}(X)$ of all subsets of X is a complete lattice for the inclusion relation.

1.1.5. For any group G let L(G) be the set of all subgroups of G. With respect to the inclusion relation, L(G) is a complete lattice. To a family { H_i, i ∈ I} of subgroups of G we associate $\bigcap_{i\in I} H_i$ as the infemum and the subgroup generated by $\bigcup_{i\in I} H_i$ as the supremum.

1.1.6. For any group G let $L_o(G)$ be the set of all normal subgroups of G (recall that a subgroup H of G is normal whenever $x\, h\, x^{-1} \in H$ for each x ∈ G, h ∈ H). It is

easily verified that $L_o(G)$ is a sublattice of $L(G)$. In particular if $H,K \in L_o(G)$ then the subgroup generated by $H \cup K$ is the set $HK =$
$\{hk,\ h \in H,\ k \in K\}$. Moreover we have $HK = KH$. The lattice $L_o(G)$ is modular. Indeed, consider $H,K,L \in L_o(G)$ with $H \subseteq K$ and then we have to verify whether $K \wedge (H \vee L) = H \vee (K \wedge L)$, i.e. whether $K \cap HL = H.(K \cap L)$. Clearly, $H.(K \cap L) \subset K \cap HL$ so we only have to establish the converse inclusion. Take $x \in K \cap HL$, $x = yz$ with $y \in H$, $z \in L$. Hence $z = y^{-1} x \in K$ and thus $z \in K \cap L$, proving the claim.

1.1.7. The lattice of all subgroups of A_4, the alternating group of four elements is not modular (because of the bad properties of the appearing non-normal subgroup).

1.1.8. If R is a ring then the set SI(R) of all subrings of R, partially ordered by inclusion, is a complete lattice.

§1.2 Noetherian and Artinian Lattices.

Let $(L,\leq)$ be a partially ordered set and consider a nonempty subset M of L; an $a \in M$ is said to be a *maximal*, resp. *minimal*, element of M if for any $x \in M$ such that $a \leq x$, resp. $x \leq a$, we have $a = x$.

1.2.1. Proposition. For any partially ordered set $(L,\leq)$ the following statements are equivalent:

1. Any ascending chain $a_1 \leq a_2 \leq \dots \leq a_n \leq \dots$ with $a_i \in L$, resp. descending chain $a_1 \geq a_2 \geq \dots \geq a_n \geq \dots$, is stationary.

2. Any nonempty subset of L has a maximal, resp. minimal, element.

Proof. $1 \Rightarrow 2$ Consider a nonempty subset A of L and suppose that A has no maximal elements. Take $a_1 \in A$; since a_1 is not maximal there exists an $a_2 \in A$ such that $a_1 \leq a_2$.
Since a_2 is not maximal we may continue this argument untill we have obtained a sequence $a_1 < a_2 < \dots < a_n < \dots$. If the latter sequence is stationary then we have to contradict the initial assumption.

2 ⇒ 1 Obvious. The dual statements in the proposition may be proved in exactly the same way by considering the ordered set L°. □

A partially ordered set (L,≤) satisfying one of the equivalent conditions mentioned in Proposition 1.2.1. is said to be Noetherian, resp. Artinian.

1.2.2. Proposition. Let (L,≤) be a partially ordered set such that each nonempty subset has a supremum, resp. infemum, then the following statements are equivalent:

1. L is a Noetherian, resp. Artinian, lattice.

2. If A is a nonempty subset of L, there exist $a_1, ..., a_n \in A$ such that sup(A) = $\sup\{a_1, a_2, ..., a_n\}$, resp. inf(A) = $\inf\{a_1, ..., a_n\}$.

Proof. Suppose that L is a Noetherian lattice (for the dual statements the same proof but in L° will yield the result). Consider the subset of L, { sup(B), B a finite subset of A }. By Proposition 2.1. there exists a maximal element, sup($B_\circ$) say, in the foregoing set. Obviously sup($B_\circ$) ≤ sup(A); put b_0 = sup($B_\circ$). If x ∈ A then $B_\circ \subset B_\circ \cup \{x\}$ and hence sup($B_\circ$) ≤ sup($B_\circ \cup \{x\}$). Consequently sup($B_\circ$) = sup($B_\circ \cup \{x\}$) and thus x ≤ $b_\circ$ or sup(A) ≤ $b_\circ$ and sup($B_\circ$) = sup(A) follows.

For the converse, let $a_1 \leq a_2 \leq ... \leq a_n \leq ...$ be an ascending chain of elements of L. The hypothesis implies that there exists an $n_0 \in \mathbb{N}$ such that $\sup\{a_i,\ i \in \mathbb{N}\}$ = $\sup\{a_1, ..., a_{n_0}\}$ and thus $a_i = a_{n_0}$ for each i ≥ n_0. □

1.2.3. Proposition. Let L be a modular lattice and a ∈ L. Then L is a Noetherian (a.c.c.), resp. Artinian (d.c.c.), lattice if and only if the lattices L' = $\{x \in L,\ x \leq a\}$ and L" = $\{x \in L, a \leq x\}$ are Noetherian, resp. Artinian.

Proof. We establish the Noetherian case, the Artinian case follows by dualizing. If L is a Noetherian lattice then the same is obviously true for L' and L". For the converse, consider an ascending chain $b_1 \leq b_2 \leq ... \leq b_n \leq ...$ with $b_i \in$ L. By the assumptions there exists an n ∈ N such that $b_n \wedge a = b_{n+1} \wedge a = u$ and $b_n \vee a = b_{n+1} \vee a = v$. We obtain $u \leq b_n \leq b_{n+1} \leq v$ and then: $b_{n+1} = b_{n+1} \wedge v = b_{n+1} \wedge (b_n \vee a) = b_n \vee (b_{n+1} \wedge a) = b_n \vee u = b_n$, i.e. $b_n = b_{n+1}$. Therefore, L is a Noetherian lattice. □

§1.3 Lattices of Finite Length

1.3.1. Proposition. Let $(L,\leq)$ be a modular lattice. For any $a,b \in L$, the intervals $[a \wedge b, a]$ and $[b, a \vee b]$ are isomorphic.

Proof. Define $f\colon [a \wedge b, a] \to [b, a \vee b]$, $f(x) = x \vee b$, $g\colon [b, a \vee b] \to [a \wedge b, a]$, $g(y) = y \wedge a$. It is not hard to check that f and g are lattice morphisms. For $x \in [a \wedge b, a]$ we have $(g \circ f)(x) = g(x \vee b) = (x \vee b) \wedge a = x \vee (b \wedge a) = x$. For $y \in [b, a \vee b]$ we have $(f \circ g)(y) = f(y \wedge a) = (y \wedge a) \vee b = y \wedge (a \vee b) = y$.
Thus f and g are inverse to each other and consequently f is a lattice isomorphism.

□

In a modular lattice $(L,\leq)$ two intervals I and J are said to be *similar* if there exist $a,b \in L$ such that I and J are written as $[a \wedge b, a]$ and $[b, a \vee b]$ or vice-versa. Two intervals I and J are said to be *projective* if there exist intervals $I_0 = I, I_1, \dots, I_n = J$ such that for each $1 \leq i \leq n$ the intervals I_{i-1} and I_i are similar.

1.3.2. Corollary. Projective intervals I and J of a modular lattice $(L,\leq)$ are isomorphic.

In a modular lattice $(L,\leq)$ two finite chains:

(1) $a = a_0 \leq a_1 \leq a_2 \leq \dots \leq a_m = b$

(2) $a = b_0 \leq b_1 \leq b_2 \leq \dots \leq b_n = b$

are said to be *equivalent* if $m = n$ and there exists a permutation π of $\{1,\dots,n\}$ such that the intervals $[a_{i-1}, a_i]$ and $[b_{\pi(i)-1}, b_{\pi(i)}]$ are projective for each
$1 \leq i \leq n$.
A *refinement* of a chain like (1) is a chain obtained from (1) by introducing a finite number of new elements.

1.3.3. Theorem (Schreier). Consider $a \leq b$ in the modular lattice $(L,\leq)$. Finite chains (1) and (2) have equivalent refinements.

Proof. We denote $(a_i \wedge b_j) \vee b_{j-1}$ by a_{ij} and $(b_j \wedge a_i) \vee a_{i-1}$ by b_{ji}, i = 1,...,m, j = 1,...,n. Put x = $a_i \wedge b_j$, y = $a_{i-1,j}$ then x ∨ y = $(a_i \wedge b_j) \vee a_{i-1,j} = (a_i \wedge b_j) \vee b_{j-1} \vee (a_{i-1} \wedge b_j) = ((a_i \wedge b_j) \vee (a_{i-1} \vee b_j)) \vee b_{j-1} = (a_i \wedge b_j) \vee b_{j-1} = a_{ij}$ (since $a_{i-1} \wedge b_j \leq a_i \wedge b_j$).
We have also, y ∧ x = $(a_i \wedge b_j) \wedge a_{i-1,j} = (a_i \wedge b_j) \wedge (a_{i-1} \vee b_{j-1}) \wedge b_j = (a_{i-1} \vee b_{j-1}) \wedge (a_i \wedge b_j)$.
Hence $[a_{i-1,j}, a_{ij}]$ = [y, x ∨ y] while [x ∧ y, x] = $[(a_{i-1} \vee b_{j-1}) \wedge (a_i \wedge b_j) , a_i \wedge b_j]$.
Putting x' = $b_j \wedge a_i$ and y' = $b_{j-1,i}$ we obtain (interchanging i and j in the foregoing) : $[b_{j-1,i} , b_{ji}]$ = [y', x' ∨ y'] and [x' ∧ y', x'] =
$[(a_{i-1} \vee b_{j-1}) \wedge (a_i \wedge b_j) , a_i \wedge b_j]$.
It follows that $[a_{i-1,j} , a_{ij}]$ and $[b_{j-1,i} , b_{ji}]$ are projective. The chains (1) and (2) now lead to refinements:

(3) a = $a_{01} \leq a_{11} \leq a_{21} \leq ... \leq a_{m1} \leq a_{12} \leq ... \leq a_{mn}$ = b
(4) a = $b_{01} \leq b_{11} \leq b_{21} \leq ... \leq b_{n1} \leq b_{12} \leq ... \leq b_{nm}$ = b

(Note: $a_i = b_{ni}$, $b_j = a_{mj}$ hence (4) refines (1), (3) refines (2)), and (3) and (4) are equivalent. □

An interval [x , y] in L is aid to be *simple* if it contains only x and y, i.e. if [x , y] = {x , y}. For a ≤ b in L we define a composition series of the interval [a , b] as a chain:

(5) a = $a_0 < a_1 < ... < a_m$ = b

such that each interval $[a_{i-1}, a_i]$ is simple, i = 1,...,m. The number m in (5) is called the length of [a , b], denoted by b[a , b] = m, and it is unambiguously defined because of:

1.3.4. Corollary. (Jordan-Hölder-Dedekind) Let a ≤ b in a modular lattice (L,≤), then only two composition series of [a , b] are equivalent.

Proof. Obvious from 1.3.3. □

If (L,≤) is a partially ordered set, a ∈ L is said to be the least, resp. largest element of L if for each x ∈ L we have a ≤ x, resp. x ≤ a. There is at most one least, resp. largest, element in L and it will be denoted by 0, resp. 1, if it exists. Clearly, any complete lattice has 0 and 1. Consider for example a partially ordered set (L,$\leq$) and an interval [a , b] in L, then a is the least (b is the largest) element in [a , b].

1.3.5. Proposition. If (L,$\leq$) is a Noetherian, resp. Artinian, lattice then L has 1, resp. 0

Proof. Suppose L is Noetherian. Then there exists a maximal element a of L. Consider x $\in$ L; since a $\leq$ a $\vee$ x we obtain a = a $\vee$ x and thus x $\leq$ a. Hence a is the largest element of L. The dual statement is proved in a similar way in L°. □

1.3.6. Corollary. Consider a modular lattice (L,$\leq$) with 0 and 1, then L has a composition series if and only if L is Noetherian and Artinian.

Proof. Assume first that $0 = a_0 < a_1 < ... < a_n = 1$ is a composition series for L. Since $[a_{i-1}, a_i]$ is simple for each $1 \leq i \leq m$ it is clear that $[a_{i-1}, a_i]$ is Noetherian and Artinian. From Proposition 1.2.3. we obtain that L = [0,1] is Noetherian and Artinian.

Conversely, if L is Artinian then there exists an $a_1 \in$ L minimal with the property of being nonzero. Again, there exists an a_2 minimal such that $a_1 \leq a_2$ and $a_1 \neq a_2$. By repetition of this argument we obtain a chain: $a_0 = 0 < a_1 < a_2 < ... < a_n < ....$ Since L is Noetherian, $a_m = a_{m+1}$ for some m $\in$ N. The construction of the chain then implies that $a_m = 1$ and that the intervals $[a_{i-1}, a_i]$ are simple for each $1 \leq i \leq$ m. This states exactly that L has a composition series.

□

A modular lattice with 0 and 1 which admits a composition series is called a *lattice of finite length.*

1.3.7. Corollary. Consider a modular lattice (L,$\leq$) and a $\leq$ b $\leq$ c in L. Then [a,c] has a composition series if and only if [a,b] and [b,c] have composition series and in this case we have: l[a,c] = l[a,b] + l[b,c].

Proof. The first assertion follows from Propositon 1.2.3. and Corollary 1.3.6.. The final claim derives from Corollary 1.3.4..

1.3.8. Corollary. Let (L,$\leq$) be a lattice of finite length; let a = $a_0 < a_1 < ... < a_m$ = b be a chain between a,b $\in$ L. The given chain may be refined to a composition series of [a,b].

Proof. Corollary 1.3.7. implies that [a,b] is a lattice of finite length and therefore it has a composition series. By Theorem 1.3.3. the chain may be refined in such a way that the result is a (refinement of a) composition series. □

§1.4 Irreducible Elements in a Lattice

An element a of a lattice (L,$\leq$) is said to be *meet-irreducible* if for any x,y $\in$ L such that a = x $\wedge$ y it follows that either a = x or a = y; a *join-irreducible element* is defined in the dual way.
A *finite meet* $\bigwedge_{i \in J} a_i$ is said to be *reduced* (or sometimes: irredundant or minimal) if for each i $\in J$ we have that $\bigwedge_{i \in J} a_i \neq \bigwedge\{a_j \ , \ j \neq i \text{ in } J\}$. Obviously, a finite meet of meet-irreducible elements may be transformed into a reduced meet of meet-irreducible elements.

1.4.1. Theorem. Pick a in the modular lattice (L,$\leq$) such that a = $x_1 \wedge ... \wedge x_m = y_1 \wedge ... \wedge y_n$, where all x_i and y_i are meet-irreducible. If $y_1 \wedge ... \wedge y_n$ is reduced then m $\geq$ n.

Proof. Write $x_i' = x_1 \wedge ... \wedge x_{i-1} \wedge x_{i+1} \wedge ... \wedge x_m$; then we have: a $\leq x_i'$ and a = $x_i \wedge x_i'$. Writing z_j for $x_i' \wedge y_j$, j = 1,...,n we obtain a $\leq z_j \leq x_i'$ and $z_j \leq y_j$, hence a $\leq \bigwedge_{j=1}^n z_j \leq \bigwedge_{j=1}^n y_j$ = a and thus a = $\bigwedge_{j=1}^n z_j$.
By Proposition 1.3.1. it follows that the interval [a,x_i'] = $[x_i \wedge x_i' \ , \ x_i']$ is isomorphic to the interval $[x_i, x_i \wedge x_i']$. Since x_i is meet-irreducible in L it is also meet-irreducible in the lattice $[x_i \ , \ x_i \vee x_i']$ and correspondingly, a is meet-irreducible in $[a \ , \ x_i']$.

The equalty a = $\bigwedge_{j=1}^{n} z_j$ now leads to the existence of an index j for which a = z_j holds. Hence, a = $x_i' \wedge y_j$ or a = $x_1 \wedge ... \wedge x_{i-1} \wedge y_j \wedge x_{i+1} \wedge ... \wedge x_m$. The latter relation entails the existence of a $j_1 \in \{1, ..., n\}$ and an $i_1 \in \{1, ..., m\}$ such that a = $y_{j_1} \wedge (\bigwedge_{i \neq i_1} x_i)$. In this way we obtain a map
$\varphi : \{1.,,,.m\} \rightarrow \{1.,,,.n\}$ such that a = $\bigwedge_{i=1}^{m} y_{\varphi(i)} = \bigwedge_{k \in Im\varphi} y_k$.
Since a = $\bigwedge_{j=1}^{m} y_j$ is reduced it follows that φ is surjective; consequently m $\geq$ n. □

1.4.2. Corollary (Kurosh, Ore). Consider a is the modular lattice (L,$\leq$) and suppose that a = $x_1 \wedge ... \wedge x_m = y_1 \wedge ... \wedge y_n$ are reduced intersections, where each x_i and y_i is meet-irreducible in L. Then m = n.

A straightforward dualization of the foregoing leads to:

1.4.3. Corollary. Consider a is the modular lattice (L,$\leq$). If a = $x_1 \vee ... \vee x_m = y_1 \vee ... \vee y_n$ are irreducible representations of a by join-irreducible elements $x_1, ..., x_m, y_1, ..., y_n$, then m = n.

1.4.4. Proposition. Each element of a Noetherian lattice (L,$\leq$) is a finite intersection of meet-irreducible elements of L.

Proof. Suppose that X = {a $\in$ L, a is not a finite intersection of meet-irreducible elements of L} is non-empty. Then X has a maximal element a_0.
Since $a_0 \in$ X it cannot be meet-irreducible, hence there are b,c $\in$ L such that $a_0 = b \wedge c$ with $a_0 < b$, $a_0 < c$. But then b,c $\neq$ X or b and c are finite intersections of meet-irreducible elements of L and then so is a_0, a contradiction. □

§1.5 Goldie dimension of a Modular Lattice

Let (L,$\leq$) be a lattice with 0 and 1 (throughout in this section) and consider a $\neq$ 0 in L. If 0 is meet-irreducible in [0,a] then a is said to be a *uniform* element (i.e. for x,y $\in$ L such that 0 < x $\leq$ a and 0 <y $\leq$ a we have x $\wedge$ y $\neq$ 0). The lattice L

is *uniform* if 1 is a uniform element. It is clear that a is meet-irreducible in L if and only if 1 is uniform in [a,1]. For a finite set $\{a_i , i \in I\}$ in L-{0} we say that the union $a = \bigvee_{i \in I} a_i$ is *direct* (or that $\{a_i , i \in I\}$ is join-independent) if and only if $a_i \wedge (\bigvee_{j \neq i} a_j) = 0$ for each $i \in I$. If I is not necessarily finite then the family $\{a_i\}_{i \in I}$ is said to be *direct* if for each finite subset $F \subset I$ the set $\{a_i , i \in F\}$ is join-independent. By dualization we obtain the notion of meet-independency and respectiveresults for this notion.

1.5.1. Proposition. Consider a modular lattice (L,≤) and a finite join-independent set $\{a_i , i \in I\}$ in L-{0}. If $I = F \cup K$ with $F \cap K = \phi$ then:
$(\bigvee_{f \in F} a_f) \wedge (\bigvee_{k \in K} a_k)$.

Proof. We may assume that $I = \{1, ..., n\}$, $F = \{1, 2, ..., p\}$, $K = \{p+1, ..., n\}$. For p=1 the assertion is clear so we proceed by induction, assuming that the assertion holds for p-1. Put $b = (\bigvee_{i=1}^{p} a_i) \wedge (\bigvee_{p+1}^{n} a_k)$, then using the induction hypothesis it follows that:

$$(\bigvee_{i=1}^{p-1} a_i) \wedge (a_p \vee b) \leq (\bigvee_{i=1}^{p-1} a_i) \wedge (\bigvee_{k=p}^{n} a_k) = 0, \text{ and}$$
$$a_p \wedge (\bigvee_{i=1}^{p-1} a_i \vee b) \leq a_p \wedge (\bigvee_{t \neq p} a_t) = 0$$

The modularity of L now entails:

$(a_p \vee b) \wedge (\bigvee_{i=1}^{p} a_i) = (a_p \vee b) \wedge (a_p \vee \bigvee_{c=1}^{p-1} a_i) = a_p \vee ((a_p \vee b) \wedge \bigvee_{i=1}^{p-1} a_i)$
$= a_p$

and $(\bigvee_{i=1}^{p-1} a_i \vee b) \wedge (\bigvee_{i=1}^{p} a_i) = (\bigvee_{i=1}^{p-1} a_i \vee b) \wedge (\bigvee_{i=1}^{p-1} a_i \vee a_p) = \bigvee_{i=1}^{p-1} a_i \vee (a_p \wedge (b \vee \bigvee_{i=1}^{p-1} a_i)) = \bigvee_{i=1}^{p-1} a_i$. On the other hand it is clear that

$b \leq (a_p \vee b) \wedge (\bigvee_{i=1}^{p} a_i)$, $b \leq (\bigvee_{i=1}^{p-1} a_i \vee b) \wedge (\bigvee_{i=1}^{p} a_i)$.

Then, $b \leq a_p \wedge (\bigvee_{i=1}^{p-1} a_i) = 0$, or $b = 0$. □

1.5.2. Proposition. Let (L,≤) be a modular lattice and A a join-independent subset of L-{0}. If $a \in L$ is a non-zero element such that for any finite subset $X \subset A$, $a \wedge \bigvee_{x \in X} x = 0$ then the set $A \vee \{a\}$ is join-independent too.

Proof. Consider a finite subset $B \subset A \cup \{a\}$; if $B \subset A$ then B is join-independent so suppose that $B = A' \cup \{a\}$ where A' is a finite subset of A. It is sufficient to show that a' $\wedge$ $(a \vee \bigvee \{x,\ x \in A' - \{a'\}\}) = 0$ for every a' $\in$ A'. We obtain the following equalities:

$\bigvee_{a' \in A'} a' \wedge (\bigvee_{x \in A'-\{a'\}} x \vee a) = \bigvee_{x \in A'-\{a'\}} x \vee (a \wedge \bigvee_{a' \in A'} a') = \bigvee_{x \in A'-\{a'\}} x,$

$a' \wedge (a \vee \bigvee_{x \in A'-\{a'\}} x) = a' \wedge (\bigvee_{a' \in A'} a' \wedge (a \vee \bigvee_{x \in A'-\{a'\}} x)) = a' \wedge \bigvee_{x \in A'-\{a'\}} x = 0$ (because A' is join-independent). □

From Proposition 1.5.2. and Zorn's lemma it follows that each join-independent subset of L-{0} is contained in a maximal join-independent subset of L-{0}. A nonzero a $\in$ L is *essential* in L if for any nonzero x $\in$ L we have a $\wedge$ x $\neq$ 0. If a $\leq$ b in L and a is essential in [0,b] then we say that b is an *essential extension* of a in L.

1.5.3. Proposition. Let (L,$\leq$) be a modular lattice and consider for each i $\in$ $\{1, ..., n\}$ elements a_i, b_i $\in$ L such that $a_i \leq b_i$ and each b_i being an essential extension of a_i. If $\{b_1, ..., b_n\}$ is join-independent then $\bigvee_{i=1}^{n} b_i$ is an essential extension of $\bigvee_{i=1}^{n} a_i$.

Proof By an easy induction argument we reduce the problem to n = 2. Since $a_1 \vee a_2 \leq a_1 \vee b_2 \leq b_1 \vee b_2$ it will suffice to establish that b $\vee$ c is an essential extension of a $\vee$ c whenever b is an essential extension of a in L and c $\in$ L is such that c $\wedge$ b = 0. To see this, consider 0 < x $\leq$ b. If x $\vee$ c $\neq$ o then x $\wedge$ (a $\vee$ c) $\neq$ 0.

If x $\wedge$ c = 0 then c < x $\vee$ c $\leq$ b $\vee$ c and since [c,b $\vee$ c] $\cong$ [0,b], by Proposition 1.3.1., it follows that (x $\vee$ c) $\wedge$ b $\neq$ 0. However, b being essential extension of a, we also have that (x $\vee$ c) $\wedge$ (b $\wedge$ a) $\neq$ 0 or (x $\vee$ c) $\wedge$ a $\neq$ 0. If we write u = (x $\vee$ c) $\wedge$ a $\neq$0 then u $\leq$ (x $\vee$ c) $\wedge$ (a $\vee$ c) = (x $\wedge$ (a $\vee$ c)) $\vee$ c. Now x $\wedge$ (a $\vee$ c) = 0 would entail u $\leq$ c and since it would also yield u $\leq$ a $\wedge$ c = 0 or u = 0, a contradiction. So we must have x $\wedge$ (a $\vee$ c) $\neq$ 0, establishing the claim. □

1.5.4. Proposition. Consider a $\leq$ b $\leq$ c in a modular lattice (L,$\leq$). If c is an essential extension of b and b is an essential extension of a then c is an essential extension of a.

Proof. If x $\neq$ 0 in [0,c] then x $\wedge$ b $\neq$ 0. Since b is an essential extension of a it follows that $(x \wedge b) \wedge a \neq 0$ and therefore $x \wedge a \neq 0$. □

1.5.5. Theorem. Consider two join-independent subsets $\{a_1, ..., a_m\}$ and $\{b_1, ..., b_n\}$ in L-$\{0\}$ for a modular lattice (L,$\leq$). If the $a_1, ..., a_m$ are uniform elements while $\bigvee_{i=1}^m a_i$ is essential in L then $m \geq n$.

Proof . Assume that $n > m$. We show inductively (up to reordering the a_j) that $\{a_1, ..., a_j, b_{j+1}, ..., b_n\}$ is join-independent for each $j \in \{0, ..., m\}$. If $j = 0$ there is nothing to prove. Put $j > 0$ and $c = a_1 \vee ... \vee a_j \vee b_{j+2} \vee ... \vee b_n$. Suppose $a_s \wedge c \neq 0$ for each s in $\{1, ..., m\}$. Uniformity of $a_1, ..., a_m$ implies that $a_s \wedge c$ is essential in a_s for each s in $\{1.,,, m\}$. By Proposition 1.5.3. we obtain that $a_1 \vee ... \vee a_m$ is an essential extension of $(a_1 \wedge c) \vee ... \vee (a_m \wedge c)$. By Proposition 1.5.4. we know that $(a_1 \wedge c) \vee ... \vee (a_m \wedge c)$ is essential in L and this implies that c is essential in L, contradicting $c \wedge b_{j+1} = 0$. So we have to have $a_s \wedge c = 0$ for some $s \in \{1, ..., m\}$. Obviously $s \notin \{1, ..., j\}$.
Putting $j+1 = s$, we may derive from Proposition 1.5.2. that $\{a_1 , ... a_{j+1} , b_{j+2} , ... , b_n\}$ is join-independent. In particular we arrive at the join- independency of $\{a_1 , ... , a_m , b_{m+1} , ... , b_n\}$ which is impossible because $a_1 \vee ... \vee a_m$ is essential in L. □

1.5.6. Corollary. If $\{a_1 , ... , a_m\}$ and $\{b_1 , ... , b_n\}$ are join-independent subsets of L-$\{0\}$ for a modular lattice (L,$\leq$) with 0 and 1, such that $a_1 , ... , a_m , b_1 , ... , b_n$ are uniform elements of L and $\bigvee_{i=1}^m a_i$ and $\bigvee_{j=1}^n b_j$ are essential in L, then $m = n$.

The *Goldie dimension* of a modular lattice (L,$\leq$) is equal to w if there exists uniform elements $a_1 , ... , a_n$ of L such that $\{a_1 , ... , a_n\}$ is join-independent and $\bigvee_{i=1}^n a_i$ is essential in L, we shall write u-dim(L) = n and we write u-dim(L) = ∞ if such a set $\{a_1 , ... , a_n\}$ does not exist.

1.5.7. Theorem. For a modular lattice (L,$\leq$) with 0 and 1 the following statements are equivalent:

(1) u-dim(L) $< \infty$.
(2) L does not contain infinite join-independent subsets.
(3) L satisfies the ascending chain condition on finite direct joins.

(4) For any chain $a_1 \leq a_2 \leq ... \leq a_m \leq ...$, in L there is an n ∈ N such that for all k ≥ n, a_k is an essential extension of a_n.

Proof. 1⇒3⇒2 is clear (using Theorem 1.5.5.) and 4⇒3 is also obvious, so it suffices to establish the implications 2⇒1 and 3⇒4.

2⇒1 Note first that for a nonzero b ∈ L there exists a nonzero uniform c ≤ b; if not, then we inductively construct a sequence c_1, c_2 ... in L-{0} such that $\{c_1, c_2, ...\}$ is join-independent but each $c_1 \vee ... \vee c_k$ being not essential in b. This construction is cristalclear when k = 1 because b is not uniform; so, assume we already constructed $c_1, ..., c_{k-1}$. Since b is not an essential extension of $c_1 \vee ... \vee c_{k-1}$, there exists a nonzero d ≠ 0, d ≤ b such that $d \wedge (c_1 \vee ... \vee c_{k-1}) = 0$. Since d is not uniform, there must exist a nonzero $c_k \leq d$ such that d is not an essential extension of c_k. We obtain in this way a join-independent $\{c_1, ..., c_k\}$ (by Proposition 1.5.2) such that b is not an essential extension of $c_1 \vee ... \vee c_k$. This would lead to an infinite join-independent set in L-{0} and a contradiction.

3⇒4 Suppose $a_1 < a_2 < ... < a_n < ...$ in L is such that for all i ≥ 1, a_i is not essential in some $a_{\varphi(i)}$ with $\varphi(i) > i$.

Define $i_1 = 1$, $i_m = \varphi(i_m - 1)$. By the foregoing there exist elements $a'_{i_m} \leq a_{i_{m+1}}$ with $a'_{i_m} \neq 0$ and $a_{i_m} \wedge a'_{i_m} = 0$. This leads to a join-independent set $\{a_{i_1}, a'_{i_2}, ..., a'_{i_m}, ...\}$, a contradiction. □

1.5.8. Corollary. Let (L,≤) be a modular lattice such that $1 = \bigvee_{i=1}^{n} a_i$ where $\{a_1, ... a_n\}$ is join-independent. The Goldie dimension of L is finite if and only if the Goldie dimensions of the lattices $[0,a_i]$ are finite for i = 1,...,n; moreover in this case:

$$\text{u-dim}(L) = \textstyle\sum_{i=1}^{n} \text{u-dim}([0, a_i]).$$

Proof. Direct from Theorem 1.5.5. and 1.5.7. □

1.5.9. Theorem. Let (L,≤) be a modular lattice with 0 and 1. If $0 = a_1 \wedge ... \wedge a_n$ is an irredundant representation of 0 by meet-irreducible elements a_i, i = 1,...,n, then u-dim(L) = n.

Proof. Write $\hat{a}_i = a_1 \wedge ... \wedge a_{i-1} \wedge a_{i+1} \wedge ... \wedge a_n$, i = 1,...,n.

Since $a_1 \wedge \hat{a}_1 = 0$, $\hat{a}_1 = a_2 \wedge ... \wedge a_n$ is an irredundant representation of $\hat{a}_1$ in $[\hat{a}_1, 1]$. The proof is now by induction on n. If n = 1 then $0 = a_1$ and L is a uniform lattice, hence u-dim(L) = 1. If n > 1 then the induction hypothesis entails that u-dim($[\hat{a}_1, 1]$) = n-1 whereas, on the other hand, $[0, \hat{a}_1] = [\hat{a}_1 \wedge a_1 , \hat{a}_1] \cong [a_1 , a_1 \vee \hat{a}_1] \subset [a_1 , 1]$. Hence $[0, \hat{a}_1]$ is uniform, or $\hat{a}_1$ is an uniform element of L.

In a similar way we derive that $\hat{a}_i$ is uniform for all $i \in \{1, ..., n\}$.

Let us assume that $\hat{a}_1 \vee \hat{a}_2 \vee ... \vee \hat{a}_n$ is not essential in L.

Then $b \wedge (\hat{a}_1 \vee ... \vee \hat{a}_n) = 0$ for some nonzero $b \in L$; consequently (invoking Proposition 1.5.2.), $\{\hat{a}_1, ..., \hat{a}_n, b\}$ is join-independent and in this case we have: $[0, \hat{a}_2 \vee ... \hat{a}_n \vee b] = [\hat{a}_1 \wedge (\hat{a}_2 \vee ... \vee \hat{a}_n \vee b), \hat{a}_2 \vee ... \vee \hat{a}_n \vee b] \cong [\hat{a}_1 , \hat{a}_1 \vee \hat{a}_2 \vee ... \vee \hat{a}_n \vee b] \subset [\hat{a}_1 , 1]$.

This implies that u-dim $([\hat{a}_1, 1]) \geq n$, a contradiction. So we must accept that $\hat{a}_1 \vee ... \vee \hat{a}_n$ is essential in L and therefore u-dim (L) = n. □

Remark. The combinations of Corollary 1.5.6. and 1.5.9. yields a new proof of the Kurosh-Ore theorem (Corrolary 1.4.2.).

A lattice (L,≤) is said to be *upper-continuous* (or it verifies condition AB5) if it is complete and such that for each directed subset A of L and any $a \in L$ we have: $a \wedge (\bigvee_{x \in A} x) = \bigvee_{x \in A}(a \wedge x)$, (where a subset A in L is aid to be *directed* if for all a,b ∈ A there is a c ∈ A such that a ≤ c and b ≤ c).

If L° is upper-continuous then we say that L is *lower-continuous.* (or that it satisfies the condition AB5*).

We say that L is *continuous* if both L and L° are upper-continuous. In any upper-continuous lattice L a nonempty subset A will be join-independent if and only if $a \wedge \bigvee\{x , x \in A - \{a\}\} = 0$ for each a ∈ A. The following results complete Theorem 1.5.9. in the upper-continuous modular case.

1.5.10. Theorem. Let (L,≤) be an upper-continuous modular lattice with 0 and 1. The following statements are equivalent:

1. u-dim(L) = n < ∞.
2. There exist $a_1, ..., a_n \in L$ which are meet-irreducible and such that $0 = a_1 \wedge ... \wedge a_n$ is an irredundant representation.

Proof. 2 ⇒ 1. cf. Theorem 1.5.9.

1 ⇒ 2. From u-dim(L) = n it follows that there exists a join-independent set $\{b_1, ..., b_n\}$ such that all b_i are uniform while $b_1 \vee ... \vee b_n$ is essential.
Let a_k be a maximal element with the property: $b_k \wedge a_k = 0$ and $\hat{b}_k \leq a_k$; where $\hat{b}_k = \bigvee_{i \neq k} b_i$, this element exists by upper-continuity and Zorn's lemma. Take u,v ∈ L such that $a_k < u$, $a_k < v$. Then $b_k \wedge u \neq 0$, $b_k \wedge v \neq o$, so $b_k \wedge (u \wedge v) \neq 0$ since b_k is uniform. Hence $a_k \neq u \wedge v$ and therefore a_k is a meet-irreducible element of L.
Next we establish that $\bigwedge_{k=1}^n a_k = 0$ and that this meet is irredundant.
Verifying that $b_l \leq \bigwedge_{k \neq l} a_k$, l = 1,...,n, one also checks that $\bigwedge_{k=1}^n a_k$ is reduced. If b $= \bigvee_{k=1}^n b_k = b_k \vee \hat{b}_k$ then we have that $b \wedge a_k = (b_k \vee \hat{b}_k) \wedge a_k = (b_k \wedge a_k) \vee \hat{b}_k$ and thus $b \wedge a_k = \hat{b}_k$. It follows that $b \wedge (\bigwedge_{k=1}^n a_k) = \bigwedge_{k=1}^n (b \wedge a_k) = \bigwedge_{k=1}^n \hat{b}_k$. The modularity of L yields $\bigwedge_{k=1}^n \hat{b}_k = 0$ and therefore $b \wedge (\bigwedge_{k=1}^n a_k) = 0$. Consequently $\bigwedge_{k=1}^n a_k = 0$, because b is essential in L. □

If (L,≤) is a modular lattice with 0 and 1, then an element a ≠ 1 in L is said to be *small* in L if for each x ∈ L such that $a \vee x = 1$, it follows that $x = 1$. We say that the lattice is *hollow* if for each element $x \neq 1$ in L is small in L.
It is easily checked that the lattice L is hollow if and only if L° is uniform and that a ∈ L is small in L if and only if a is essential in L°.
We say that the *dual Goldie dimension* of L is n if there exists a finite set $\{a_1, ..., a_n\}$ in L such that $a_1 \wedge ... \wedge a_n$ is small in L while the lattices $[a_i, 1]$ are hollow for i = 1,...,n; we shall write h-dim(L) = n in this case and h-dim(L) = ∞ if such an n does not exist. Obviously h-dim(L) = u-dim(L°).
A straightforward dualization of Theorem 1.5.7. yields:

1.5.11. Theorem. The following conditions are equivalent

1. h-dim(L) < ∞.
2. L does not contain infinite meet-independent subsets.
3. L satisfies the descending chain condition for direct finite meets.
4. For a sequence $a_1 \geq a_2 \geq ... \geq a_n \geq ...$ in L, there exists an n ∈ ℕ such that for all $k \geq n$, a_n is small in $[a_k, 1]$.

1.5.12. Corollary. Let (L,≤) be a modular lattice such that $0 = \bigwedge_{i=1}^n a_i$ for a meet-independent $\{a_1, ..., a_n\}$. The dual Goldie dimension of L is finite if and only if the dual Goldie dimensions of $[a_i, 1]$ are finite for all i = 1,...,n.

Moreover, h-dim(L) = $\sum_{i=1}^{n}$ h-dim($[a_i, 1]$).

1.5.13. Theorem. If $1 = a_1 \vee ... \vee a_n$ is an irredundant representation in a modular lattice (L,$\leq$) with 0 and 1 such that each $[0, a_i]$ is hollow then h-dim(L) = n.

1.5.14. Theorem. Let (L,$\leq$) be an upper-continuous modular lattice with 0 and then the following assertions are equivalent:

1. h-dim(L) = n $< \infty$.
2. There exists $a_1, ..., a_n \in$ L such that $1 = a_1 \vee ... \vee a_n$ is an irredundant representation and the lattices $[0, a_i]$ are hollow.

For completeness sake we mention some results concerning infinite Goldie dimension of modular lattices. A nonempty set I in a lattice (L,$\leq$) is an *ideal* of L if for all x,y $\in$ I we have that $[0, x \vee y] \subset$ I. The set $\mathcal{I}$(L) of all ideals of L is itself an upper-continuous lattice with respect to the intersection (for $\wedge$) and the operation of taking the ideal generated by ideals (for $\vee$).
The map p: L$\rightarrow \mathcal{I}$(L) defined by p(a) = [0,a] is an embedding of lattices; this allows to treat L as a sublattice of $\mathcal{I}$(L).
The following lemma contains some of the evident properties of the lattice I(L):

1.5.15 Lemma. Let (L,$\leq$) be a modular lattice with 0, then the following properties hold:

1. For any subset S of $\mathcal{I}$(L) and every I $\in \mathcal{I}$(L) such that I $\wedge$ ($\bigvee_{s \in S} s$) is nonzero, there exists a finite subset $S_1 \subset$ S such that I $\wedge$ ($\bigvee_{s \in S_1} s$) $\neq$ 0.
2. An x $\in$ L is uniform in L if and only if [0,x] is uniform in $\mathcal{I}$(L).
3. S is a join-independent subset of L if and only if the set $S^p = \{[0, x],\ x \in S\}$ is join-independent in $\mathcal{I}$(L).
4. S is a maximal join-independent set in L if and only if S^p is a maximal join-independent subset of $\mathcal{I}$(L), if and only if S^p is join-independent and $\bigvee_{\bar{s} \in S^p} \bar{s}$ is essential in $\mathcal{I}$(L).

1.5.16. Proposition. If M is a maximal join-independent set in L such that all elements of M are uniform then for any join-independent set N in L, $|N| \leq |M|$ (where $|\ |$ denotes the cardinality).

Proof. By the lemma it will suffice to establish the claim for $\mathcal{I}(L)$. If M or N is finite then Theorem 1.5.5. and consequent results prove the assertion. Therefore, assume $|M| \geq \mathcal{X}_o, |N| \geq \mathcal{X}_o$. Then $|M|$ is also the cardinality of the set of all finite subsets of M, $S_\omega(M)$.
Because of Lemma 1.5.15, 4, $a = \bigvee_{m\in M} m$ is essential in $\mathcal{I}(L)$; thus for $b \in N$ we have $b \wedge a \neq 0$. Again by the Lemma (1.) we may select a finite subset M_b in M such that $b \wedge (\bigvee_{x\in M_b} x) \neq 0$ for given $b \in N$. If we define $f: N \to S_\omega(M)$ by $f(b) = M_b$ and pick $a_1 \neq \ldots \neq a_k \in f^{-1}(M_b)$ then $0 \neq c_i = a_i \wedge (\bigvee_{x\in M_b} x) \leq a_i$, for $i = 1,\ldots,k$, and $\{c_1, \ldots, c_k\}$ is a join-independent subset of $[0; \bigvee_{x\in M_b} x]$.
Since M_b is join-independent and consisting of uniform elements we have u-dim$([0, \bigvee_{x\in M_b} x]) = |M_b| \geq k$. Consequently, $|f^{-1}(M_b)| \leq |M_b|$ and when we combine this with $|N| \geq \mathcal{X}_o$ we obtain $|N| = |f(N)| \leq |S_\omega(M)| = |M|$ □

A subset B of L-$\{0\}$ is said to be a *basis* of L if B is a maximal join-independent subset of L such that all of its elements are uniform in L.
The foregoing proposition entails that the cardinality of a basis is independent of the chosen basis, therefore we can extend the notion of Goldie dimension to the infinite case if we define it to be the cardinality of a basis of L whenever it exists.
It is noteworthy to point out that the Goldie dimension does generalize the dimension of vector spaces; indeed, B is a basis of the vectorspace V over a field K if and only if $\{Kb,\ b \in B\}$ is a basis for the lattice L(V) of all linear K-subspaces of V.

1.5.17. Proposition. With notations and assumptions as before, the following assertions are equivalent:

1. L has a basis.
2. $\mathcal{I}(L)$ has a basis.
3. Any nonzero $\overline{x} \in \mathcal{I}(L)$ contain a uniform element in $\mathcal{I}(L)$.
4. For every nonzero x in L there is a uniform $y \in L$ such that $y \leq x$.

Proof. 1⇒2. follows from Lemma 1.5.15, 2 and 4.
2⇒3. For a basis B of $\mathcal{I}(L)$, $\overline{x} = \bigvee_{\overline{b}\in B} \overline{b}$ is essential in $\mathcal{I}(L)$.

For some $\overline{y} \in \mathcal{I}(L)$ we then have $\overline{y} \wedge \overline{x} \neq 0$, hence $\overline{y} \wedge (\overline{b}_1 \vee ... \vee \overline{b}_n)$ is nonzero for some finite subset $\{\overline{b}_1, ..., \overline{b}_n\}$ in B. But then :
u-dim($[0, \overline{y} \wedge (\overline{b}_1 \vee .. \vee \overline{b}_n)]$) $\leq$ u-dim($[0, \overline{b}_1 \vee ... \vee \overline{b}_n]$) $= n < \infty$.
By Theorem 1.5.5. there exists a uniform element $\overline{z}$, $\overline{z} \leq \overline{y} \wedge (\overline{b}_1 \vee ... \vee \overline{b}_n) \leq \overline{y}$.
3⇒4. Consider $x \neq 0$ in L. By the assumptions there is a uniform $\overline{y} \in \mathcal{I}(L)$ such that $\overline{y} \leq [0, x]$. For any nonzero $z \leq x$ it is clear that [0,z] is uniform in $\mathcal{J}(L)$. By the lemma, z is uniform in L.
4⇒1. Choose (Zorn's Lemma) a maximal join-independent subset B consisting of uniform elements in L. By 4., B is a maximal join-independent subset of L and thus B is a basis of L. □

Remark. A sublattice of a lattice with a basis need not have a basis but an ideal of a lattice with a basis does have a basis.

As a consequence of Lemma 1.5.15. we also have:

1.5.18. Proposition. If S is a maximal join-independent subset of L and B_s is a basis for [0,s] then $\bigcup_{s \in S} B_s$ is a basis of L.

Proof. Exercise. □

§1.6 Goldie dimension and Chain Conditions for Modular Lattices with Finite Group Actions.

In this chapter we follow P. Grzesczuk and E. Puczylowski, cf. [3], and extend some results of J. Fisher, cf. [1], concerning chain conditions for modular lattices with finite group actions to the case of the Goldie dimension.

Let L be a lattice containing 0 and 1. An isomorphism L→L is an automorphism of L and the automorphisms of L form a group Aut(L). We consider a finite group in Aut(L) and we write L^G for $\{a \in L \, , \, a^g = a \text{ for all } y \in G\}$.

1.6.1. Proposition. 1. If L is complete then $(\bigvee_{s\in S} s)^g = \bigvee_{s\in S} s^g$ holds for every subset of L.
2. If L is upper-continuous then $Z = \{a \in L, \bigwedge_{g\in G} a^g = 0\}$ is closed under taking joins of chains.
3. $\mathcal{I}(L^G) \cong \mathcal{I}(L)^G$, where the actions of G on $\mathcal{I}(L)$ is extended from L in the obvious way: $[0,a]^g = [0,a^g]$.

Proof. 1. Immediate from the fact that g as well as g^{-1} is increasing.
2. Immediate from upper-continuity and 1.
3. Define f: $\mathcal{I}(L^G) \to \mathcal{I}(L)^G$, g: $\mathcal{I}(L)^G \to \mathcal{I}(L^G)$ as follows:

for $\bar{x} \in \mathcal{I}(L^G)$, $f(\bar{x})$ is the L-ideal generated by $\bar{x}$; for $\bar{y} \in \mathcal{I}(L)^G$, $g(\bar{y}) = \bar{y} \cap L^G$. Both f and g are lattice homomorphisms and they obviously satisfy fog = $id(\mathcal{I}(L)^G)$, gof = $id(I(L^G))$. □

For the sequel of this section we let L be a modular lattice with 0 and 1. First we study the ascending and descending chain conditions but we only include proofs for the ascending chain conditions because the equivalent properties concerning the descending chain condition will follow by the usual straightforward dualization argument.
Recall (see 1.2.) that (L,≤) satisfies a.c.c., resp. d.c.c., if and only if every chain $a_1 \le a_2 \le ...$, resp. $a_1 \ge a_2 \ge ...$, in L terminates. Modularity of the lattice L (Proposition 1.3.1. in particular) entails that lattices $[a_1, 1], ..., [a_n, 1]$ with $a_1, ..., a_n \in L$ satisfy a.c.c., resp. d.c.c., if and only if $[a_1 \wedge ... \wedge a_n, 1]$ satisfies a.c.c., resp. d.c.c..

1.6.2. Lemma. L satisfies a.c.c., resp. d.c.c., if and only if $\mathcal{I}(L)$ satisfies a.c.c., resp. d.c.c.

Proof. Since p: $L \Rightarrow \mathcal{I}(L)$ is an embedding of Lattices, the chain conditions on $\mathcal{I}(L)$ imply those on L. Conversely, let $\bar{x}_1 < \bar{x}_2 < ...$ be a strictly ascending chain of ideals of L then we may select $a_i \in \bar{x}_{i+1} - \bar{x}_i$ for each i and we write $b_i = \bigvee_{k=1}^{i} a_k$. Clearly $b_1 \le b_2 \le ...$ with $b_i \in \bar{x}_{i+1}$. Since $a_1 \le b_i$ and $a_i \ne \bar{x}_i$ we have $b_i \ne \bar{x}_i$ since $\bar{x}_i$ is an ideal; therefore the sequence $b_1 \le b_2 \le ...$ is strictly ascending. The fact that the latter sequence must terminate then yields the a.c.c. for $\mathcal{I}(L)$. □

1.6.3. Theorem. The modular lattice $(L,\leq)$ with 0 and 1 satisfies a.c.c., resp. d.c.c., if and only if L^G satisfies a.c.c., resp. d.c.c.

Proof. Since L^G embeds in L one implication is obvious. Now assume L^G satisfies a.c.c.

By Proposition 1.6.1. (3) and Lemma 1.6.2. we may pass to $\mathcal{I}(L)$ i.e. we may assume that L is upper-continuous.

If L does not satisfy a.c.c. let $a \in L^G$ be maximal with the property that [a,1] does not satisfy a.c.c. It is clear that [a,1] is upper-continuous and that G acts on it as a group of automorphisms.

By Proposition 1.6.1. (2), the set $Z(a) = \{x \in [a,1] , \bigwedge x^g = a\}$ contains a maximal element, x_0 say. Now for some $g \in G$ we have that $[x_0^g, 1]$ does not satisfy a.c.c.. But $[x_0^g, 1] \cong [x_0, 1]$ for all $g \in G$, hence $[x_0, 1]$ cannot satisfy a.c.c.. Repetition of the argumentation leeds to a strictly ascending $x_0 < x_1 < x_2 < ....$ The choice of x_0 yields $a < \bigwedge_{g\in G} x_1^g$ but then the choice of a entails that $[\bigwedge_{g\in G} x_1^g , 1]$ satisfies a.c.c.. However, $\bigwedge_{g\in G} x_1^g < m_1 < m_2$ ... leads to a contradiction. □

Before dealing with the Goldie dimension we need some additional lemmas.

1.6.4. Lemma. For each $a \in L$, u-dim(L) $\leq$ u-dim([o,a])+u-dim([a,1])

Proof. If $\{a, x_1, ..., x_r\}$ is join-independent in L then $\{a \vee x_1, ... , a \vee x_r\}$ is join-independent in [a,1] (by the modularity of L). The statement is trivial if one of the terms on the right hand side is infinite, so we may assume that both terms are finite.

By the foregoing remark there is a finite bound on the cardinalities of families $x_1, ..., x_n$ in L such that $\{a, x_1, ..., x_r\}$ is join-independent. Assume that $x_1, ..., x_r$ are chosen in such a way as to make r maximal. Then it follows that each x_i is uniform, r $\leq$ u-dim([a,1]), and $a \vee x_1 \vee ... \vee x_r$ is essential in [a,1]. Since we assumed that d = u-dim[o,a] is finite, there exists a join-independent $\{u_1, ..., u_d\}$ consisting of uniform elements of [o,a] and then $u_1 \vee ... \vee u_d$ is essential in [o,a]. Furthermore, $u_1 \vee ... \vee u_d \vee x_1 \vee ... \vee x_r$ is essential in L (a version of Proposition 1.5.3. using that $a \vee x_1 \vee ... \vee x_r$ is essential in [a,1] while $u_1 \vee ... \vee u_d$ is essential in [o,a]).

Finally, u-dim(L) = d+r $\leq$ u-dim([o,a]) + u-dim([a,1]) follows. □

1.6.5. Corollary. If $a_1 \wedge ... \wedge a_n = 0$ in L then u-dim(L) $\leq \sum_{i=1}^{n}$ u-dim($[a_i,1]$)

Proof.

If n = 1 there is nothing to prove; proceed by induction on n.
If n $\geq$ 2 put $\hat{a}_1 = a_2 \wedge ... \wedge a_n$. By the induction hypothesis we have: u-dim($[\hat{a}_1,1]$) $\leq \sum_{i=2}^{n}$ u-dim($[a_i,1]$). From $a_1 \wedge \hat{a}_1 = 0$ we derive: $[0,a_1] = [a_1 \wedge \hat{a}_1 , a_1] \cong [\hat{a}_1 , a_1 \vee \hat{a}_1] = [\hat{a}_1,1]$. Hence, u-dim($[0,a_1]$) $\leq \sum_{i=2}^{n}$ u-dim($[a_i,1]$. The lemma finishes the proof. □

1.6.6. Lemma. u-dim(L) = u-dim($\mathcal{I}$(L)).

Proof. That u-dim(L) $\leq$ u-dim($\mathcal{I}$(L)) is obvious.
For the converse, let $\overline{x}_1, ..., \overline{x}_n$ be a join-independent subset of $\mathcal{I}$(L) and consider nonzero $y_i \in \overline{x}_i$ for c = 1,...,n. For all j $\in \{1, ..., n\}$,
$y_j \wedge (y_1 \vee ... \vee y_{j-1} \vee y_{j+1} ... \vee y_n) \in$
$\overline{x}_j \wedge (\overline{x}_1 \vee ... \overline{x}_{j-1} \vee \overline{x}_{j+1} \vee ... \vee \overline{x}_n)$ and the latter is the zero interval. Consequently, $\{y_1, ..., y_n\}$ is a join-independent subset of L and u-dim($\mathcal{I}$(L)) $\leq$ u-dim(L) follows. □

1.6.7. Theorem. Let L be a modular lattice with 0 and 1 and let G be a finite group of lattice isomorphisms acting on L.

1. u-dim(L) is finite if and only if u-dim(L^G) is finite and in this case u-dim(L^G) $\leq$ u-dim(L) $\leq |$ G $|$. u-dim(L^G)
2. For an infinite cardinal α, u-dim(L) = α if and only if u-dim(L^G) = α.

Proof. 1. By Lemma 1.6.6. we may pass to $\mathcal{I}$(L) i.e. we may assume that L is upper-continuous. Then we may select a maximal element x_0 in $\{x \in L, \bigwedge_{g\in G} x^g = 0\}$ and by Corollary 1.5.6. it will be sufficient to establish that u-dim($[x_0,1]$) $\leq$ u-dim(L^G) and for the latter inequality it will suffice to show that for every set of join-independent $x_1, ..., x_m \in [x_0,1]$ we have that $\{y_i = \bigwedge_{g\in G} x_i^g$, i=1,...,m$\}$ is join-independent in L^G. The choice of the element x_0 entails $x_i \neq 0$ for all i $\in \{1, ..., m\}$, hence we have to show that for each i $z_i = y_i \wedge (y_1 \vee ... \vee y_{i-1} \vee y_{i+1} \vee ... \vee y_m)$ = 0. However, for every i = 1,...,m, $x_0 < x_0 \vee y_i \leq x_i$ and $\{x_1, ..., x_m\}$ is

join-independent in $[x_0,1]$, hence $\{x_0 \vee y_1, \ldots, x_0 \vee y_m\}$ is join-independent in $[x_0,1]$. Consequently, for all i = 1,...,m, $x_0 = (x_0 \vee y_i) \wedge (\bigvee_{j \neq i} x_0 \vee y_i) \geq y_i \wedge (y_1 \vee \ldots \vee y_{i-1} \vee y_{i+1} \vee \ldots \vee y_m)$.
The choice of x_0 and the fact that $z_i \in L^G$ yields:
$z_i = \bigwedge_{g \in G} z_i^g \leq \bigwedge_{g \in G} x_0^g = 0$, proving the claims.
2. Using a basis B with $|B| = \alpha$ and argumentation similar to the argumentation of the proof of 1. leads to a proof in the infinite case. □

Remark. Some specific applications, e.g. those mentioned by P. Grzesczuk and E. Puczylowski in [3] have been included in consequent sections. Invariants of finite group actions reappear in 2.17., 4.2.4.

§1.7 Complements and Pseudo-Complements

Let $(L,\leq)$ be a lattice with 0 and 1. If $a \in L$ then $b \in L$ is said to be a *complement* of a (in L) if $a \vee b = 1$ and $a \wedge b = 0$. A lattice is said to be *complemented* if each element of L has a complement.
Any $b \in L$ such that b is maximal with the property $b \wedge a = 0$ is called a *pseudo-complement* of a in L.

1.7.1. Lemma. Let $(L,\leq)$ be a modular lattice with 0 and 1. Every complement of $a \in L$ is a pseudo-complement of a in L.

Proof. Let b be a complement of a in L and suppose $b \leq b'$ for some $b' \in L$ such that $b' \wedge a = 0$. By modularity of L we obtain: $b' = b' \wedge 1 = b' \wedge (a \vee b) = (b' \wedge a) \vee b = b$. □

1.7.2. Lemma. If $(L,\leq)$ is upper-continuous and modular then each $a \in L$ has a pseudo-complement.

Proof. For $a \in L$ put $A = \{x \in L, a \wedge x = 0\}$. By the upper-continuity of L, A is inductively ordered hence we may find a maximal element in A which is then clearly a pseudo-complement of a. □

1.7.3. Lemma If b is a pseudo-complement of a in the modular lattice $(L,\leq)$ then a. $\vee$ b is essential in L.

Proof. If $x \neq 0$ in L such that $(a \vee b) \wedge x = 0$ then $a \wedge x = 0$ and $b \wedge x = 0$. By modularity of L: $a \wedge (b \vee x) \leq (a \vee b) \wedge (b \vee x) = b \vee ((a \vee b) \wedge x) = b$; thus $a \wedge (b \vee x) \leq b$. From $a \wedge b = 0$ we derive: $a \wedge (b \vee x) = a \wedge a \wedge (b \vee x) \leq a \wedge b = 0$, hence $a \wedge (b \vee x) = 0$. Maximality of b yields $b = b \vee x$ or $x \leq b$ and $x = 0$ follows, a contradiction. □

1.7.4. Lemma. Let $(L,\leq)$ be a modular upper-continuous lattice, let b be a pseudo-complement of $a \in L$ and suppose that $c \in L$ is maximal with the properties $a \leq c$ and $b \wedge c = 0$, then c is a maximal essential extenion of a in L.

Proof. Note that c exists by the assumptions on L. Let $0 < x \leq c$; if $a \wedge x = 0$, consider $b \vee x$. Since $b \neq b \vee x$ we obtain $a \wedge (b \vee x) \neq 0$; but $a \wedge (b \vee x) = a \wedge c \wedge (b \vee x) = a \wedge (c \wedge (b \vee x)) = a \wedge (x \vee (b \wedge c)) = a \wedge x$. So $a \wedge x \neq 0$, a contradiction. Therefore c is an essential extension of a.
Consider another essential extension $a \leq c'$ such that $c \leq c'$. If $c \neq c'$ then $b \wedge c' \neq 0$ and hence $a \wedge (b \wedge c') \neq 0$. Since $a \wedge b = 0$ it follows that $(a \wedge b) \wedge c = 0$, a contradiction. □

1.7.5. Corollary. Let $(L,\leq)$ be a modular upper-continuous lattice. An $a \in L$ is a pseudo-complement in L if and only if a does not have proper essential extensions in L.

Proof. If a is a pseudo-complement, of b say, in L and $a < c$ is essential then $b \wedge c \neq 0$ and thus $a \wedge (b \wedge c) \neq 0$. But $a \wedge (b \wedge c) = (a \wedge b) \wedge c = 0$, a contradiction. Conversely, if a does not have proper essential extensions in L then $a = c$ if c is an element as in Lemma 1.7.4. But c is a pseudo-complement of b in L, hence a is a pseudo-complement. □

Remark. The above results may be used to give a proof of Lemma 1.6.4. (up to going over to $\mathcal{I}(L)$ in order to apply the above to the upper-continuous case). Consider a pseudo-complement b of a in L, then $a \vee b$ is essential. From u-dim(L) = u-dim([0, $a \vee b$]) + u-dim([0,b]) and $[0,b] = [a \wedge b, b] \cong [a, a \vee b] \subset [a,1]$ it follows that u-dim[0,b] $\leq$ u-dim([a,1]) and thus the statement of Lemma 1.6.4. follows from this elegant argument.

§1.8 Semiatomic Lattices and Compactly Generated Lattices

An $a \in L$ is an *atom* if $b \leq a$ entails either $b = 0$ or $b = a$, i.e. if and only if [0,a] is simple. An upper-continuous lattice with 0 and 1 is *semiatomic* if 1 is a join of atoms. An a in an upper-continuous lattice is said to be *compact* if for each directed subset A of L such that $a \leq \bigvee_{x \in A} x$ there exists an $x_0 \in A$ such that $a \leq x_0$.An upper-continuous lattice is *compact* (or finitely generated) if 1 is compact in L and L is *compactly generated* if each element of L is a join of compact elements.

1.8.1. Lemma. In an upper-continuous lattice $(L,\leq)$ each atom is also compact.

Proof. Consider an atom a of L and let A be a directed subset of L such that $a \leq \bigvee_{x \in A} x$.
Upper-continuity of L entails: $a = a \wedge (\bigvee_{x \in A} x) = \bigvee_{x \in A} (a \wedge x)$. Clearly, there exists $x_0 \in A$ such that $a \wedge x_0 \neq 0$. Then we obtain $a \wedge x_0 \leq a$, i.e. $a \wedge x_0 = a$ or $a \leq x_0$. □

1.8.2. Theorem. Any semiatomic lattice L is complemented.

Proof. By upper-continuity (and Zorn's lemma) there exists a directed set of atoms $\{s, s \in \mathcal{V}\}$ which is maximal with the property: $a \wedge (\bigvee_{s \in \mathcal{V}} s) = 0$. Put b for the join of $\mathcal{V}$. If $a \vee b \neq 1$ then there exists an atom s_0 such that
$s_0 \not\leq a \vee b$, i.e. such that $s_0 \wedge (a \vee b) = 0$.
By the modularity of L we have:

$a \wedge (b \vee s_0) \leq (a \vee b) \wedge (b \wedge s_0) = b \vee ((a \vee b) \vee s_0) = b$, hence,
$a \wedge (b \vee s_0) = a \wedge a \wedge (b \vee s_0) \leq a \wedge b = 0$, or $a \wedge (b \vee s_0) = 0$.
From $s_0 \wedge b = 0$, $s_0 \notin \mathcal{V}$ follows and therefore we arrive at $b \vee s_0 = \bigvee \{s, s \in \mathcal{V} \cup \{s_0\}\}$, contradicting the maximality of $\mathcal{V}$.
Consequently $a \vee b = 1$ and the proof is complete. □

1.8.3. Corollary. In a semiatomic lattice $(L,\leq)$ there always exists a join-independent $\{s_i, i \in \mathcal{J}\}$ consisting of atoms such that $1 = \bigvee_{i\in J} s_i$.

1.8.4. Corollary. In a semiatomic lattice $(L,\leq)$, the lattices $[0,a]$ and $[a,1]$ are semiatomic for each $a \in L$.

Proof. Let $\mathcal{A}$ be the set of atoms of L, then $1 = \bigvee \mathcal{A}$ and also $1 = 1 \vee a = \bigvee \{s \vee a, s \in \mathcal{A}\}$. For each $s \in \mathcal{A}$, $s \vee a = a$ or $s \vee a$ is an atom in $[a,1]$. Indeed if $s \vee a \neq a$ and $a \leq x \leq s \vee a$ then $x = x \wedge (s \vee a) = a \vee (x \wedge s)$ where $x \wedge s = 0$ or $x \wedge s = s$. If $x \wedge s = 0$ then $x = a$ follows, if $x \wedge s = s$ then $x = a \vee s$.
Put $\mathcal{A}' = \{s \in \mathcal{A}, a \vee s \neq a\}$. Clearly $1 = \bigvee \{s \vee a, s \in \mathcal{A}'\}$ or $[a,1]$ is semiatomic.
By Theorem 1.8.2. we may pick a complement b for a. Then $[0,a] = [a \wedge b, a] \cong [b, a \vee b] = [b,1]$. Since $[b,1]$ is semiatomic in view of the foregoing, it follows that $[0,a]$ is semiatomic too. □

1.8.5. Lemma (Krull). Suppose $(L,\leq)$ is a compact lattice and consider $a \neq 1$ in L, then there exists an $a_0 \neq 1$ which is maximal with the property that $a \leq a_0$.

Proof. Consider $A = \{b \in L, a \leq b < 1\}$. It is clear that $A \neq \phi$ and A is inductive because L is compact. Applying Zorn's lemma we arrive at the fact that A posesses a maximal element. □

1.8.6. Lemma. A complete Noetherian lattice is compact.

Proof. Let A be a directed subset of L. There is an $a_0 \in A$ which is a maximal element and it is also the largest element of A since this set is directed, i.e. $a_0 =$

$\bigvee\{x, x \in A\}$. For each a $\in$ L we obtain: $a \wedge \bigvee A = a \wedge a_0$. On the other hand it follows from $a \wedge x \leq a \wedge a_0$ for all x $\in$ A, that $\bigvee_{x\in A}(a \wedge x) = a \wedge a_0$. Therefore L is upper-continuous and one easily checks compactness. □

1.8.7. Theorem. A compactly generated complemented lattice (L,$\leq$) is semiatomic.

Proof. Take a nonzero a in L; we aim to establish that a contains an atom. Since L is compactly generated we may assume a to be compact. Lemma 1.8.5. applied to [0,a] leads to the existence of b which is maximal with the properties: $b \leq a$ and $b \neq a$. Since L is complemented we may select b' $\in$ L for which $b \vee b' = 1$,

$b \wedge b' = 0$. Modularity of L now entails $b \vee (a \wedge b') = a \wedge (b \vee b') = a \wedge 1 = a$,

$b \wedge (a \wedge b') = a \wedge (b \wedge b') = 0$. This yields that $c = a \wedge b'$ is a complement of b in [0,a]. If $0 \leq x \leq c$ then: $x = x \vee 0 = x \vee (b \wedge c) = c \wedge (b \vee x)$.

If $x \neq o$ then the maximality of b yields $b \vee x = a$ (because $b \wedge x = 0$) and then $c \wedge a = x$ or $c = x$, proving that c is an atom contained in a. Write a_0 for the join of all atoms of L. In case $a_0 \neq 1$ there must exist a complement b_0 of a_0 with $b_0 \neq 0$. Knowing now that b_0 must contain an atom s we derive easily from $s \leq a_0$,

$a_0 \wedge b_0 = 0$, that $s = 0$, a contradiction. Therefore $a_0 = 1$ follows. □

If L is a complete lattice, let r_L be the meet of all proper maximal elements, putting $r_L = 1$ if no such maximal elements exist. We refer to r_L as the *radical* of L. Lemma 1.8.5. entails that $r_L \neq 1$ for a compact lattice.

1.8.8. Lemma. Let L be a compact lattice. If a $\in$ L and $a \vee r_L = 1$ then $a = 1$ (any element of L with this property is said to be *small*). In particular, r_L is the largest small element of L.

Proof. If $a \neq 1$ then Lemma 1.8.5. entails that there exist a maximal m in L such that $a \leq m < 1$. Since $r_L \leq m$ we obtain $a \vee r_L \leq m$ and the contradiction $m = 1$. Let now a $\in$ L be a small element. If a is not smaller than r_L then there is a maximal element m in L such that $a \not\leq m$.

But $a \vee m = 1$ then implies $m = 1$, a contradiction. □

1.8.9. Lemma. If (L,$\leq$) is a semiatomic and compactly generated lattice then r_L = 0.

Proof. Let s $\in$ L be an atom and let m be a complement of s in L. Let m < a. If s $\wedge$ a $\neq$ 0 then s $\wedge$ a = s or s $\leq$ a, and then m $\vee$ s $\leq$ a and a = 1 follows. If s $\wedge$ a = 0 we obtain a contradicton because m < a. So m is a maximal proper element of L. Thus s $\wedge$ r_L = 0 for each atom s of L but if r_L $\neq$ 0 then r_L contains an atom since L is semiatomic, hence r_L = 0. □

1.8.10. Theorem. If L is a compact Artinian lattice then [r_L,1] is semiatomic.

Proof. Up to replacing L by [r_L,1] we may assume that r_L = 0. By Lemma 1.8.6., L° is compact and upper-continuous, but L° is also semiatomic since r_L = 0. By Theorem 1.8.2., L° is complemented and thus L is complemented, but then L is semiatomic in view of Theorem 1.8.7. □

§1.9 Semiartinian Lattices

In this section we assume throughout that *L is a modular upper-continuous lattice with 0 and 1.* The *socle* of the lattice is the join of all atoms of L, we denote it by s(L). It is not hard to verify that [0,s(L)] is semiatomic.

By transfinite recursion we define an ascending chain in L as follows. Let α be an ordinal number. If $\alpha = 0$, put $s_0(L) = 0$; if $\alpha = 1$, put $s_1(L) = s(L)$; if $\alpha = \beta + 1$, put $s_\alpha(L) = s([s_\beta(L),1])$ and if α is a limit ordinal, put $s_\alpha(L) = \bigvee_{\beta < \alpha} s_\beta(L)$. So we obtain

(*) $$0 = s_0(L) \leq s_1(L) \leq ... \leq s_\alpha(L) \leq ...$$

We say that (*) is the *Loewy* series of L. The semiatomic $[s_\alpha(L),s_{\alpha+1}(L)]$ are called the factors of the series (*).

Since L is a set there exists a least ordinal α such that $s_\alpha(L) = s_{\alpha+1}(L) = ...$ This ordinal is called the *Loewy-length* of L, denoted $\Lambda(L)$. With assumptions as before:

1.9.1. Lemma. If $(L,\leq)$ is compactly generated then $s(L)$ is the meet of all essential elements of L.

Proof. Put $d = \bigwedge \{x, x \text{ essential in } L\}$. If $s \in L$ is an atom and x is essential in L then $s \wedge x \neq 0$, so $s \wedge x = s$ or $s \leq x$. Thus $s(L) \leq x$ and $s(L) \leq d$. The lattice $[0,d]$ is again compactly generated. Consider $b \in [0,d]$ and let c be a pseudo-complement of b in L. Lemma 1.7.3. yields that $b \vee c$ is essential in L, hence $d \leq b \vee c$.
Since L is modular we calculate: $d = d \wedge (b \vee c) = b \vee (d \wedge c)$, showing that b has a complement in the lattice $[0,d]$. Theorem 1.8.7. yields that $[0,d]$ is semiatomic, thus $d \leq s(L)$ and $s(L) = d$ follows. □

The lattice $(L,\leq)$ is *semiartinian* if for all $a \neq 1$ in L, $[a,1]$ contains an atom.

1.9.2. Lemma. In a semiartinian lattice $(L,\leq)$, $s(L)$ is essential in L.

Proof. Consider $a \neq o$ in L and let b be a pseudo-complement of a. Clearly $b \neq 1$ and thus $[b,1]$ contains an atom, c say. Since $b < c$ it follows that $a \wedge c \neq 0$.
Put $d = a \wedge c$ and consider $0 < x \leq d$. Since $x \leq a$ and $x \leq c$ we deduce that $b < b \vee x \leq c$. Since c is an atom in $[b,1]$ it follows that $b \vee x = c$, and hence $d \wedge (x \vee b) = x \vee (d \wedge b) = x \vee (a \wedge c \wedge b) = x \vee (a \wedge b \wedge c) = x \vee 0 = x$.
Therefore $d \wedge c = x$ or $d = x$ and it follows that d is an atom.
From $d \leq a$ it follows that a contains an atom. Thus $a \wedge s(L) \neq o$. □

1.9.3. Proposition. $(L,\leq)$ is semiartinian if and only if $[0,a]$ and $[a,1]$ are semiartinian for any $a \in L$.

Proof. If $(L,\leq)$ is semiartinian then $[a,1]$ is semiartinian for any $a \in L$. If $b < a$ then $[b,1]$ is semiartinian and by Lemma 1.9.2. $c \leq a$ for some atom c of $[b,1]$. It is easily verified that c is also an atom in $[b,a]$, hence $[0,a]$ is semiartinian.
Conversely, let $[0,a]$ and $[a,1]$ be semiartinian and let $b \neq 1$ in L. If $a \leq b$ then the fact that $[a,1]$ is semiartinian implies that $[b,1]$ contains an atom. If $a \not\leq b$ let $c = a \wedge b$. Then $c < a$ and the semiartinian property of $[0,a]$ yields that $[c,a]$ contains an atom d. Put $d' = d \vee b$. If $d' = b$ then $d \leq b$ and $d \leq b \wedge a = c$, a contradiction.

Thus b < d' and we now proceed to prove that d' is an atom of [b,1].
Consider $b \leq x \leq d'$, then:
$c = b \wedge a \leq x \wedge a \leq d' \wedge a = a \wedge (d \wedge b) = d \vee (a \wedge b) = d \vee c = d$.
But d is an atom in [0,a], hence $x \wedge a = c$ and $x \wedge c = d$. If $x \wedge a = c = b \wedge a$ then we have: $b \vee (x \wedge a) = b \vee (b \wedge a) = b$. Modularity of L entails that $b \vee (x \wedge a) = x \wedge (b \vee a)$. From $d \leq a$, $d' \leq a \vee b$ follows and then $x = x \wedge d' \leq x \wedge (b \vee a) = b$, i.e. $x = b$. Suppose now that $x \wedge a = d$. Then $d' = d \vee b = b \vee (x \wedge a) = x \wedge (a \vee b)$ and consequently $d' \leq x$. Together with $x \leq d'$ this leads to $x = d'$. In other words the lattice [b,1] contains an atom, proving that L is semiartinian. □

1.9.4. Proposition. If $1 = \bigvee_{i \in I} a_i$ in $(L,\leq)$ and $[0,a_i]$ is semiartinian for all $i \in I$, then L is semiartinian.

Proof. If $a \neq 1$ in L then $a_i \not\leq a$ for some $i \in I$. Thus $a \wedge a_i < a_i$. Since $[0,a_i]$ is semiartinian there exists an atom c of $[a \wedge a_i, a_i]$. As in the proof of Proposition 1.9.3. it follows that $a \vee c$ is an atom of [a,1]. Thus L is a semiartinian lattice. □

1.9.5. Corollary. With assumptions on $(L,\leq)$ as before we have: L is semiartinian if and only if there exists an ordinal α such that $1 = s_\alpha(L)$.

Proof. One implication is obvious and the other follows by transfinite recursion on the Loewy length of L. □

§1.10 Indecomposable Elements in a Lattice

Let $(L,\leq)$ be a modular lattice with 0 and 1. An $a \in L$ which has a complement is called a *direct factor* of L. If $b \leq a$ in L are such that b is a direct factor of [0,a] then we say that b is a *direct factor of a*. We say that an $a \neq 0$ in L is *indecomposable* if 0 and a are the only direct factors of a.

Note that atoms of L are also indecomposable elements.

1.10.1. Proposition. If (L,$\leq$) is Noetherian or Artinian then each a $\in$ L is a finite direct join of indecomposable elements.

Proof. Suppose a does not satisfy the statement. There exist a' $\neq$ 0 and b' $\neq$ 0 such that a = a' $\vee$ b', a' $\wedge$ b' = 0 and a' is not a direct join of indecomposable elements. Then a' = a" $\vee$ b" with a" $\wedge$ b" = 0, a" $\neq$ 0 and b" $\neq$ 0, and a" is not a direct join of indecomposable elements.
Repeating this argument we eventually obtain infinite chains b' < b' $\vee$ b" < ... , a' > a" > a"' > ... , a contradiction. □

Finally let us include the following theorem, the proof of which may be found in Birkhoff, [1]:

1.10.2. Theorem (Ore). Consider a modular lattice (L,$\leq$) and suppose that L has finite length. If a $\in$ L may be written as a direct join of indecomposable elements of L in two ways, say a = $\bigvee_{j=1}^{n}\ b_j$, then we have the equality m = n.

§1.11 Exercises.

(1) The natural numbers ordered by divisibility form a lattice which is not complete.

(2) Let C[0,1] be the continuous functions $[0,1] \to \mathbb{R}$ For f,g $\in$ C[0,1] we define $f \leq g$ by $f(x) \leq g(x)$ for all $x \in [0,1]$. Show that (C[0,1],$\leq$) is a lattice.

(3) Consider a complete upper-continuous lattice and an essential extension $a_i \leq b_i$ in L for each $i \in I$, some index set. If $\{b_i,\ i \in I\}$ is join-independent then $\bigvee_{i \in I} a_i$ is essential in $\bigvee_{i \in I} b_i$.

(4) Let L be a modular upper-continuous lattice with 0 and 1. If $b \leq a$ in L then b is meet-irreducible in [0,a] if and only if there exists a meet-irreducible c of L such that $b = a \wedge c$.

(5) A lattice (L,$\leq$) is *distributive* if $a \wedge (b \vee c) = (a \wedge b) \vee (a \wedge c)$ for a,b,c $\in$ L. Show that (L,$\leq$) is distributive if and only if $a \vee (b \wedge c) = (a \vee b) \wedge (a \vee c)$ for all a,b,c $\in$ L.

(6) Show that a totally ordered set is distributive.

(7) Show that the lattice of Exercise (1) is distributive.

(8) Show that a distributive lattice is modular.

(9) If L is a distributive lattice with 0 and 1 then each $a \in L$ has at most one complement.

(10) A *Boolean* lattice is a distributive lattice with 0 and 1 in which every element has a complement (also called: Boolean algebra). If (L,$\leq$) is a complete Boolean algebra then, for any $\{a_i,\ i \in L\}$ in L and $b \in L$ we have: $(\bigvee_{i \in I} a_i) \wedge b = \bigvee_{i \in I} (a_i \wedge b)$ and $(\bigwedge_{i \in I} a_i) \vee b = \bigwedge_{i \in I} (a_i \vee b)$.

(11) Let (B,$\leq$) be a complete Boolean algebra. If every element of B contains an atom then $B \cong \mathcal{P}(S)$, the latter being the Boolean algebra of subsets of a set S. (Hint: apply (10) to S = { a $\in$ B, a is an atom of B}.)

(12) The lattice of ideals of the ring $\mathbb{Z}$ is distributive.

(13) Let R be a ring. An $e \in R$ is said to be idempotent if $e^2 = e$. If for all $a \in R$, ae = ea then e is a central idempotent. Let B(R) denote the set of all central idempotents of R. Show that B(R) is a Boolean algebra with ordering defined by $e \leq f$ if and only if e = ef. Note the following relations in B(R): $e \wedge f = ef$, $e \vee f = e + f - ef$, $e^* = 1 - e$, where e^* is the complement of e in B(R).

(14) A ring R such that $a^2 = a$ holds for all $a \in R$ is called a Boolean ring. Let $(B,\leq)$ be a Boolean algebra and define $a.b = a \wedge b$ and $a + b = (a \wedge b^*) \vee (b \wedge a^*)$ (where x^* is the complement of $x \in B$) for a and b in B. Show that $(B,+,.)$ is a Boolean ring. Derive from Exercise (13) that the correspondence $(B,\leq) \to (B,+,.)$ is a bijection between the classes of Boolean algebras and Boolean rings.

(15) Let G_1 and G_2 be finite groups with coprime orders. Put $G = G_1 \times G_2$. Show that $L(G) \cong L(G_1) \times L(G_2)$.

(16) For a modular complemented lattice $(L,\leq)$ show that L is Noetherian if and only if L is Artinian.

(17) Consider a topological space X and let Open(X), Cl(X) resp., be the set of open, resp. closed, subsets of X.

a. Open(X), Cl(X) ordered by inclusion are distributive lattices.
b. Show that Open(X) $\cong Cl(X)^\circ$.
c. Open(X) $\wedge$ Cl(X) is a Boolean algebra.

(18) Consider a modular lattice $(L,\leq)$ which is compactly generated. We say that L is *finitely co-generated* if $0 = \bigwedge_{i\in I} a_i$ with $a_i \in L$ implies that $0 = \bigwedge_{i\in J} a_j$ for some finite subset J of I.

a. Show that L is finitely co-generated if and only if s(L) is essential in L and [0,s(L)] is finitely co-generated, if and only if s(L) is essential and compact in L.
b. Show that L is Artinian if and only if [a,1] is finitely co-generated for all $a \in L$.

(19) Let $\mathcal{I}(L)$ be the lattice of ideals of L as in Section 1.6. Verify that $\mathcal{I}(L)$ is complete and modular if L is. Is $\mathcal{I}(L)$ distributive?

(20) Give a proof for Lemma 1.5.15.

(21) Prove that $\mathcal{I}(L)$ is upper-continuous.

(22) Let L be $\mathcal{P}(\mathbb{N})$. Check that L has a basis. Let $N_1, N_2 \subset \mathbb{N}$ be infinite subsets such that $N_1 \cup N_2 = \mathbb{N}$, $N_1 \cap N_2 = \phi$. Define N_{11}, N_{12} and N_{21}, N_{22} (infinite again) such that $N_{11} \cup N_{12} = N_1$, $N_{11} \cap N_{12} = \phi$ and $N_{21} \cup N_{22} = N_2$, $N_{21} \cap N_{22} = \phi$ etc... Finally we obtain a sublattice L' of L, L' $= \{\phi, N_1, N_2, N_{12}, N_{11}, N_{21}, N_{22}, ...\}$. Verify that L' does not have a basis. (cf. Greszczuk and Puczylowski [2])

(23) If R is a ring with units, let I(R) be the lattice of all twosided ideals of R. Show that the Goldie dimension of $I(R)^\circ$ equals the cardinality of the set of maximal ideals of R.

(24) Consider the lattices considered in Exercises (1), (2), (11), (12) and (19) and construct automorphisms of these latices. Calculate L^G in each of these cases for a

finite group of automorphisms at your choice.

(25) Let G be a finite group of automorphisms of a modular upper-continuous lattice $(L,\leq)$ with 0 and 1. Prove that $\text{h-dim}(L^G) \leq \text{h-dim}(L) \leq |G|\ \text{h-dim}(L^G)$.

Bibliographical Comments to Chapter 1

In this chapter we provided the basic notions, needed in subsequent chapters, expressed in the general language of Lattice Theory.

Perhaps the reader with a ring theoretic interest will decide that this generality is not desirable. Nevertheless we hope to prove that the lattice theoretical formulations of some properties of certain dimensions are not only elegant but also avoid duplication of proofs and results in situations which are only seemingly different e.g. when graded dimensions have to be defined and compared to the ungraded equivalents, but also in several other situations.

We have included some of the basic theorems of Lattice Theory such as: Schreier 's theorem, the Jordan-Hölder-Dedekind theorem, the Kurosh-Ore theorem etc... Some specific references for this part of the chapter are: G. Birkhoff [1], G. Michler [1], J. Fisher [1], B. Stenström [1] and [2].

The Goldie dimension of a lattice is the transcription to the situation of modular lattices of the notion of Goldie rank of a module introduced by A. Goldie in 1957. For the later theory we made use of the papers and preprints by P. Grezsczuk, E. Puczylowski, [1], [2], [3], K. Varadarajan [1] and the book of C. Năstăsescu [10].

Chapter 2

FINITENESS CONDITIONS FOR MODULES.

§2.1 Modules

In this book all rings considered are associative rings with unit unless otherwise stated. We write R for such a ring, $1 \in R$ is its unit element. A *left R-module* is an abelian group (the group law is denoted additively) with a scalar multiplication $R \times M \to M$, $(a,x) \mapsto ax$ satisfying the following properties:

(1.) $(a+b)x = ax + bx$ for $x \in M$, $a, \in R$
(2.) $a(x+y) = ax + ay$ for $x,y \in M$, $a \in R$
(3.) $(ab)x = a(bx)$ for $x \in M$, $a,b \in R$
(4.) $1.x = x$ for $x \in M$

The notion of a *right R-module* is defined in a similar way in terms of a scalar mutiplication $M \times R \to M$. If $R = K$ is a field (not necessarily commutative) then a module over K is called a vectorspace or a K-space. Whenever necessary we write ${}_RM$ or resp. M_R to make clear whether M is considered as a left R-module or as a right R-module. Throughout this book module will mean left module.

2.1.1. Observations. 1. By R° we denote the opposite ring associated to R (it is the abelian group of R with multiplication operation $*$ defined by $a*b = ba$ for all $a,b \in R$. For a left R-module M, the aplication $(x,a) \mapsto ax$, $M \times R^\circ \to M$, defines on the abelian group the structure of a right R°-module. This module is called the

opposite of the module M and it will be denoted by M°. For commutative rings we do not distinguish between a module M and its opposite M.
2. In a left R-module M, the following equalities hold: $a.0 = 0$, $0.x = 0$, $(-a)x = a(-x) = -(ax)$ for $a \in R$, $x \in M$ and where we have written 0 for both the zero elements in R and the zero in M but this will not lead us into confusion.
3. Any abelian group may be viewed as a $\mathbb{Z}$ -module

An *R-submodule* L of M is a nonempty subset L of M such that $x+y \in L$ and $ax \in L$ for all $x,y \in L$, $a \in R$.

2.1.2. Examples.

1. If M is an R-module then $\{0\}$ and M are the trivial submodules of M; the submodule $\{0\}$ is called the *zero-submodule* and it is denoted simply by 0.

2. If $X \in M$, $Rx = \{ax,\ a \in R\}$ is a submodule of M and it is called the *cyclic submodule* generated by x.

3. The ring R itself may be viewed as a left (also right) module over R; using the multiplication of R as the scalar multiplication. The submodules of $_RR$ are the *left ideals* of R and the submodules of R_R are the *right ideals*. A left ideal which is simultaneously a right ideal will be refered to as an *ideal* of R.

Let L be a submodule of M, then we may define an operation on the factor group M/L by $R \times M/L \to M/L$, $(a,\overline{x}) \mapsto \overline{ax}$, for $a \in R$, $\overline{x} \in M/L$ (one easily checks that this operation is well-defined).
With the operation introduced above, M/L becomes an R-module and it is called the *factor module* (sometimes: *quotient module*) of M by L. For two R-modules M,N we may consider functions f: $M \to N$ satisfying $f(x+y) = f(x) + f(y)$, $f(ax) = af(x)$ for all $x,y \in M$, $a \in R$; such functions are called *module homomorphisms* or *R-linear maps.*

2.1.3. Examples.

1. Consider R-modules M,N. The function $f{:}M \to N$ given by $f(x) = 0$ for all $x \in M$ is called the *zero homomorphism.*

2. If L is a submodule of M then the canonical surjection π: $M \to M/L$, $\pi(x) = \overline{x}$, is

a module homomorphism (epimorphism). In particular if L = 0 then π becomes the identity function.

3. If R is a commutative ring and M is an R-module then to any a in R we may associate a homomorphism φ_a: M $\to$ M, x $\mapsto$ ax.

4. For any m $\in$ M, M an R-module and R any ring, we may define φ_m: R $\to$ M, a $\mapsto$ am.

If f: M $\to$ N, g: N $\to$ P are R-linear maps then so is g $\circ$ f: M $\to$ P. If f: M $\to$ N is an R-linear map then Ker f = {x $\in$ M, f(x) = 0}, Im f = {f(x), x $\in$ M} are R-modules and submodules of M resp. N.
The module Ker f is called the *kernel of f* and Im f is called the *image of f*.
Module homomorphisms which are injective, resp. surjective will be called *monomorphisms*, resp. *epimorphisms*. It is clear that an R-linear map f: M $\to$ N is monomorphic, resp. epimorphic, if and only if Ker f = 0, resp. Im f = N.
An R-linear map f: M $\to$ N is said to be an *isomorphism* if there exists an R-linear map g: N $\to$ M such that g $\circ$ f = 1_M and f $\circ$ g = 1_N (we use 1_X for the identity map of a module X). A trivial verification learns that an R-linear f: M $\to$ N is an isomorphism if and only if f is bijective.
We say that the modules M and N are *isomorphic*, written M $\cong$ N, if and only if there exists an isomorphism f: M $\to$ N. The relation defined by $\cong$ is an equivalence relation in the class of R-modules.

2.1.4. The Fundamental Isomorphism Theorem. If f: M $\to$ N is a module homomorphism then $\overline{f}$: M/Ker f $\to$ Im f, $\overline{x} \mapsto$ f(x), is an isomorphism.

Proof. If y $\in$ Im f then there exists an x $\in$ M such that f(x) = y. Define $\overline{g}$: Im f $\to$ M/Ker f, y $\mapsto$ $\overline{x}$. It is elementary to verify that $\overline{g}$ is well-defined and that $\overline{f}$ and $\overline{g}$ are invere to each other. □

2.1.5. Noether's First Isomorphism Theorem. Let M be an R-module and consider submodules L $\subset$ N $\subset$ M, then (M/L)/(N/L) $\cong$ M/N.

Proof. Define f: M/L $\to$ M/N, $\overline{x} \to \overline{\overline{x}}$ ($\overline{x}$ being the class of x in M/L, $\overline{\overline{x}}$ the class of x in M/L); this is an epimorphism such that Ker f = N/L and the claim follows from Theorem 2.1.4. □

If L and N are submodules of M we put L + M = {x+y, x ∈ L, y ∈ N}. Clearly L + N is a submodule of M; similarly it is checked that L ∩ N is also a submodule of M.

2.1.6. Noether's Second Isomorphism Theorem. For submodules L and N of M we have L + N/N ≅ L/L∩N.

Proof. Define f: L → L+N/N by f(x) = $\overline{x}$, $\overline{x}$ the class of x in M/N. Obviously f is an epimorphism and Ker f = L ∩ N. Again, by Theorem .1.4. the proof is complete. □

A sequence of modules and R-linear maps:

$$\cdots \longrightarrow M_{n-1} \underset{fn-1}{\longrightarrow} M_n \underset{fn}{\longrightarrow} M_{n+1} \underset{fn+1}{\longrightarrow} \cdots$$

is said to be *exact at* M_n if Im f_{n-1} = Ker f_n; the sequence is said to be *exact* if it is exact at each M_i in the sequence.

An exact sequence:

$$o \longrightarrow M, \underset{f}{\longrightarrow} M \underset{g}{\longrightarrow} M'' \longrightarrow o$$

is called a *short exact sequence*. The exactness of such a short sequence is equivalent to: Im f = Ker g, f is a monomorphism and g is an epimorphism. By the isomorphism theorems it follows that Im f ≅ M', M" ≅ M/Im f. Consequently, if N is a submodule of M then the sequence:

$$o \longrightarrow N \xrightarrow{i} M \xrightarrow{\pi} M/N \longrightarrow o$$

where i is the canonical injection, π the canonical surjection, is a short exact one.

Consider a family of R-modules $\{M_i$, i ∈ I}. The cartesian product set $X_{i\in I}M_i$ = $\{(x_i)_{i\in I},\ x_i\ \in\ M_i\}$ may be equipped with an R-module structure in the following way: if $\underline{x}\ =\ (x_i)_{i\in I}$ and $\underline{y}\ =\ (y_i)_{i\in I}$ then $\underline{x}+\underline{y} = (x_i\ +\ y_i)_{i\in I}$, and $a\underline{x}\ =\ (ax_i)_{i\in I}$. The R-module we just defined will be denotedby $\Pi_{i\in I}\ M_i$, it is called the *direct*

product of the family $\{M_i,\ i \in I\}$. For each j ∈ I there exist canonical surjective homomorphisms $p_j : \Pi_{i\in I} M_i \rightarrow M_j$, $(x_i)_{i\in I} \mapsto x_j$, and canonical injective homomorphisms $i_j : M_j \rightarrow \Pi_{i\in I} M_i$, $x \rightarrow (x_i)_{i\in I}$, where $x_j = x$ and $x_i = 0$ for all $i \neq j$.
Within the direct product we consider the subset $\bigoplus_{i\in I} M_i$ given as $\{(x_i)_{i\in I} \in \Pi_{i\in I} M_i,\ x_i \in M_i$ and $x_i = 0$ for almost all i ∈ I}. It is easily checked that $\bigoplus_{i\in I} M_i$ is a submodule of $\Pi_{i\in I} M_i$, it will be called the *(external) direct sum* of the family $\{M_i, i \in I\}$.
The image of the canonocal injections i_j is contained in $\bigoplus_{i\in I} M_i$ and the restriction of each p_i to $\bigoplus_{i\in I} M_i$ is again surjective.
For a finite set I we obtain the quality $\Pi_{i\in I} M_i = \bigoplus_{i\in I} M_i$ and we usually write $\bigoplus_{i=1}^n M_i = M_1 \oplus M_2 \oplus \ldots \oplus M_n$. In the special case where $M_i = M$ for all i ∈ I we will write $M^I = \Pi_{i\in I} M_i$, $M^{(I)} = \bigoplus_{i\in I} M_i$.
Given a family of module homomorphisms $f_i : M_i \rightarrow N$, resp. $f_i : N \rightarrow M_i$, then there exists a unique homomorphism f: $\bigoplus_{i\in I} M_i \rightarrow N$, resp. f: $N \rightarrow \Pi_{i\in I} M_i$, such that $f o i_j = f_j$, resp $p_j \circ f = f_j$ for all j ∈ I It is clear how to define f, e.g.
f: $\bigoplus_{i\in I} M_i \rightarrow N$, $(x_i)_{i\in I} \mapsto \sum_{i\in I} f_i(x_i)$ (the sum is finite!), resp.
f: $N \rightarrow \Pi_{i\in I} M_i$, x ↦ $(f_i(x))_{i\in I}$, and it is also clear that f is the unique homomorphism with the predescribed properties.

A set $\{x_i,\ i \in I\}$ in an R-module M is said to be a system of *generators of* M (or the x_i are said to *generate M over R*) if every x ∈ M may be expressed as a linear combination of the elements x_i in the sense that there exists a finite number of elements $a_j \in$ R such that x = $\sum a_{j_i} x_{j_i}$. An R-module M with a finite set of generators is called a *finitely generated module* or M is said to be of *finite type*. A set of elements x_i, $i \in I$, in M is said to be *linearly independent* over R if
$\sum_{i\in I} a_i x_i = 0$ for a finite number of possibly nonzero elements $a_i \in$ R entails that $a_i = 0$ for each i.
A system of generators which happens to be linearly independent is called a *basis of* M over R. Any R-module allowing a basis is said to be a *free (left) R-module.* An example of a free R-module may be obtained as follows: let I ≠ Φ, then $_R(R^{(I)})$ is free with basis $(e_i)_{i\in I}$ where $e_i = (\delta_{ij})_{j\in I}$ and $\delta_{ij} = 0$ if j ≠ i, $\delta_{ii} = 1$.

2.1.7. Proposition. If M is a free R-module then there exists an index set I ≠ Φ such that M ≅ $R^{(I)}$.

Proof. Let $\{x_i,\ i \in I\}$ with I ≠ Φ be a basis for M over R. Define f: $R^{(I)} \rightarrow M$, $\sum a_i e_i \rightarrow \sum a_i x_i$ where only finitely many of the a_i are possibly nonzero. From the definition of the concept of a basis one immediately derives that f has to be an isomorphism. □

2.1.8. Proposition. If M is an R-module, resp. a finitely generated module, then there exists a free module, resp. a free finitely generated module, L say, and a submodule $K \subset L$ such that $M \cong L/K$.

Proof. Since M always has a set of generators e.g. take all nonzero elements of M, we may pick such a set $\{x_i,\ i \in I\}$. Consider $L = R^{(I)}$ and define f: $L \to M$, $\sum'_{i\in I} a_i e_i \mapsto \sum'_{i\in I} a_i x_i$. Since the x_i generate M it is clear that f is surjective. Put K = Ker f, then the fundamental isomorphism theorem yields $M \cong L/K$. □

For any two R-modules M and N we let Hom_R (M,N) denote the set of module homomorphisms from M to N. This set is an abelian group with respect to the operation defined by $(f+g)(x) = f(x) + g(x)$, for all $x \in M$, $f,g \in \mathrm{Hom}_R(M,N)$.
If f: $M \to N$ is a module homomorphism and X is an R-module then there is a morphism of abelian groups $f_* : \mathrm{Hom}_R(X,M) \to \mathrm{Hom}_R(X,N)$ defined by $f_*(\alpha) = f \circ \alpha$, $\alpha \in \mathrm{Hom}_R(X,M)$, and also a morphism of abelian groups $f^*: \mathrm{Hom}_R(N,X) \to \mathrm{Hom}_R(M,X)$ defined by $f^*(\beta) = \beta \circ f$, $\beta \in \mathrm{Hom}_R(N,X)$. If we have an exact sequence:

$$0 \longrightarrow M' \xrightarrow[f]{} M \xrightarrow[g]{} M'' \longrightarrow 0$$

then the following sequences are exact:

$$0 \longrightarrow \mathrm{Hom}_R(X,M') \xrightarrow[f_*]{} \mathrm{Hom}_R(X,M) \xrightarrow[g_*]{} \mathrm{Hom}_R(X,M'')$$
$$0 \longrightarrow \mathrm{Hom}_R(M'',Y) \xrightarrow[g^*]{} \mathrm{Hom}_R(M,Y) \xrightarrow[f^*]{} \mathrm{Hom}_R(M',Y).$$

In case N = M, Hom_R (M,N) is a ring with respect to the multiplication, $f.g = g \circ f$ for all $f,g \in \mathrm{Hom}_R(M,N)$; this ring will then be denoted by End_R (M) and it is called the *endomorphism* ring of the module M. We may now view M as a right module over End_R (M) in the following natural way: $x.f = f(x)$ for all $x \in M$ and $f \in \mathrm{End}_R(M)$. We refer to the right End_R (M)-module associated to $_RM$ as the *contramodule* module associated to $_RM$. Observe that for a right R-module M_R the multiplication in $\mathrm{End}_R(M_R)$ is given by $f.g = f \circ g$ for all $f,g \in \mathrm{End}_R(M_R)$.

2.1.9. Example. The abelian group $\mathrm{Hom}_R(_RR, M) \cong M$, because $\varphi: M \to$

$\mathrm{Hom}_R\ ({}_RR, M)$, $m \mapsto \varphi(m)$, $\varphi(m)(x) = xm$ for all $x \in R$, $m \in M$, is easily seen to be a group isomorphism. When $M = {}_RR$, φ becomes an isomorphism of rings $R \cong \mathrm{End}_R\ ({}_RR)$.

Let φ: $R \to S$ be a ring morphism (with $\varphi(1) = 1$ as always). A left S-module M may be considered as a left R-module by "restriction of scalars" i.e.
$a.x = \varphi(a)$ for $a \in R$, $x \in M$.
It is customary to denote the R-module obtained by restriction of scalars over φ by $\varphi_*(M)$. It is straightforward to verify that any $f \in \mathrm{Hom}_S\ (M,N)$ for S-modules M and N satisfies $f \subset \mathrm{Hom}_R(\varphi_*(M), \varphi_*(N))$.

Now consider an R-module M and a nonempty $X \subset M$; the *left annihilator* of X in R is defined to be the set $l_R(X)$, $l_R(X) = \{r \in R, rx = 0 \text{ for all } x \in X\}$. On the other hand, if $Y \subset R$ is nonempty then $r_M(Y) = \{x \in M, rx = 0 \text{ for all } r \in Y\}$ is the *right annihilator* of Y in M. If $X = \{x\}$ and $Y = \{a\}$ then $l_R(\{x\})$ and $l_R(\{a\})$ will simply be written as $l_R(x)$, resp. $r_M(a)$. If X is a submodule of M then $l_R(X)$ is an ideal of R. In particular, the ideal $l_R(M) = \mathrm{Ann}_R(M)$ is defined to be the *annihilator of the module* M. On the other hand, $r_M(Y)$ is a submodule in the contramodule associated to M. The module M is said to be faithful if $\mathrm{Ann}_R(M) = 0$. Observe that $r_M(Y) = \bigcap_{a\in Y} r_M(a)$.

2.1.10. Proposition. Let X,Y be nonempty subsets of the R-module M and let I,J be nonempty subsets of R, then:

1. $X \subset Y$ implies $l_R(Y) \subset l_R(X)$; $I \subset J$ implies $r_M(J) \subset r_M(I)$.
2. $X \subset r_M(l_R(X))$ and $I \subset l_R(r_M(I))$.
3. $l_R(X) = l_R(r_M(l_R(X)))$ and $r_M(I) = r_M(l_R(r_M(I)))$.

Proof. Only 3. needs a proof. In view of 2. we have: $l_R(r_M(l_R(X))) \subset l_R(X)$. Take $a \in l_R(X)$ and $x \in r_M(l_R(X))$. From $ax = 0$, $x \in l_R(r_M(l_R(X)))$ and the inclusion $l_R(X) \subset l_R(r_M(l_R(X)))$ follows.
The second equality in 3. may be derived in a similar manner. □

Consider a submodule N of M and let π: $M \to M/N$ be the canonical epimorphism. Write $l_R(\pi(X)) = (N{:}X)$ and $(N{:}Y) = \pi^{-1}(r_{M/N}(Y))$, then it is immediately

seen that
$(N{:}X) = \{r \in R,\ rx \in N \text{ for all } x \in X\}$.
$(N{:}Y) = \{X \in M,\ yx \in N \text{ for all } y \in Y\}$.
For $X = \{x\}$, $Y = \{y\}$ we write simply $(N{:}X) = (N{:}x)$, $(N{:}Y) = (N{:}y)$.

In conclusion of this section we introduce bimodules.
Let R and S be rings. An abelian group M (written additively) which happens to be a left R-module and also a right S-module such that $(ax)b = a(xb)$ for all $a \in R$, $b \in S$, $x \in M$, is said to be an *R-S-bimodule*.
We may now define the notions of subbimodule, factorbimodule, homomorphism of bimodules, isomorphism of bimodules; isomorphism theorems for bimodules hold and one may also define exact sequences, direct products and direct sums of bimodules. All of this is formally similar to the module theory developed before.
Perhaps the most obvious example of a bimodule is obtained by considering a ring R as an R-R-bimodule (R is associative !); this bimodule will be denoted by ${}_RR_R$, the subbimodules of ${}_RR_R$ are the ideals of R.

§2.2 The Lattice of Submodules of a Module

Look at a family L_i, $i \in I$, of submodules of a left R-module M. The *sum* of the L_i, $i \in I$, is the submodule of M given as $\sum_{i\in I} L_i = \{\sum'_{i\in I} x_i,\ x_i \in L_i$ and $\sum'$ denoting a finite sum $\}$. When I is a finite set, $\{1, \dots, n\}$ say, then we also write $L_1 + \dots + L_n$ for the sum of the family $\{L_i, i \in I\}$. So $\{x_i, i \in I\}$ is a set of generators for M if and only if $M = \sum_{i\in I} Rx_i$.
We will write $L_R(M)$ (or just L(M) if this cannot lead to confusion) for the set of all submodules of M.

2.2.1. Proposition. The set L(M), partially ordered by inclusion, is a modular complete lattice which is upper-continuous and compactly generated (and it has 0 and 1).

Proof. We have 0 for the zero module and 1 for M. For any family $\{L_i, i \in I\}$ we have $\bigwedge_{i\in I} L_i = \bigcap_{i\in I} L_i$, $\bigvee_{i\in I} L_i = \sum_{i\in I} L_i$. In order to establish modularity, consider submodules L, K, N of M such that $K \subset L$, we have to show that $L \cap (K+N) =$

K + (L∩N). We only have to prove L ∩ (K+N) ⊂ K + (L∩N), so consider

x ∈ L ∩ (K+N). Then x ∈ L and x = y+z with y ∈ K and z ∈ N. From x = x-y and y ∈ L we derive that z ∈ L hence z ∈ L ∩ N. Because y ∈ K we also have x = y+z ∈ K + (L∩N). In order to establish upper-continuity consider $(L_i)_{i\in I}$; a directed family of elements of L(M), and look at K ∈ L(M). For any i,j ∈ I there exists k ∈ I such that $L_i \subset L_k$, $L_j \subset L_k$, therefore $\sum_{i\in I} L_i = \bigcup_{i\in I} L_i$ follows, and hence: K ∩ $(\sum_{i\in I} L_i)$ = K ∩ $(\bigcup_{i\in I} L_i) = \bigcup_{i\in I} (K \cap L_i) = \sum_{i\in I} (K \cap L_i)$. The latter establishes upper-continuity of L(M). Now consider a finitely generated submodule N of M and suppose that N ⊂ $\sum_{i\in I} L_i$ where $(L_i)_{i\in I}$ is a directed family of elements of L(M). Again $\sum_{i\in I} L_i = \bigcup_{i\in I} L_i$. Choose $x_1, ..., x_n$ ∈ N such that N = Rx_1 + ... + Rx_n. For each x_k there exists an i_k *in* I such that $x_k \in L_{i_k}$ and since the family $(L_i)_{i\in I}$ is directed we find some i_0 such that $L_{i_k} \subset L_{i_0}$ for all k, i.e.

N ⊂ L_{i_0}.

We proved that N is a compact element in L(M). Finally, the fact that any module is the union of all its finitely generated submodules entails that L(M) is compactly generated. □

2.2.2. Remark. The lattice L(M) is not necessarily lower-continuous. For example put R = $\mathbb{Z}$ and consider the family of ideals $2^n\mathbb{Z}$, n ≥ 1. From $2^n\mathbb{Z} + 3\mathbb{Z} = \mathbb{Z}$ we derive that $\bigcap_{n\geq 1} (2^n\mathbb{Z} + 3\mathbb{Z}) = \mathbb{Z}$ whilst, on the other hand, $\bigcap_{n\geq 1} 2^n\mathbb{Z} = 0$ or $3\mathbb{Z} + \bigcap_{n\geq 1} 2^n\mathbb{Z} = 3\mathbb{Z}$ and thus $3\mathbb{Z} + \bigcap_{n\geq 1} 2^n\mathbb{Z} \neq \bigcap_{n\geq 1} (2^n\mathbb{Z} + 3\mathbb{Z})$.

2.2.3. Proposition. Consider L,K ∈ L(M) such that K ⊂ L. The interval [K,L] is isomorphic to L(L/K).

Proof. If N ∈ [K,L] define φ: [K,L] → L(L/K), N ↦ N/K. The reader will readily check that φ is indeed a lattice isomorphism. □

2.2.4. Corollary. If M is an R-module, resp. a finitely generated R-module, the there exists a free module, resp. a finitely generated module, L say, together with a submodule K of L such that L(M) ≅ [K,L].

Proof. Envoke Proposition 2.1.8. and apply the above. □

If the family $(N_i)_{i\in I}$ is join-independent then we say that the family is an independent family of submodules of M; in this case we will have that $\sum_{i\in I} N_i \cong \bigoplus_{i\in I} N_i$. Indeed, define f: $\bigoplus_{i\in I} N_i \to \sum_{i\in I} N_i$, $(x_i)_{i\in I} \mapsto \sum_{i\in I} x_i$, where $x_i \in N_i$. Clearly, f is surjective. Moreover if x = $(x_i)_{i\in I}$ ∈ Ker f then $\sum_{i\in I} x_i = 0$ but the independency of the family $(N_i)_{i\in I}$ then leads to $x_i = 0$ for all i ∈ I, or Ker f = 0 and f has to be an isomorphism. Whenever M = $\sum_{i\in I} N_i$ for some independent family then M is said to be the *internal direct sum* of that family and we write M = $\bigoplus_{i\in I} N_i$.A submodule N of M such that N has a complement in L(M) is said to be a *direct summand* of M. Direct summands may be related to idempotent elements in certain rings. An element e ∈ R is said to be *idempotent* if e^2 = e. If e is idempotent then 1-e is idempotent and e(1-e) = (1-e)e = 0.

2.2.5. Proposition. For an R-module M and N ∈ L(M), the following propositions are equivalent:

1. N has a complement in L(M).
2. The canonical inclusion i: N → M has a left inverse F ∈ Hom_R (M,N) such that foi = 1_N.
3. The canonical epimorphism π: M → M/N has a right inverse g ∈ Hom_R (M/N,M) such that πog = $1_{M/N}$.
4. There exits an idempotent element u in End_R (M) such that Im (u) = N.

Proof. 1⇒2. If M = N ⊕ P and x ∈ M is decomposed as x = y+z with y ∈ N, z ∈ P, let f: M → N, x ↦ y. Clearly: foi = 1_N.
2⇒4. If we put u = iof then Im u = Im f = N and u^2 = iofoiof = iof = u or u is an idempotent element of End_R (M).
4⇒1. Given u as in 4. put P = Ker u. If x ∈ M then u^2(x) = u(x) yields u(u(x)-x) = 0 or u(x)-x ∈ P and then x ∈ P + N follows. Where x in P ⋂ N then u(x) and x = u(y) for some y ∈ N, but then 0 = u(x) = u^2(y) = u(y) = x. Consequently P ⋂ N = 0 and P is a complement for N.
1⇒3. From M = N ⊕ P, x ∈ M, x = y+z with y ∈ N, z ∈ P, it is clear that πog= $1_{M/N}$ if we define g: M/N → M, $\overline{x}$ → z.
3⇒1. If we are given a morphism g: M/N → M such that πog = $1_{M/N}$ then we may take P = Im g and there remains no problem in showing that M = N ⊕ P. □

We say that a short exact sequence

$$0 \longrightarrow M' \xrightarrow[f]{} M \xrightarrow[g]{} M'' \longrightarrow 0$$

splits if there exists an isomorphism α: M $\rightarrow$ M' $\oplus$ M" making the following diagram with exact rows commutative:

$$\begin{array}{ccccccccc} 0 & \longrightarrow & M' & \xrightarrow{f} & M & \xrightarrow{g} & M'' & \longrightarrow & 0 \\ & & \downarrow{\scriptstyle 1_{M'}} & & \downarrow{\scriptstyle \alpha} & & \downarrow{\scriptstyle 1_{M''}} & & \\ 0 & \longrightarrow & M' & \xrightarrow[i]{} & M' \oplus M'' & \xrightarrow[p]{} & M'' & \longrightarrow & 0 \end{array}$$

where p and i are the canonical maps. By the foregoing proposition it follows that the short exact sequence 0 $\longrightarrow$ M' $\longrightarrow$ M $\longrightarrow$ M" $\longrightarrow$ 0 is split if and only if there exists a homomorphism u: M $\rightarrow$ M' such that u$\circ$f $= 1_{M'}$ or if and only if there exists a homomorphism v: M" $\rightarrow$ M such that g$\circ$v $= 1_{M''}$.

2.2.6. Remarks. For a ring R, L($_RR$), resp. L(R_R), is the lattice of left ideals,resp. right ideals, of R. We denote these lattices by L_l(R),rsp. L_r(R).
If R and S are two rings and $_RM_S$ is an R-S-bimodule, then we can consider the lattice L($_RM_S$) of submodules of $_RM_S$. Exactly like in the module case we prove that the lattice L($_RM_S$) is a complete, modular, upper-continuous and compactly generated lattice. More specificely, if $(N_i)_{i\in I}$ is a family in L($_RM_S$) then $\bigvee_{i\in I} N_i = \sum_{i\in I} N_i$, $\bigwedge_{i\in I} N_i = \bigcap_{i\in I} N_i$. The compact elements of this lattice are the finitely generated submodules (N $\in$ L($_RM_S$) is finitely generated if there exists $x_1, ..., x_n \in$ N such that N $= \sum_{i=1}^{n} Rx_iS$). In particular we consider the lattice of ideals of R, L($_RR_R$) and we will write L(R) for this lattice when this cannot lead to ambiguity. The lattice L(R) is a sublattice of L_l(R) and of L_r(R).

We now proceed in the theory of modules following the path set out in Chapter 1 on lattices, in particular we first study Noetherian and Artinian modules and modules of finite length.

§2.3 Noetherian and Artinian Modules

An R-module M is said to be *(left) Noetherian*, resp. *(left) Artinian* if L(M) is Noetherian, resp. Artinian. Directly from the results of Section 1.2 we now derive:

2.3.1. Proposition. For an R-module M, the following statements are equivalent:

1. M is Noetherian, resp. Artinian.
2. Any ascending, resp. descending, chain of submodules of M is stationary.
3. Submodules of M are finitely generated, resp. for any family $(N_i)_{i\in I}$ of submodules of M there exists a finite J in I such that $\bigcap_{i\in I} N_i = \bigcap_{j\in J} N_j$.

2.3.2. Proposition. Consider a short exact sequence

$$0 \longrightarrow M' \xrightarrow[f]{} M \xrightarrow[g]{} M" \longrightarrow 0$$

of R-modules. Then M is Noetherian, resp. Artinian, if and only if M' and M" are Noetherian resp. Artinian.

Proof. Clearly M' ≅ Im f, M" ≅ M/Im f and L(M") ≅ [Im f, M], then apply Proposition 1.2.3. □

2.3.3. Proposition. Consider a finite family M_i, i = 1,...,n of R-modules. Then $M_1 \oplus ... \oplus M_n$ is Noetherian, resp. Artinian, if and only if each M_i is Noetherian, resp. Artinian.

Proof. Using induction on n we may restrict attention to the case n = 2. Then we have the short exact sequence

$$0 \longrightarrow M_1 \longrightarrow M_1 \oplus M_2 \longrightarrow M_2 \longrightarrow 0$$

and the result is again a consequence of Proposition 1.3.2.

2.3.4. Remark. An infinite direct sum of nonzero modules is not Noetherian, not Artinian.

If an R-module M and N ∈ L(M) is irreducible, resp.: uniforn, indecomposable, essential, a direct factor, then N is called an *irreducible*, resp. *uniform*, *indecomposable*, *essential*, a *direct summand*, submodule of M.

2.3.5. Proposition. Let M be a Noetherian or an Artinian R-module.

1. M has finite Goldie dimension.
2. Each submodule of M is a finite intersection (reduced in the Artinian case) of irreducible submodules.
3. M is a finite direct sum of indecomposable submodules.

Proof. For 1. and 2., apply Theorem 1.5.5. and for 3. apply Proposition 1.10.1. □

2.3.6. Proposition. For a ring R the following statements are equivalent:

1. R is left Noetherian resp. left Artinian.
2. A finitely generated R-module is Noetherian, resp. Artinian.

Proof. 2⇒1. Obvious because ${}_RR$ is finitely generated.
1⇒2. Apply Proposition 2.3.2. and Proposition 2.3.3. □

§2.4 Modules of Finite Length

We say that a module $M \neq 0$ is *simple* if M is an atom in the lattice L(M). A propre submodule N of M is said to be *maximal* whenever N is maximal element of L(M) - {M}.
From Proposition 2.2.3. it follows that N is maximal in M if and only if M/N is a simple R-module.

2.4.1. Lemma. For a nonzero R-module M, the following assertions are equivalent:

1. M is simple.
2. For all nonzero x in M, $Rx = M$.
3. $M \cong R/W$ where W is a maximal left ideal of R.

Proof. 1⇒2. Evident.

2⇒1. Let N be a submodule of M. If N ≠ 0 then pick x ≠ 0 in N.
From M = Rx ⊂ N we obtain N = M, so M is simple.
1⇔3. By the fundamental isomorphism theorem 1⇒3 follows, the converse is trivial
□

For an R-module M we say that a chain of submodules $0 = M_0 \subset M_1 \subset ... \subset M_n = M$ is a *composition sequence* if M_i/M_{i-1} is simple for i = 1,...,n. The number n as above is called the *length* of the module M and it is denoted by l(M) = n.

2.4.2. Theorem (Jordan-Hölder-Dedekind). If the R-module M allows two composition series, say $0 = M_0 \subset M_1 \subset ... \subset M_n = M$, $0 = N_0 \subset N_1 \subset ... \subset N_n = M$, then m = n and there exists a permutation ε of {1,2,..,n} such that $M_i/M_{i-1} \cong N_{\varepsilon(i)}/N_{\varepsilon(i)-1}$ for all i = 1,2,...,n.

Proof. Corollary 1.3.4. yields m = n and there exists a permutation ε such that the intervals $[M_{i-1}, M_i]$ and $[N_{\varepsilon(i)-1}, N_{\varepsilon(i)}]$ are projective in L(M). Using Noether's second isomorphism theorem, $M_i/M_{i-1} \cong N_{\varepsilon(i)}/N_{\varepsilon(i)-1}$.

2.4.3. Corollary. M is of finite length if and only if it is Noetherian and Artinian.

Proof. Apply Corollary 1.3.6. □

2.4.4. Corollaries. 1. Let N be a submodule of M then M has finite length if and only if N and M/N are of finite length. Moreover: l(M) = l(N) + l(M/N).
2. Consider a ring isomorphism φ: R → S and an S-module M. Then ${}_SM$ is Noetherian, resp. Artinian, resp. of finite length, if and only if $\varphi_*(M)$ is Noetherian resp. of finite length.

Proof. 1. Apply Corollary 1.3.7. and Proposition 2.2.1.
2.The map X ↦ $\varphi_*(X)$ defines a lattice isomorphism $L(\varphi)$, $L(\varphi)$: $L_s(M) \to L_R(\varphi_*(M)$. □

§2.5 Semisimple Modules

An R-module is said to be *semisimple* if the lattice L(M) is semiatomic. By Theorem 1.8.2. and 1.8.7. we obtain:

2.5.1. Proposition. The following assertions are equivalent:

1. The R-module M is semisimple.
2. M is a direct sum of simple modules.
3. Any submodule of M is a direct summand.
4. M contains no essential submodules different from M.

2.5.2. Corollary. For a vectorspace V over K we have:

1. V is a free K-module.
2. The lattice L(V) is complemented.

Proof. Put $V = \sum_{x \in V}$ Kx. For $x \neq 0$ we have $Kx \cong K$ and Kx is a simple K-module, henc L(V) is semiatomic. The foregoing proposition then finishes the proof. □

A further consequence of Corollary 1.8.4. is:

2.5.3. Proposition. If M is a semisimple R-module then each submodule N of M is semisimple too; moreover, M/N is semisimple as well.

Let Ω denote the set of isomorphism classes of simple R-modules. If M is a semisimple R-module and $\omega \in \Omega$ then we let M_ω be the sum of all simple submodules of M which are in the class ω. We say that M_ω is the *isotypic ω-component* of the module M; a semisimple module M is called *isotypic* if there exists a $\omega \in \Omega$ such that $M = M_\omega$.

2.5.4. Proposition. If M is a semisimple R-module then $M = \bigoplus_{w \in \Omega} M_\omega$.

Proof. We have to establish: $M_\omega \cap (\sum_{\omega\neq\omega} M_{\omega'}) = 0$.
If x ∈ $M_\omega \cap (\sum_{\omega\neq\omega} M_{\omega'})$ is nonzero then Rx ⊂ M_ω and Rx ⊂ $\sum_{\omega\neq\omega} M_{\omega'}$ imply that there exists a simple R-module S in the class ω, together with simple R-modules $S_1, ..., S_n$ in some class $\omega' \neq \omega$, such that $S \subset S_1 \oplus S_2 ... \oplus S_n$. By induction on n we may conclude that S ≅ S_i for some i ∈ {1,...,n}, a contradiction. □

The socle of L(M) will also be called the *socle of M*, it will be denoted by s(M). From the results of Section 1.9. it follows that s(M) is the sum of all simple submodules of M. In particular (cf. Lemma 1.9.1.) s(M) equals the intersection of all essential submodules of M. The Loewy series of L(M) is now called the *Loewy series of the module* M and it will be denoted by: $(s_\alpha(M))_\alpha$, α an ordinal. The length of the Loewy series will be denoted by λ(M).

2.5.5. Proposition. Consider R-modules M and N and f ∈ Hom_R(M,N).

1. f(s(M)) ⊂ s(N).
2. $f(s(M)_\omega) \subset s(N)_\omega$ for all $\omega \in \Omega$.
3. $s({}_RR)$ and $s({}_RR)_\omega$ are ideals, for all $\omega \in \Omega$.

Proof. From Proposition 2.5.3., 1. and 2. follows.
3. That $s({}_RR)$ is a left ideal is clear. Take a ∈ R and consider the R-linear map φ_a: ${}_RR \rightarrow {}_RR$, x ↦ xa. From 1. we conlude: $\varphi_a(s({}_RR)) \subset s({}_RR)$, thus $s({}_RR)a \subset s({}_RR)$ and it is now clear that $s({}_RR)$ is an ideal of R. From 2. we conclude that $s({}_RR)_w$ is an ideal for all w ∈ Ω. □

If the lattice L(M) is semiartinian then M is said to be *semiartinian* i.e. for any submodule N of M, N ≠ M, M/N contains a simple module.

2.5.6. Proposition. 1. M is semiartinian if and only if N and M/N are semiartinian for each R-submodule N of M.
2. A direct sum of semiartinian modules is a semiartinian module.
3. M is semiartinian if and only if there is an ordinal α such that M = s_α(M).

Proof. 1. cf. Proposition 1.9.3., Proposition 2.2.2.

2. cf. Proposition 1.9.4.
3. cf. Corollary 1.9.5. □

The ring R is called *left semiartinian* if the R-module $_RR$ is semiartinian It is now a direct consequence of the foregoing proposition that R is left semiartinian if and only if any left R-module is semiartinian.

§2.6 Semisimple and Simple Artinian Rings

A ring R is said to be *semisimple* if $_RR$ is semisimple.

2.6.1. Proposition. The following statements are equivalent:

1. R is (left) semisimple.
2. Each R-module is semisimple.
3. Each left ideal of R is a direct summand of R.
4. Each essential left ideal of R coincides with R.

Proof. 1⇒2. follows from Proposition 2.5.3. and 2.5.1. and by Proposition 2.1.8. 2⇒3⇒4⇒1 follow from Proposition 2.5.1. □

If R is a semisimple ring then we may write R $\cong \bigoplus_{i\in F} S_i$ where the S_i are simple R-modules. Because R is Artinian, F is a finite set. By the Jordan-Hölder theorem $|F| = l(_RR)$.

Any simple R-module S is isomorphic to R/I where I is a maximal left ideal of R. If R is semisimple, I is a direct summand of R, hence there exists $i_0 \in F$ such that R $= S_{i_0} \oplus$ I and so S$\cong S_{i_0}$. We may conclude that Ω is finite.

2.6.2. Proposition. The following statements are equivalent:

1. The ring R is semisimple and Ω is a singleton

2. The ring R is semisimple and 0 and R are the only ideals of R.
3. The ring R is left Artinian and 0 and R are the only ideals of R.

Proof. 1⇒2. We may write R = $S_1 \oplus ... \oplus S_n$, where $S_1, ..., S_n$, are minimal ideals of R. If I is a nonzero ideal of R then I contains some minimal ideal S of R. Because Ω consists of a unique element, S $\cong$ S_i for all i = 1,...,n; let f_i: S $\to$ S_i be an isomorphism. Now S is a direct summand of R hence there may be found a $g_i \in \mathrm{End}_R({}_RR)$ extending f_i, i = 1,...,n. Hence $g_i(S) = f_i(S) = S_i$. On the other hand, there exists an $a_i \in$ R such that $g_i(x) = xa_i$ for all x $\in$ R. Consequently $Sa_i = S_i$. Since S $\subset$ I and I being an ideal, $S_i \subset$ I follows for all i = 1,...,n. Therefore I = R and 2. follows.
2⇒3. Trivial.
3⇒1. Since R is left Artinian, $s({}_RR) \neq 0$. By 3. it follows that R = $s({}_RR)$ and hence R is semisimple. By Proposition 2.5.5.b. it follows that Ω is a singleton. □

A ring R having no nontrivial ideals is said to be a *simple ring*. If R satisfies one of the conditions of Proposition 2.6.2. then R is called an *Artinian simple* ring.

2.6.3. Lemma (Schur). If S is simple R-module then the ring $\mathrm{End}_R(S)$ is a field (not necessarily commutative).

Proof. Pick $\alpha \in \mathrm{End}_R(S)$, $\alpha \neq 0$. Then Ker $\alpha = 0$ and Im α = S. Hence α is an isomorphism and there exists an $\alpha^{-1} \in \mathrm{End}_R(S)$. □

2.6.4. Lemma. Let M be an R-module. For any number n $\geq$ 1 we have a ring isomorphism: $\mathrm{End}_R(M^n) \cong M_n(\mathrm{End}_R(M))$.

Proof. Consider the canonical morphisms $i_k : M \to M^n$, $p_k : M^n \to M$ for k = 1,...,n, then $1_{M^n} = \sum_{k=1}^n i_k \circ p_k$.
Define $\varphi : \mathrm{End}_R(M^n) \to M_n(\mathrm{End}_R(M))$, $\alpha \mapsto (\alpha_{ij})$, where $\alpha_{ij} = p_j \circ \alpha \circ i_i$. Let $\alpha, \beta \in \mathrm{End}_R(M^n)$ and write $\nu = \beta \circ \alpha$. Then $\varphi(\alpha\beta) = \varphi(\beta \circ \alpha) = (\nu_{ij})$ where $\nu_{ij} = p_j \circ \beta \circ \alpha \circ i_i$, i,j = 1,...,n. On the other hand:
$\varphi(\alpha)\varphi(\beta) = \sum_{k=1}^n \alpha_{ik}\beta_{kj} = \sum_{k=1}^n \beta_{kj} \circ \alpha_{jk} = \sum_{k=1}^n p_j \circ \beta \circ c_k \circ p_k \circ \alpha \circ i_i$

$= p_j \circ \beta \circ \sum_{k=1}^{n} i_k \circ p_k \circ \alpha \circ i_i = p_j \circ \beta \circ \alpha \circ i_i$.
This shows that $\varphi(\alpha\beta) = \varphi(\alpha)\varphi(\beta)$ and hence φ is a ring homomorphism. If $\varphi(\alpha) = 0$ then $\alpha_{ij} = 0$ for all ij = 1,...,n and hence $i_k \circ \alpha_{jk} = 0$ for all k= 1,..n. Consequently $\sum_{k=1}^{n} i_k \circ \alpha_{ik} = 0$ or:
$o = \sum_{k=1}^{n} i_k \circ p_k \circ \alpha \circ i_i = (\sum_{k=1}^{n} i_k \circ p_k) \circ \alpha \circ i_i = \alpha \circ i_i$.
Hence, $\alpha \circ i_k = 0$ for all k $\in$ {1..,n} and we obtain that $0 = \sum_{k=21}^{n} \alpha \circ i_k \circ p_k = \alpha \circ \sum_{k=1}^{n} i_k \circ p_k = \alpha$ provided that φ is injective.
The universal property of the direct sum and product entails that φ is surjective as well. □

2.6.5. Corollary. If n $\geq$ 1 is a natural number then $M_n(R) \cong \mathrm{End}_R({}_RR^n)$.

2.6.6. Lemma. If D is a field and V is a D-vector space of dimension n then $\mathrm{End}_D(V) \cong M_n(D)$ is left and right Artinian.

Proof. A direct proof is possible but we prefer to refer to Corollary 2.10.6.

2.6.7. Theorem (Wedderburn). Let R be a simple Artinian ring with $l({}_RR) =$ n. If S is a simple module, let D = $\mathrm{End}_R(S)$. Then R = $M_n(D)$.

Proof. From $l({}_RR)$ = n it follows that ${}_RR \cong S^n$ and hence R $\cong$ $\mathrm{End}_R({}_RR) \cong \mathrm{End}_R(S^n)$. By Lemma 2.6.4. we obtain that $\mathrm{End}_R(S^n) \cong M_n(D)$ and the theorem follows. □

2.6.8. Corollary (Wedderburn-Artin). Let R be a semisimple ring. There exist fields $D_1, ..., D_s$ and $n_1, n_2 ..., n_s \geq 1$ such that R $\cong M_{n_1}(D_1) \times .. \times M_{n_s}(D_s)$.

Proof. There are simple nonisomorphic R-modules $S_1.., S_s$ such that: ${}_RR \cong S_1^{n_1} \oplus S_2^{n_2} \oplus ... \oplus S_s^{n_s}$ = M. Writing D_i for $\mathrm{End}_R(S_i)$ we infer from Proposition 2.5.5.2. that $\mathrm{End}_R(M) = \mathrm{End}_R(S_1^{n_1}) \times ... \times \mathrm{End}_R(S_s^{n_s}) \cong M_{n_1}(D_1) \times .. \times M_{n_s}(D_s)$
Since R $\cong \mathrm{End}_R(M)$ the claim follows. □

2.6.9. Corollary. If R is left semisimple then the opposite R^{o} is a semisimple ring.

Proof. Direct from Lemma 2.6.6. and Corollary 2.6.8. □

§2.7 The Jacobson Radical and the Prime Radical of a Ring

Let M be a R-module. The radical of $L_R(M)$ (cf. Chapter 1) is now called the *Jacobson radical of M* and it is denoted by $J_R(M)$ (sometimes simply J(M)). By definition J(M) is the intersection of all maximal submodules of M and if M does not contain a maximal submodule different from M we put J(M) = M.
The definition states that $J(M) = \bigcap \{\text{Ker } f, f \in \text{Hom}_R(M,S)$, for S a simple R-module$\}$. The following properties of the Jacobson radical follow directly from the lattice theory in Section 1.8.

Properties. 1. If M is finitely generated then $J(M) \neq M$. 2. $J(M/J(M)) = 0$. 3. If M is semisimple then $J(M) = 0$. 4. If M is finitely generated and N is a submodule of M such that N+J(M)=M then M = N.

2.7.1. Proposition. If $f \in \text{Hom}_R(M,N)$ then $f(J(M)) \subset J(N)$.

Proof. Consider an R-linear map g: $N \to S$ where S is a simple R-module. If $x \in J(M)$ then $x \in \text{Ker } g \circ f$, hence $g(f(x)) = 0$ or $f(J(M)) \subset J(N)$. □

The Jacobson radical of $_RR$ is called *Jacobson radical of R* and we denote it by J(R). Since right multiplication by some $a \in R$ defines an R-linear map
φ_a: $_RR \to {}_RR$, we derive from the foregoing that $\varphi_a(J(R)) \subset J(R)$ i.e.
$J(R)a \subset J(R)$ and this proves that J(R) is an ideal of R.

2.7.2. Proposition. $J(R) = \{a \in R$, 1-xa is invertible in R for all $x \in R\}$

Proof. Pick a $\in$ J(R) and x $\in$ R. Obviously 1-xa cannot be in a left maximal ideal, hence R(1-xa) = R or 1 = μ(1-xa) for some $\mu \in$ R. Now μ= 1 + μxa with μxa $\in$ J(R) yields that 1 = μ'(1+μxa) for some μ' $\in$ R. From μ(1-xa) = 1 and μ'μ = 1 it follows that 1-xa = μ' and thus also (1-xa)μ = 1, proving that 1-xa has (left and right) inverse μ.
Conversely, if 1-xa is invertible for all x $\in$ R and a $\notin$ J(R) then a $\notin$ ω for some maximal left ideal ω of R. Hence, Ra + ω = R yields 1 = ra + s where r $\in$ R, s $\in \omega$. Then s = 1-ra is invertible by assumption but this implies ω = R, a contradiction. □

2.7.3. Proposition. J(R) is the largest ideal such that for a $\in$ J(R), 1-a is invertible.

Proof. If a $\in$ J(R) then 1-a is invertible in view of the foregoing. Let I $\neq$ R be an ideal with the property that for all b $\in$ I, 1-b is invertible. Were I not contained in J(R) then I + ω = R for some maximal left ideal ω of R. Write 1 = a+x with a $\in$ I, x $\in$ ω. Then x = 1-a must be invertible and consequently ω = R a contradiction. □

2.7.4. Corollary J(R) = J(R°).

2.7.5. Proposition. For an R-module M, J(R)M $\subset$ J(M).

Proof. Let f: M $\to$ S be an R-linear map to a simple R-module. If a $\in$ J(R) and x $\in$ M then f(ax) = af(x) = 0, hence ax $\in$ J(M) and J(R)M $\subset$ J(M). □

2.7.6. Corollary (Nakayama's lemma) Let M be a finitely generated module and N an arbitrary submodule of M. If N + J(R)M = M then N = M.

Proof. Follows from Proposition 2.7.5. and the property 4 of J(M). □

Consider the lattice L(R) of ideals of R. The radical of this lattice is the *simplectic radical* of R, denoted by $J_s(R)$. By definition $J_s(R)$ is the intersection of the maximal ideals ($\neq$R).

Note that an ideal I is maximal if and only if R/I is simple.

2.7.7. Proposition. $J(R) \subset J_s(R)$.

Proof. Let $I \neq R$ be a maximal ideal of R and let π: $R \to R/I$ be the canonical epimorphism. If J(R) is not contained in I then $\pi(J(R))$ is a nonzero ideal of R/I. The latter ring is simple, hence $\pi(J(R)) = R/I$ and this implies $I + J(R) = R$. Nakayama's lemma then entails $I = R$, a contradiction. Therefore we must have $J(R) \subset I$ and thus $J(R) \subset J_s(R)$. □

2.7.8. Remark. In case R is commutative, $J(R) = J_s(R)$.

A ring R with a unique maximal left (right) ideal is called a *local* ring. It is easily verified that the ring R is local if and only if R/J(R) is a field and in this case $J_s(R) = J(R)$.

Another very important radical is obtained by studying prime ideals of a ring rather than maximal ideals and then it will be related to nilpotent elements and nilideals. An $a \in R$ is *nilpotent* if $a^n = 0$ for some positive integer n; a left ideal such that each of its elements is nilpotent is said to be *nil* or to be a *nilideal.*

2.7.9. Proposition. If I is a left nilideal of R then $I \subset J(R)$.

Proof. Pick $a \in I$ and $x \in R$.Since $y = xa \in I$ we have $(xa)^n = 0$ for some n From $(1-y)(1+y+...+y^{n-1}) = 1-y^n = 1$, we retain that 1-xa is invertible in R. By Proposition 2.7.2, $I \subset J(R)$. □

An ideal P of R, $P \neq R$, is said to be *prime* if $IJ \subset P$ entails $I \subset P$ or $J \subset P$ for any two ideals I and J of R. An ideal P is prime if and only if for all $a,b \in R$ such that $aRb \subset P$ we have either $a \in P$ or $b \in P$. The set of all prime ideals of R is denoted by Spec(R).
By rad(R) we denote $\bigcap \{P \; P \in Spec(R)\}$. This ideal is called the *prime radical* of

the ring R (sometimes also called the *nilradical*). An a$\in$ R is strongly nilpotent if for any sequence a_0 , a_1 , ... , a_m , ... in R, such that a = a_0 , ... , $a_{n+1} \in a_n R a_n$, ... , there exists a positive integer k such that $a_k = 0$.

2.7.10. Theorem. In a ring R, rad(R) coincides with the set of all strongly nilpotent elements of the ring.

Proof. If a $\in$ R is strongly nilpotent and a $\notin$ P for some prime ideal P of R then aRa $\not\subset$ P hence there is an $a_1 \in$ aRa such that $a_1 \notin$ P. Put a = a_0, repeat the foregoing argument with a_1 and so on untill we have obtained the sequence a_0 , a_1 , a_2 , ..., with $a_{n+1} \notin$ P for all n $\geq$ 0. Since a is strongly nilpotent we arrive at $a_k = 0$ for some k but this leads to a contradiction. Hence a $\in$ P for all prime ideals P of R. Conversely, consider a $\in$ rad(R) and suppose that a is not strongly nilpotent. Then there exists a sequence a_0 , a_1 , .. , a_n , . , with a_0 = a, $a_{+1} \in a_n R a_n$ and $a_n \neq 0$ for any n $\geq$ 0. Put S = $\{a_n,\ n \geq 0\}$. Hence 0 $\notin$ S. By Zorn's lemma there exists an ideal P, maximal with the property P $\cap$ S = ϕ, P $\neq$ R.
If I and J are ideals not contained in P then (I+P) $\cap$ S $\neq \phi$ and (J+P) $\cap$ S $\neq \phi$. Pick $a_i \in$ I+P, $a_j \in$ J+P; suppose i $\leq$ j. From $a_j \in$ I+P we derive $a_{j+1} \in a_j R a_j \subset$ (I+P)(J+P) $\subset$ IJ + P and from $a_{j+1} \notin$ P it follows that IJ $\not\subset$ P and hence P is a prime ideal. □

2.7.11. Corollary. rad(R) is a nilideal.

Proof. Pick a $\in$ rad(R) and consider the sequence $a_0 = a$, ... , $a_n = a^{2^n}$, ... Obviously, $a_{n+1} \in a_n R a_n$. Since a is strongly nilpotent there exists a k $\geq$0 such that $a_k = 0$, hence $a^{2^k} = 0$ or a is nilpotent. □

2.7.12. Corollary. rad(R) $\subset$ J(R).

Proof. By the foregoing corollary and Proposition 2.7.9. □

We call R a *prime ring* if the zero ideal is a prime ideal. Clearly, if P is a prime

ideal then R/P is a prime ring.
We say that R is a *semiprime* ring if rad(R) = 0; it is clear that R/rad(R) is semiprime

2.7.13. Proposition The following statements are equivalent:

1. R is semiprime.
2. If aRa = 0 for all a ∈ R then a = 0.
3. There are no nonzero nilpotent left ideals in R.

Proof. 1⇒3. Suppose that I is a left ideal of R such that $I^n = 0$. Since $I^n \subset$ P for every P ∈ Spec(R) it follows that I ⊂ P for every P ∈ Spec(R) hence I ⊂ rad(R) = 0.
3⇒2. If aRa = 0 for some a ∈ R then $(RaR)^2 = 0$ and hence RaR = 0 and consequently a = 0.
2⇒1. Suppose a ∈ rad(R) and a ≠ 0, then aRa ≠ 0 and we may select an $a_1 \neq 0$ in aRa such that $a_1Ra_1 \neq 0$. By repetition of this argument we obtain a sequence $a_0 = a$, a_1 , .. a_n , ..., such that $a_{n+1} \in a_nRa_n$ and such that $a_n \neq 0$ for all n ≥ 0. It follows that a is not strongly nilpotent, contradiction. □

2.7.14. Corollary. If the lattice L(R) of ideals of R is Noetherian then rad(R) is nilpotent.

Proof. Let I be a maximal nilpotent ideal of R. Obviously I ⊂ rad(R). By the foregoing proposition it follows that R/I is a semiprime ring. Evidently rad(R) ⊂ I and rad(R) = I follows. □

§2.8 Rings of Fractions. Goldie's Theorems.

A nonempty subset S of R is said to be a *multiplicatively closed set* (or *multiplicative system*) if 1 ∈ S and for all s,s' ∈ S we also have ss' ∈ S.

2.8.1. Definition. A *left ring of fractions of R with respect to S* is a ring B together with a ring homomorphism φ: R $\rightarrow$ B such that the pair (B,φ) satisfies the following conditions:

1. For all s $\in$ S, $\varphi(s)$ is invertible in B
2. For every b $\in$B, there is an a $\in$ R and an s $\in$ S such that b = $\varphi(s)^{-1}\varphi(a)$.
3. If $\varphi(a) = 0$ for a $\in$ R then sa = 0 for some s $\in$ S

2.8.2. Proposition (Universal Property of the Ring of Fractions).
Let (B,φ) be a left ring of fractions of R relative to S and let Ψ: R $\rightarrow$ T be any homomorphism of rings such that $\Psi(s)$ is invertible for all s $\in$ S There exists a unique homomorphism of rings $\overline{\Psi}$: B $\rightarrow$ T such that $\overline{\Psi} \circ \varphi = \Psi$

Proof. Take b $\in$ B, say b $= \varphi(s)^{-1}\ \varphi(a)$ for a $\in$ R, s $\in$ S. Define $\overline{\Psi}(b) = \Psi(s)^{-1}\ \Psi(a)$ and let us verify that $\overline{\Psi}$ is well defined.
If b $= \varphi(s')^{-1}\ \varphi(a')$ then $\varphi(a) = \varphi(s)\ \varphi(s')^{-1}\ \varphi(a') = \varphi(s'')^{-1}\ \varphi(a'')\ \varphi(a')$ where s" $\in$ S, a" $\in$ R. But we have: $\varphi(s'')\ \varphi(s) = \varphi(a'')\ \varphi(s')$ and $\varphi(s'')\ \varphi(a) = \varphi(a'')\ \varphi(a')$ and thus: $\varphi(s''s - a''s') = 0$ and $\varphi(s''a - a''a') = 0$. Therefore there exists $t_1, t_2 \in$ S such that $t_1\ (s''s - a''s') = 0$ and $t_2\ (s''a - a''a') = 0$.

Applying Ψ to these equalties and taking into account that $\Psi(t_1)$ and $\Psi(t_2)$ are invertible it follows that: $\Psi(s'')\Psi(s) = \Psi(a'')\Psi(s')$ and $\Psi(s'')\Psi(a) = \Psi(a'')\Psi(a')$. Substituting $\Psi(a'') = \Psi(s'')\Psi(s)\ \Psi(s')^{-1}$ in the latter equality we obtain
$\Psi(s'')\ \Psi(s)^{-1}\ \overline{\Psi}(a) = \Psi(s')^{-1}\ \Psi(a')$. From this it is easily derived that $\overline{\Psi}$ is the unique map with the property that $\overline{\Psi} \circ \varphi = \Psi$. □

From the above proposition we conclude that if a left ring of fractions exists, then it is unique up to isomorphism and we denote this "unique" ring of left fractions by S^{-1}R.
The map φ: R $\rightarrow$ S^{-1}R is called the *canonical morphism.*

2.8.3. Corollary. If S^{-1}R and RS^{-1} exist (the latter being defined as S^{-1}R by left-right symmetry) then they are canonically isomorphic.

2.8.4. Theorem. Let S be a multiplicative system in R, then the following conditions are equivalent:

1. The left ring of fractions $S^{-1}R$ exists.
2. S satisfies the following conditions:

* For all a $\in$ R, s $\in$ S there exist t $\in$ S, b $\in$ R such that ta = bs (then S is said to be *left permutable*

** For all a $\in$ R, if s $\in$ S is such that as = 0 then there exists a t $\in$ S such that ta = 0 (S is said to be *left reversible*).

Proof. 1⇒2. The element $\varphi(a)\ \varphi(s)^{-1} \in S^{-1}$R is of the form $\varphi(t)^{-1}\ \varphi(b)$ for some t $\in$ S, b $\in$ R, whence it follows that $\varphi(ta - bs) = 0$ and thus r(ta-bs) = 0 for some r $\in$ S. This estblishes (*).
Moreover, if as = 0 then $\varphi(a)\ \varphi(s) = 0$ or $\varphi(a) = 0$. This implies the existence of some t $\in$ S such that ta = 0, i.e. S also satisfies (**).
2⇒1. Consider the set S $\times$ R and define the equivalence relation: (s,x) $\sim$ (t,y) if and only if there exist a and b in R for which as = bt and ax = by. The equivalence class of (s,x) will be denoted by $\frac{x}{s}$. On S $\times$ R/$\sim$ we define the following operations: $\frac{x}{s} + \frac{y}{t} = \frac{ax+by}{u}$ where a,b $\in$ R are chosen such that u = as = bt $\subset$ S; $\frac{x}{s} \cdot \frac{y}{t} = \frac{x_1 y}{t_1 s}$ where $t_1 \in$ S, $x_1 \in$ R such that t_1x = x_1t.
The properties (*) and (**) imply that these definitions do not depend on the choice of the representatives. In this way the set S $\times$ R/$\sim$ turns out to be ring and the function φ: R $\to$ S $\times$ R/$\sim$, a $\mapsto$ $\frac{a}{1}$, is a ring homomorphism. Clearly, (S $\times$ R/$\sim$, φ) is a left ring of fractions of R with respect to S. □

The properties (*) and (**) of S are called the *left Ore conditions* and R is said to *satisfy the left Ore conditions with respect to S.*

An s $\in$ R is said to be *regular* or a *nondivisor of zero* if $l_R(s) = r_R(s) = 0$, and we write S_{reg} = {s $\in$ R, s regular in R}. Obviously S_{reg} satisfies (**). If S_{reg} is also satisfying (*) then we write Q_{cl}(R) for S_{reg}^{-1}R and this ring is then called the *left classical ring of fractions.*
Assume that R satisfies the left Ore conditions with respect to a multiplicative system S and consider a left R-module M. On the set S $\times$ M we may again define an equivalence relation: (s,m) $\sim$ (t,n) if and only if there exist a and b in R such that as = bt and am = bn. Again the equivalence class of (s,m) will be denoted by $\frac{m}{s}$. Addition on S $\times$ M/$\sim$ is defined by $\frac{m}{s} + \frac{n}{t} = \frac{am+bn}{u}$ where a,b $\in$ R satisfy u = as

$=$ bt $\in$ S and we define a scalar multiplication for elements of S^{-1}R by $\frac{r}{s} \cdot \frac{n}{t} = \frac{rn}{t_1 s}$ where t_1r = rt, $t_1 \in$ S.
One easily verifies that S^{-1}M is an S^{-1}R-module. There is a canonical map α_M: M $\to$ S^{-1}M, m $\mapsto$ $\frac{m}{1}$ which is R-linear and it is also clear that $\frac{r}{1} \cdot \frac{m}{1} = \frac{rm}{1}$.
To a given R-linear map f: M $\to$ N we may associate a map $[S^{-1}]$f: S^{-1}M $\to$ S^{-1}N, $\frac{m}{s} \mapsto \frac{f(m)}{s}$ which is obviously a homomorphism of S^{-1}R-modules.

The diagram:

$$\begin{array}{ccc} M & \xrightarrow{\quad f \quad} & N \\ {\scriptstyle \alpha_M}\downarrow & & \downarrow{\scriptstyle \alpha_N} \\ S^{-1}M & \xrightarrow[\ [S^{-1}]f\ ]{} & S^{-1}N \end{array}$$

is commutative and if we consider S^{-1}R-modules as R-modules by restriction of scalars over the ring morphism φ: R $\to$ S^{-1}R then all maps in the diagram are R-linear.

A submodule N of M is said to be *S-saturated* (or *S-closed*) if sx $\in$ N with s $\in$ S, x $\in$ M entails that x $\in$ N. Let L_{sat}(M) be the set of all S-saturated R-submodules of M.

2.8.5. Proposition. Let R satisfy the left Ore conditions with respect to S and let M be an R-module.

1. Ker α_M = t(M) where t(M) = {x $\in$ M, sx = 0 for some s $\in$ S}.
2. If o $\longrightarrow$ M' $\xrightarrow[f]{}$ M $\xrightarrow[g]{}$ M" $\longrightarrow$ 0 is a short exact sequence then the following sequence is also exact:
0 $\longrightarrow$ S^{-1}M' $\xrightarrow[[S^{-1}]f]{}$ S^{-1}M $\xrightarrow[[S^{-1}]g]{}$ S^{-1}M" $\longrightarrow$ 0
3. To $\overline{N} \in$ L(S^{-1}M) we associate $\alpha_M^{-1}(\overline{N})$, this defines a strictly ascending bijection: L(S^{-1}M) $\to$ L_{sat}(M). If N = $\alpha_M^{-1}(\overline{N})$ then we have $\overline{N}$ = S^{-1}N = S^{-1}R . α_M(N).
4. If M is a Noetherian R-module, resp. Artinian, resp. semisimple, then S^{-1}M is a Noetherian, resp. Artinian, resp. semisimple,

Proof. 1. By definition of the equivalence relation ~.

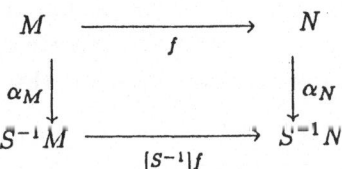

2. Clearly follows from 1. 4 obviously follows from 3.
3. It is immediately seen that $N = \alpha_M^{-1}(\overline{N})$ is S-saturated. Let $\frac{x}{s} \in S^{-1}N$. Because $\frac{x}{s} = \frac{1}{s} \cdot \frac{x}{1}$ and $x \in N$ it follows that $\frac{x}{1} \in \overline{N}$ or $\frac{x}{s} \in \overline{N}$. Conversely, if $\frac{x}{s} \in \overline{N}$ then $\frac{s}{1} \cdot \frac{x}{s} = \frac{x}{1} \in \overline{N}$ and thus $x \in N$. This establishes $\frac{x}{s} \in S^{-1}N$ and thus the eqality $S^{-1}N = \overline{N}$ is evident. □

When $M = t(M)$, M is said to be *S-torsion* and M is said to be *S-torsion free* whenever $t(M) = 0$. In particular $M/t(M)$ is always S-torsion free. If M is S_{reg}-torsion, resp. S_{reg}-torsion free then we simply say that M is torsion, resp. torsion free.

We now turn to the proof of Goldie's theorems describing which rings have a semisimple Artinian classical ring of fractions.

2.8.6. Lemma. If ${}_RR$ has finite Goldie dimension and if $l_R(a) = 0$ for $a \in R$ then Ra is essential.

Proof. Suppose that $I \neq 0$ is a left ideal of R such that $Ra \cap I = 0$. Then it is clear that $I \oplus Ia \oplus Ia^2 \oplus \ldots \oplus Ia^n \oplus \ldots$ is a direct sum, a contradiction. □

For an R-module M, $Z(M) = \{x \in M, l_R(x) \text{ is essential in } R\}$ is an R-submodule of M called the *singular submodule*. In particular $Z({}_RR)$ is a left ideal but it is easy to verify that $Z({}_RR)$ is actually an ideal of R; we write $Z(R)$ for this ideal and call it the *singular ideal* of R.

2.8.7. Proposition. Suppose that R is a semisimple ring satisfying the ascending chain condition for left annihilator ideals of R, then:
1. $Z(R) = 0$
2. There are no nonzero left nilideals.
3. If ${}_RR$ has finite Goldie dimension then any $a \in R$ with $l_R(a) = 0$ is a regular element.

Proof. 1. Put $Z(R) = A$. If $A \neq 0$ then $A^n \neq 0$ for all $n \geq 0$. There exists an $n \geq 0$ such that $l_R(A^n) = l_R(A^{n+1})$ because we have an ascending chain of left annihilator

ideals $l_R(A) \subset l_R(A^2) \subset \dots$. Pick a $\in$ A such that a $A^n \neq 0$ and $l_R(a)$ being maximal with respect to this property. Pick b $\in$ A. Since $l_R(b)$ is essential, $l_R(b) \cap$ Ra $\neq 0$ and thus we may select a y $\in$ R such that
ya $\neq 0$ and yab $\neq 0$. Hence $l_R(a) \subsetneq l_R(ab)$ and therefore ab$A^n = 0$. But the latter means that a$A^{n+1} = 0$ or a $\in l_R(A^{n+1}) = l_R(A^n)$, a contradiction. So we must have A = 0.

2. Suppose that I is a nonzero left nilideal and pick a $\neq 0$ in I. Then Ra is nil and aR is nil (because R is semisimple !). Pick b $\in$ aR, b $\neq 0$ and such that $l_R(b)$ is maximal. If r $\in$ R then br is nilpotent, so let n be the last natural number such that $(br)^n = 0$,hence $(br)^{n-1} \neq 0$. Since $l_R(b) \subset l_R((br)^{n-1})$ it follows that $l_r(b) = l_R((br)^{n-1}$ and therefore br $\in l_R(b)$ because brb = 0. Now r was arbitrary, so bRb = 0 and as R is semiprime b = 0, a contradiction. Therefore I = 0 is the only possibility.

3. Suppose that ab = 0 for some b $\in$ R. Then Ra $\subset l_R(b)$. By Lemma 2.8.6. we know that $l_R(b)$ is essential and then b $\in$ Z(R) or b = 0. □

A ring R is said to be a *left Goldie* ring if${}_R R$ has finite Goldie dimension and if it verifies the ascending chain conditions for left annihilator ideals. If R is a subring of Q such that Q is (isomorphic to) the classical ring of left fractions of R then we say that R is a *left order* of Q.

2.8.8. Theorem (A. Goldie). For a ring R, the following statements are equivalent:

1. R is a left order in a semisimple ring.
2. R is a semiprime left Goldie ring.
3. A left ideal of R is essential if and only if it contains a regular element of R.

Proof. 1⇒2. Let Q be a classical ring of left frations of R. Since Q is semisimple, left ideals I and J of R will have the property I $\cap$ J = 0 if and only if QI $\cap$ QJ = 0. Consequently, ${}_R R$ has finite Goldie dimension. For any finite subset X $\subset$ R it follows from $l_R(X) = R \cap l_Q(X)$ that R verifies the ascending chain condition for left annihilator ideals.
Consider a nilpotent ideal I; then $r_R(I)$ is essential as a left ideal. Indeed, if J is a nonzero left ideal, we may choose n ≥ 0 maximal with respect to $I^n J \neq 0$. From $I^{n+1}J = 0$ it then follows that $I^n J \subset r_R(I) \cap J$, establishing $r_R(I) \cap J \neq 0$ and hence $r_R(I)$ is essential in R. Then $Qr_R(I)$ is essential in Q and the semisimplicity of

Q entails $Qr_Q(I) = Q$. From Proposition 2.8.5. we retain that $1 = \frac{a}{s}$ with $a \in r_R(I)$ and $s \in S_{reg}$, consequently $I \cap S_{reg} \neq \phi$. It follows that $I = 0$ and therefore R is semiprime.

2⇒3. Let I be an essential left ideal. By Proposition 2.8.7. R does not contain nonzero nilideals, so there must exist $a_1 \in I$, $a_1 \neq 0$ such that $l_R(a_1) = l_R(a_1^2)$. If $l_R(a_1) \neq 0$ then $I \cap l_R(a_1) \neq 0$ and there exists an $a_2 \in I \cap l_R(a_1)$, $a_2 \neq 0$ such that $l_R(a_1) = l_R(a_2^2)$. If $l_R(a_1) \cap l_R(a_2) \neq 0$ then $I \cap l_R(a_1) \cap l_R(a_2) \neq 0$ and there exists an $a_3 \in I \cap l_R(a_1) \cap l_R(a_2)$ such that $a_3 \neq 0$ and $l_R(a_3) = l_R(a_3^2)$.
Continuing this argument we arrive at a sequence of nonzero elements a_1, a_2, ..., a_n,...such that $l_R(a_i) = l_R(a_i^2)$ and for all $i \in \{1,...,n\}$, $a_i \in I \cap l_R(a_1) \cap ... \cap l_R(a_{i-1})$. As a consequence of this we obtain that $Ra_1 + ... + Ra_n$ is direct. The finiteness of the Goldie dimension implies that $l_R(a_1) \cap ... \cap l_R(a_m) = 0$ for some m where the direct sum $Ra_1 \oplus ... \oplus Ra_m$ is maximal as such.
Put $s = a_1 + a_2 + ... + a_m$. Then $l_R(s) = 0$. By Proposition 2.8.7. it follows that s is a regular element of S. Finally, since $s \in I$ we have proved 3.

3⇒1. Consider $s \in S_{reg}$, $a \in R$. Because Rs is essential in R, $(Rs{:}a) = \{r \in R, ra \in Rs\}$ is essential too. Therefore, there exists $t \in (Rs{:}a) \cap S_{reg}$ and this entails that $bs = ta$ for some $b \in R$, proving the left Ore condition for R with respect to S_{reg}.
For an essential left ideal $\overline{I} \subset Q^l_{cl}(R)$, $\overline{I} \cap R$ is essential in R hence:
$(\overline{I} \cap R) \cap S_{reg} \neq \phi$. The latter implies that $\overline{I} = Q^l_{cl}(R)\ (\overline{I} \cap R) = Q^l_{cl}(R)$. □

2.8.9. Corollary. R is an order in a simple Artinian ring Q if and only if R is a prime left Goldie ring.

Proof. Assuming that Q is simple we have to prove that R is prime. If $aRb = 0$ for some a and b in R then it straightforward to check that $\frac{a}{1}Q\frac{b}{1} = 0$, hence $\frac{a}{1} = 0$ or $\frac{b}{1} = 0$, say $\frac{a}{1} = 0$. There exists an $s \in S_{reg}$ such that $sa = 0$ but this implies $a = 0$ and consequently $a = 0$.
Conversely, if I and J are ideals of Q such that $IJ = 0$ then $(I \cap R)(J \cap R) = 0$ implies $I \cap R = 0$ or $J \cap R = 0$, say $I \cap R = 0$. From Proposition 2.8.5. it follows that $I = 0$ and so it follows that Q is prime. Because Q is semisimple and prime it is clear that Q has to be simple. □

2.8.10. Corollary. A domain R is a left Ore domain if and only if R has finite Goldie dimension.

2.8.11. Corollary. A semisimple left Noetherian ring is an order in a semisimple Artinian ring.

§2.9 Artinian Modules which are Noetherian.

2.9.1. Lemma. For R-modules U and M, the following statements are equivalent:

1. For each submodule M' of M, M' $\neq$ M, there exists an f $\in$ Hom_R(U,M) such that Im f $\not\subset$ M'.
2. There exists a set $\Lambda \neq \phi$ and an epimorphism $U^{(\Lambda)} \to M \to 0$

Proof. 1$\Rightarrow$2. Put $\Lambda = \mathrm{Hom}_R(U,M)$ and define φ: $U^{(\Lambda)} \to M$ by $(x_f)_{f\in\Lambda} \mapsto \sum_{f\in\Lambda} f(x_f)$ (of course this sum is finite). If Im f $\neq$ M then there exists a g $\in$ Λ such that Im g $\not\subset$ Im φ. By the definition of φ it is obvious that Im g $\subset$ Im φ, a contradiction. Hence Im φ = M or φ is an epimorphism.
2$\Rightarrow$1. Let φ: $U^{(\Lambda)} \to M$ be an epimorphism and M' $\subset$ M a submodule such that M' $\neq$ M. Put for $\lambda \in \Lambda$, i_λ: $U \to U^{(\Lambda)}$, where i_λ is the canonical inclusion. Since M $\neq$ M' we find a $\lambda \in \Lambda$ such that Im $(\varphi \circ i_\lambda) \not\subset$ M'. Denoting: f = $\varphi \circ i_\lambda$, then we obtain that f $\in$ Hom_R(U,M) and Im f $\not\subset$ M'. □

A module M satisfying one of the conditions of Lemma 2.9.1. is said to be *U-generated.* If in addition all submodules of M are U-generated then we say that M is *U-strongly generated.* If all R-modules are U-generated then we say that U is a *generator* for the class of all left R-modules. For example ${}_R R$ is a generator. Indeed, let M be an R-module and M $\neq$ M' a submodule; pick x $\in$ M - M' and define f: R $\to$ M, r $\mapsto$ rx, then f(1) = x entails that Im f $\not\subset$ M'.

For an R-module M we may define the following ascending chain of submodules (σ): $0 = M_0 \subset M_1 \subset \ldots \subset M_n \subset \ldots$, where M_1 = s(M) is the socle of M, M_2/M_1 = s (M/M_1) , ... , M_{n+1}/M_n = s (M/M_n), ... From Proposition 2.5.6. we retain that each M_n in the sequence is a semiartinian module.

2.9.2. Theorem. If U is an Artinian R-module and M is strongly U-generated then the chain (σ) is stationary.

Proof. Suppose (σ) is strictly ascending and let f $\in$ Hom_R(U,M) be such that Im f $\subset$ M_n. The assumption on U yields that Im f is an Artinian module.

Consider the induced chain:

$$(\sigma_f)\ 0 = M_0 \cap \mathrm{Im}\, f \subset M_1 \cap \mathrm{Im}\, f \subset \ldots \subset M_n \cap \mathrm{Im}\, f,$$

where each $M_i \cap$ Im f$/M_{i-1} \cap$ Im f is an Artinian semisimple module and thus of finite length. Consequently Im f has finite length. Now we prove inductively that l(Im f) $\geq$ n if f $\in$ Hom_R(U,M) is such that Im f $\subset$ M_n, Im f $\not\subset$ M_{n-1}. If n=1 then Im f $\neq$ 0 and Im f$\subset$ M_1 yields l(Im f) $\geq$ 1. Let now f: U$\rightarrow$M be such that Im f$\subset$ M_{n+1}, Im f$\not\subset$ M_n. Hence Im f+$M_{n-1}/M_{n-1} \neq 0$. Since M_{n+1}/M_{n-1} is a semiartinian module, s(M_{n+1}/M_{n-1}) is essential in M_{n+1}/M_{n-1}. From s$(M_{n+1}/M_{n-1}) = M_n/M_{n-1}$ we obtain that $(M_n/M_{n-1}) \bigcap$ (Im f + $M_{n-1}/M_{n-1}) \neq 0$. Let us write K = $M_n \cap$ (Im f + M_{n-1}) = M_{n-1} + ($M_n \cap$ Im f), and L = $M_n \cap$ Im f. From M_{n-1} $\subset$ K $\subset$ M_n we derive L $\not\subset$ M_{n-1}.

The fact that M is strongly U-generated entails the existence of a morphism

g: U $\rightarrow$ L such that Im g $\not\subset$ M_{n-1} and thus Im g $\subset$ L $\subset$ M_n. By the induction hypothesis l(Im g) $\geq$ n. Since M_n does not contain Im f it follows that L $\subset$ Im f and hence Im g $\underset{\neq}{\subset}$ Im f. The latter inequality implies l(Im f) > l(Im g) or l(Im f) $\geq$ n+1.

Now put G = {U', U' is an R-submodule of U such that U/U' has finite length}. If U_1,U_2 are in G then there exists a monomorphism

o $\rightarrow$ U/$U_1 \cap U_2$ $\rightarrow$ U/U_1 $\bigoplus$ U/U_2 and hence $U_1 \cap U_2$ $\in$ G. The Artinian property of U yields the existence of a minimal element U_0 in G and it follows from the foregoing that U_0 is the least element of G.

Write μ = l(U/U_0). Since M is strongly U-generated and taking into account that (σ) is supposed to be strictly ascending, there exist R-linear maps f_n: U $\rightarrow$ M such that Im f_n $\subset$ M_n and Im f_n $\not\subset$ M_{n-1} for any n $\geq$ 1.

On the other hand each Im f is an R-module of finite length, so Ker f_n $\in$ G and consequently U_0 $\subset$ Ker f_n for all n $\geq$ 1.

Therefore l(Im f_n) = l(U/Ker f_n) $\leq$ l(U/U_0) = μ.

From l(Im f_n) $\geq$ n we now derive that μ $\geq$ n for each n, a contradiction. □

2.9.3. Corollary. Suppose U and M are Artinian R-modules. If M is strongly U-generated then M is Noetherian.

Proof. By the theorem, there is an n such that $M_n = M_{n+1} = \ldots$. The Artinian property for M yields M_n = M. But every M_n is a module of finite length, so M has finite length too. □

2.9.4. Corollary. If R is left Artinian then every Artinian left R-module is left Noetherian.

2.9.5. Corollary (Hopkins). If R is left Artinian then R is left Noetherian.

Note that the converse of Corollary 2.9.5. does not hold (e.g. consider the ring of integers $\mathbb{Z}$).

2.9.6. Corollary. For a left Artinian ring R, R/J(R) is semisimple Artinian and J(R) is nilpotent (i.e. $J(R)^n = 0$ for some natural number n).

Proof. From Theorem 1.8.10. it follows that R/J(R) is semisimple. Also it has been noted that ${}_RR$ has finite length. Let $0 = I_0 \subset I_1 \subset \ldots \subset I_n = R$ be a composition series of ${}_RR$. Since each I_k/I_{k-1} is simple we have $J(R)\ I_k \subset I_{k-1}$ and therefore: $J(R) = J(R)\ I_n \subset I_{n-1}$. The latter leads to: $J(R)^2 \subset J(R)I_{n-1} \subset I_{n-2}, \ldots,$ $J(R)^n \subset I_0 = 0$, and thus J(R) is nilpotent (of index n). □

§2.10 Projective and Injective Modules.

We say that the left R-module P is *projective* if for every surjective R-linear map $M \xrightarrow{\alpha} M'' \longrightarrow 0$ and for every R-linear map f: $P \to M''$ there exists an R-linear g: $P \to M$ such that $f = \alpha \circ g$.

2.10.1. Lemma. A free R-module is projective.

Proof. Let L be a free R-module with basis $(e_i)_{i \in I}$ and let α: $L \to M''$ be given. By the surjectivity of α we may select $y_i \in M$ such that $\alpha(y_i) = e_i$ for all $i \in I$. Define g: $L \to M$ by mapping $\sum_{i \in I} a_i e_i$ to $\sum_{i \in I} a_i y_i$, for all $a_i \in R$. It is obvious that this R-linear map is well-defined and that $f = \alpha \circ g$ holds. □

2.10.2. Lemma. If $(P_i)_{i\in I}$ is any family of modules then P = $\bigoplus_{i\in I} P_i$ is projective if and only if P_i is projective for all i ∈ I.

Proof. If P is projective, consider the diagram:

$$\begin{array}{ccccc} P & \xrightarrow[p_j]{} & P_j & & \\ \downarrow g & & \downarrow f & & \\ M & \xrightarrow[\alpha]{} & M'' & \rightarrow & 0 \end{array}$$

where p_j in the canonical projection, and g such that foP_j = αog exists by the projectivity of P. Write i_j: $P_j \rightarrow P$ for the canonical injection and put h = goi_j. Then αoh = αogoi_j = fop_joi_j = fo1_{P_j} = f and therefore P_j is projective. Conversely, let P_i be projective for all j ∈ I. Consider the diagram:

$$\begin{array}{ccccc} P_j & \xrightarrow[i_j]{} & P & & \\ \downarrow & & \downarrow f & & \\ M & \longrightarrow & M'' & \rightarrow & 0 \end{array}$$

where g_j exists by the projectivity of P_j, fo$i_j = \alpha \circ g_j$.

The universal property of the direct sum learns that there is a unique R-linear map g: P → M such that goi_j = g_j for all j ∈ I. One calculates: fo$i_j = \alpha \circ g_j = \alpha \circ (g \circ i_j)$ $= (\alpha \circ g) \circ i_j$ and again using the universal property of the direct sum we arrive at f = αog proving that P is projective. □

2.10.3. Proposition. For an R-module P the following assertions are equivalent:

1. P is projective.
2. P is a direct summand of a free module.
3. Every short exact sequence $0 \longrightarrow L \xrightarrow[f]{} M \xrightarrow[g]{} P \longrightarrow 0$ is split.

Proof. 2⇒1. follows from the foregoing lemmas.
1⇒3. The projectivity of P entails the existence of h: P → M such that goh = 1_P i.e. the sequence in 3. is split.
3⇒2. There eists a free module L and an epimorphism f: L → P → 0. By 3. the sequence 0 → Ker f → L → P → 0 is split, consequently L ≅ P ⊕ Ker f □

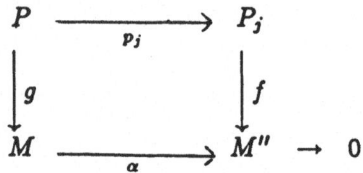

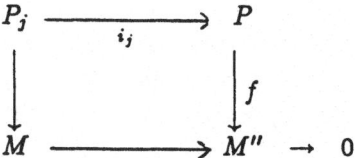

For an R-module M we let M^* be the abelian group $\mathrm{Hom}_R(P,M)$ for some projective R-module P. We let A denote the ring $\mathrm{End}_R(P)$. Then M^* is a left A-module in the following way: for $\alpha \in A$, $f \in M^*$ put $\alpha.f = f \circ \alpha$. For an R-submodule N of M it is easily verified that the set $X_N = \{f \in M^*, \mathrm{Im}\, f \subset N\}$ is an A-submodule of M^*.

2.10.4. Lemma. For a finitely generated projective R-module P every A-submodule of M^* is of the form X_N for some submodule N of H.

Proof. Let Y be an A-submodule of M^* and put $N = \sum_{f \in Y} \mathrm{Im}\, f$. Obviously $Y \subset X_N$. Conversely, consider $g \in X_N$ and the diagram

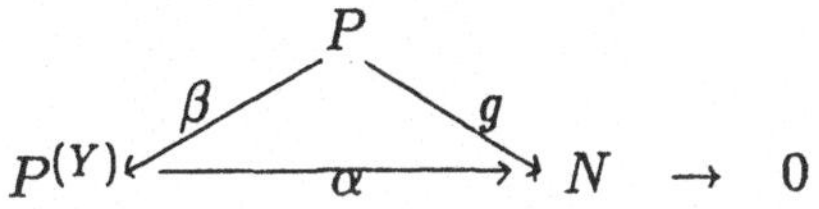

where α is the canonical map: $(x_f)_{f \in Y} \mapsto \sum_{f \in Y} f(x_f)$. Projectivity of P yields that β with $g = \alpha \circ \beta$ exists.
Because P is finitely generated, $\mathrm{Im}\, \beta \subset P^{(Y')}$ for some finite subset Y' of Y. This leads to a commutative diaram:

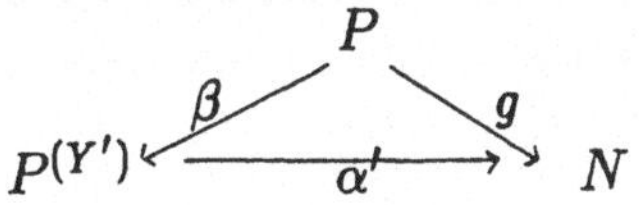

where α' is the restriction $\alpha \mid P^{(Y')}$. Write $Y' = \{f_1, ..., f_n\}$. So we consider α': $P^n \to N$, $(x_1, ..., x_n) \mapsto \sum_{i1}^n f_i(x_i)$. Letting p_j: $P^n \to P_j$ denote the canonical projections we obtain for $x \in P$, $\beta(x) = (\beta_i(x))_{i=1,...,n}$, where $\beta_i = p_i \circ \beta \in A$.
The equality $g = \alpha' \circ \beta$ now entails: $g(x) = \sum_{i=1}^n (f_i \circ \beta_i)(x) = \sum_{i=1}^n (\beta_i f_i)\,(x)$ or $g = \sum_{i=1}^n \beta_{i_i}$.
Since Y is a submodule of M^* we conclude that $g \in Y$ and so we have established that $Y = X_N$. □

2.10.5. Theorem. Let P be a finitely generated projective R-module. If M is Noetherian, resp. Artinian resp. semisimple, then M^* is Noetherian, resp. Artinian, resp. semisimple as an A-module.

Proof. Look at an ascending chain of A-submodules of M^*: $Y_1 \subset Y_2 \subset ... \subset Y_n \subset ...$. Putting $N_i = \sum_{f \in Y} \mathrm{Im}\, f$ we obtain an ascending chain of R-submodules of M: $N_1 \subset N_2 \subset \ldots \subset N_n \subset \ldots$. The Noetherian condition of M yields that $N_n = N_{n+1} = \ldots$. By the lemma we obtain $Y_n = X_{N_n} = X_{N_{n+1}} = Y_{n+1}$ and so it follows that M* is a Noetherian A-module.

A similar proof is valid in the Artinian case.

If M is semisimple, consider an A-submodule Y $\subset$ M* By the lemma we obtain that Y = X_n, N = $\sum_{f \in Y}$ Im f in M. Since N is a direct summand of M, M = N $\bigoplus$ P for some R-submodule P of M. One checks that M* = $X_N \bigoplus X_P = Y \bigoplus Y_P$ and thus Y is a direct summand of M*. Hence M* is semisimple as claimed. □

2.10.6. Corollary. Let P be a projective finitely generated R-module. If P is Noetherian, resp. Artinian, semisimple then the ring A is left Noetherian, resp. Artinian, semisimple.

2.10.7. Corollary. If R is a Noetherian ring, resp. Artinian, semisimple then M_n(R) has he same properties.

For any idempotent e of R, eRe = {exe, x $\in$ R} is a ring having e as the identity element. From R = Re + R(1-e) it follows that Re is a projective R-module.

2.10.8. Lemma. The rings eRe and End_R(Re) are isomorphic.

Proof. Take $\alpha \in \text{End}_R(\text{Re})$, put α(e) = ae for some a $\in$ R. From $\alpha(e) = \alpha(e^2) = e\alpha(e)$ it follows that α(e) = eae so we may define φ: End_R(Re) $\mapsto$ eRe, $\alpha \mapsto$ eae. If $\alpha, \beta \in \text{End}_R(\text{Re})$ then $\varphi(\alpha\beta) = \varphi(\beta \circ \alpha) = (\beta \circ \alpha)(e) = \beta(\alpha(e)) = \beta(\text{eae}) = \text{ea}\beta(\text{e})$ $= \text{ea}\beta(e^2) = \text{eae}\beta(\text{e}) = (\text{eae})(\text{ebe})$ where we put $\varphi(\beta) = \beta(\text{e}) = \text{ebe}$, some b $\in$ R. Evidently, φ is a ring homomorphism.

We have $\varphi(\alpha) = 0$ when eae = 0 hence when α(e) = 0, but then $\alpha(\text{xe}) = \text{x}\alpha(\text{e}) = 0$ or $\alpha = 0$, and φ is injective. On the other hnd, for aea $\in$ eRe we may define α: Re $\rightarrow$ Re, x e $\mapsto$ xeae. Clearly $\varphi(\alpha)$ = eae and thus φ is surjective as well. □

2.10.9. Corollary. If R is a left Noetherian ring, resp. Artinian, semisimple, and e is an idempotent element of R then the ring eRe is left Noetherian, resp. Artinian, semisimple.

Let us now focus on the class of injective modules. An R-module Q is said to be *injective* if for every monomorphism: $0 \longrightarrow M' \underset{\alpha}{\longrightarrow} M$, and for every homomorphism f: $M' \to Q$, there exists a homomorphism g: $M \to Q$ such that $g \circ \alpha = f$.

2.10.10. Proposition. If $(Q_i)_{i \in I}$ is a family of R-modules then $Q = \Pi_{i \in I} Q_i$ is injective if and only if every Q_i is injective.

2.10.11. Proposition. The following assertions are equivalent: 1. The R-module Q is injective.
2. Any exact sequence $0 \to Q \to M \to N$ splits.

Proof. Propositions 2.10.10. and 2.10.11. may be proved as Propositions 2.10.2 and 2.10.3. by reversing the arrows □

2.10.12. Theorem (Baer). An R-module Q is injective if and only if every R-linear f: $I \to Q$ where I is a left ideal of R extends to an R-linear f_o: $R \to Q$.

Proof. Assume that Q is an injective R-module, since I is a left R-submodule of R one implication follows immediately from the definition of injectivity. Now assume the second statement of the theorem and consider the situation:

$$\begin{array}{ccccc} 0 & \longrightarrow & N & \longrightarrow & M \\ & & \big\downarrow f & & \\ & & Q & & \end{array}$$

where N is an R-submodule of M. Let F consist of all pairs (N',f") where $N \subset N' \subset M$, N' is an R-submodule of M and f': $N' \to Q$ restricts to f. Since (N,f) $\in$ F, $F \neq \phi$. We define a partial order on F by defining (N',f") $\leq$ (N",f") if $N' \subset N"$ and $f"|N' = f'$.
Zorn's lemma yields the existence of a maximal pair $(N_0, f_o) \in F$. If $N_0 = M$ the statement follows. If $N_0 \neq M$ let $x \in M - N_0$ and consider the left ideal $I = \{a \in R, ax \in N_0\}$ of R. Define the R-linear map g: $I \to Q$, $a \mapsto f_0(ax)$ and let h: $R \to Q$ be the extension of g to R. Put $N_1 = N_0 + Rx$, f_1: $N_1 \to Q$, $f_1(x_0 + rx) = f_0(x_0) + r\,h(1)$ for $x_0 \in N_0$, $r \in R$. Whenever $x_0 + rx = x_0' + r'x'$ then

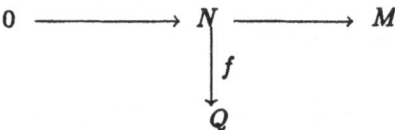

(r-r') x $\in N_0$ and we apply f_0 to $x_0 - x'_0$ and obtain: $f_0(x'_0 - x_0) = f_0((r-r')x) =$ g(r-r') = h(r-r') = (r-r')h(1). Consequently, $f_0(x_0)$ + rh(1) = $f_0(x'_0)$ + r'h(1) and thus f_1 is well-defined. It is also clear that $f_1 \mid N_0 = f_0$ hence $(N_1, f_1) \in$ F and $(N_0, f_0) \leq (N_1, f_1)$, a contradiction. □

2.10.13. Theorem. For an R-module M there exists an injective R-module Q such that M is an R-submodule of Q.

Proof. Hint: construct an R-module containing M maximal with the property of being an essential extension and apply Theorem 2.10.12. □

For an R-module M, a pair (Q,i) where i:M → Q is a monomorphism, Q is injective and Im i is an essential submodule of M, is said to be an *injective envelope* of M. It is easily seen that such an injective envelope is unique up to isomorphism if it exists; we write E(M) for "the" injective envelope (or *hull*) of M.

2.10.14. Theorem (Eckmann-Schopf). An R-module always has an injective envelope. (cf. hint in Theorem 2.10.13.).

2.10.15. Corollary. For an injective R-module the following statements are equivalent:

1. Q is an indecomposable module.
2. Q is uniform.
3. Any nonzero submodule of Q is coirreducible.
4. The ring $\text{End}_R(Q)$ is local.

Proof. 1⇒2. If there exist R-submodules M and N of Q, both nonzero such that M ∩ N = 0, let E(M) be the injective envelope of M. Then E(M) is isomorphic to a submodule, again denoted E(M), of Q and hence Q = E(M) because Q is indecomposable. Then Q is an essential extension of M and M ∩ N = 0 is impossible
2⇒1. Obvious
2⇔3. Easy.
2⇒4. Consider f,g $\in \text{End}_R(Q)$ such that 1 = f + g. From Ker f ∩ Ker g = 0,

Ker f = 0 or Ker g = 0 follows. If Ker f = 0 then Im f is an injective R-module but then Im f = Q and f is invertible. It follows that $\mathrm{End}_R(Q)$ is a local ring.
4⇒1. If Q were decomposable, say $Q = Q_1 \oplus Q_2$ with $Q_i \neq 0$, i = 1,2, then we may define f_1, f_2: Q → Q such that $f_1(x_1 + x_2) = x_1$, $f_2(x_1 + x_2) = x_2$ for each $x_1 \in Q_1$, $x_2 \in Q_2$. Evidently $1 = f_1 + f_2$, so the fact that $\mathrm{End}_R(Q)$ is local yields that either f_1 or f_2 is invertible. If f_1 is invertible then $Q_2 = 0$ follows. □

2.10.16. Corollary. For an R-module M of finite Goldie dimension E(M) is a finite direct sum of indecomposable injective R-modules.

Proof. Since E(M) is essential over M it is clear that E(M) has finite Goldie dimension and the corollary follows easily. □

We now define a notion dual to the notion of a generator module. A left R-module C is said to be a *cogenerator* for the class of all left R-modules if for each left R-module M and a submodule N, M ≠ N, there exists a morphism f: M → C such that f(N) ≠ 0.
The following lemma is just the dual of Lemma 2.9.1.

2.10.17. Lemma. The following statements are equivalent:

1. ${}_RC$ is a cogenerator.
2. For each left R-module M there exists a set $I \neq \phi$ and a monomorphism
   $0 \to M \to C^I$.

2.10.18. Proposition. Let C be an injective R-module, then C is a cogenerator if and only if each simple (left) R-module S is isomorphic to a submodule ofx_RC.

Proof. One implication is evident. Conversely, assume that C is not a cogenerator, hence there exists an R-module M and a submodule N of M, N ≠ M, such that $\mathrm{Hom}_R(M/N,C) = 0$. We may reduce the problem to N = 0 and M ≠ 0. Since M is the union of all its finitely generated submodules and since C is injective we have $\mathrm{Hom}_R(M,C) = 0$ if and only if $\mathrm{Hom}_R(M',C) = 0$ for all finitely generated R-submodules M' of M. Therefore we may assume that M is finitely generated. Since

$M \neq 0$ we can select a maximal submodule P of M. Injectivity of C combined with $\mathrm{Hom}_R(M,C) = 0$ yields $\mathrm{Hom}_R(M/P,C) = 0$ but the latter is a contradiction because M/P is a simple R-module. □

2.10.19. Corollary. The class of left R-modules has an injective cogenerator.

Proof. Let $\mathcal{I}$ be the set of isomorphism classes of simple R-modules then $C = E\,(\bigoplus_{s\in\mathcal{I}} S)$ is an injective cogenerator in view of the foregoing proposition. □

2.10.20. Example. The abelian group $Q/\mathbb{Z}$ is an injective cogenerator for the class of all abelian groups.

If M is an R-module then the abelian group $M^* = \mathrm{Hom}_{\mathbb{Z}}(M,Q/\mathbb{Z})$ is a right R-module if we define $(f.a)(x) = f(ax)$, $f \in M^*$, $a \in R$. The right R-module M^* is called the *character module* of M.

2.10.21. Proposition. The sequence of R-modules:

$$0 \longrightarrow M' \xrightarrow[\alpha]{} M \xrightarrow[\beta]{} M'' \longrightarrow 0$$

is exact if and only if the sequence of character modules:

$$0 \longrightarrow M''^* \xrightarrow[\beta^*]{} M^* \xrightarrow[\alpha^*]{} M'^* \longrightarrow 0,$$

is exact.

Proof. Derives from the fact that $Q/\mathbb{Z}$ is a cogenerator for abelian groups. □

§2.11 Tensor Product and Flat Modules.

We are given an arbitrary ring R, a left R-module N and a right R-module M. A map φ: $M \times N \to G$, where G is an abelian group, is said to be *R-balanced* if and only if $\varphi(xa,y) = \varphi(x,ay)$ for all $x \in M$, $y \in N$ and $a \in R$. Let $L = \mathbb{Z}^{(M\times N)}$

be the free $\mathbb{Z}$ -module with basis $\{(x,y), x \in M, y \in N\}$ and let K be the subgroup generated by the elements of the form: $(x_1 + x_2,y)$ - (x_1,y) - (x_2,y) or $(x,y_1 + y_2)$ - (x,y_1)- (x,y_2), or (xa,y) - (x,ay) for $x, x_1, x_2 \in M$, $y, y_1, y_2 \in N$, $a \in R$. We write $M \bigotimes_R N$ for L/K and call this the *tensor product of M over R*. Let π: L $\to$ L/K be the canonical surjection and τ: $M \times N \to L$ the canonical injection. Put $\varphi = \pi \circ \tau$. Then φ: $M \times N \to M \bigotimes_R N$ will be denoted by $\varphi(x,y) = x \otimes y$. By definition, each element of $M \bigotimes_R N$ may be written as a finite sum $\sum_i' x_i \bigotimes y_i$, $x_i \in M$, $y_i \in N$, but such expression need not be unique.

The definition of $M \bigotimes_R N$ also forces the following relations: $(x_1 + x_2) \otimes y = x_1 \otimes y + x_2 \otimes y$, $x \otimes (y_1 + y_2) = x \otimes y_1 + x \otimes y_2$, $xa \otimes y = x \otimes ay$ The map φ is a $\mathbb{Z}$ -bilinear and R-balanced map.

Our first aim now is to establish that the tensor product is a sort of universal object with respect to the properties we mentioned. First, for M and N as above, we say that a pair (G,Ψ) where G is an abelian group and Ψ: $M \times N \to G$ a $\mathbb{Z}$ -linear R-balanced map, is *a tensor product of M and N* if for each abelian group H and $\mathbb{Z}$ -bilinear R-balanced θ: $M \times N \to H$ there exists a unique group homomorphism u: $G \to H$ such that $u \circ \Psi = \theta$.

Obviously $(M \otimes_R N, \varphi)$ is a tensor product of M and N and it is easily verified that a tensor product is in fact unique up to isomorphism of abelian groups.

Consider R-linear maps f: $M \to M'$, g: $N \to N'$ where f is right linear g is left linear. We may define Ψ: $M \times N \to M' \bigotimes N'$ by putting $\Psi(x,y) = f(x) \otimes g(y)$ for all $x \in M$, $y \in N$. By the foregoing it follows that there exists a morphism of abelian groups, $f \otimes g$: $M \bigotimes_R N \to M' \bigotimes_R N'$ such that $(f \otimes g)(x \otimes y) =$
$f(x) \otimes g(y)$. The map $f \otimes g$ is called the tensor product of f and g. It is easy enough to verify the following equalities:

$$(f_1 + f_2) \otimes g = f_1 \otimes g + f_2 \otimes g$$
$$f \otimes (g_1 + g_2) = f \otimes g_1 + f \otimes g_2$$
$$f \otimes 0 = 0 \otimes g = 0$$
$$1_M \otimes 1_N = 1_{M \otimes N}.$$

By switching to the opposite ring and the opposite modules when necessary we may establish the following properties:

2.11.1. Proposition. For a left R-module N and a right R-module M there is a unique isomorphism of abelian groups:

$$\varphi_{M,N}: M \bigotimes_R N \to N^0 \bigotimes_{R^0} M^0$$

such that $\varphi_{M,N}(x \otimes y) = y \otimes x$ for $x \in M$, $y \in N$.

2.11.2. Proposition. Let R and S be rings and consider the modules M_R, ${}_RN_S$, ${}_SP$. Then there exists a unique isomorphism of abelian groups

$\varphi_{M,N,P}$: $(M \otimes_R N) \otimes_S P \to M \otimes_R (N \otimes_S P)$,

such that $\varphi_{M,N,P}((x \otimes y) \otimes z) = x \otimes (y \otimes z)$ for $x \in M$, $y \in N$, $z \in P$.

It follows from Proposition 2.11.1. that for commutative R, $M \otimes_R N$ and $N \otimes_R M$ are isomorphic in a natural way; we say then that the tensor product is commutative.

If N is a left R-module then there is an R-linear isomorphism u_N: $R \otimes_R N \to N$, $a \otimes x \mapsto ax$. Moreover, if $f \in \mathrm{Hom}_R(N,N')$ then we obtain a commutative diagram:

$$\begin{array}{ccc} R \otimes_R N & \xrightarrow{1_R \otimes f} & R \otimes_R N' \\ \downarrow u_N & & \downarrow u_{N'} \\ N & \xrightarrow{f} & N' \end{array}$$

If R,S and T are rings, while ${}_SM_R$ and ${}_RN_T$ are bimodules then $M \otimes_R N$ will be an S-T-bimodule with operations defined by $s(x \otimes y) = sx \otimes y$, $(x \otimes y)t = x \otimes yt$ for $s \in S$, $t \in T$.

2.11.3. Proposition. Let $(N_i)_{i\in\Lambda}$ be a family of left R-modules. If M is a left S-module and a right R-(bi)module, then there exists an S-isomorphism η_M: $M \otimes_R N \to \bigoplus_{i\in\Lambda} (M \otimes_R N_i)$, such that $\eta_M(x \otimes y) = (x \otimes y_i)_{i\in\Lambda}$ where $x \in M$, $y = (y_i)_{i\in\Lambda} \in N$. Moreover, if f:$x_S M_R \to x_S M'_R$ is a morphism of bimodules then we obtain the following diagram of S-modules which is commutative:

$$\begin{array}{ccc} M \otimes_R N & \xrightarrow{\eta_M} & \bigoplus_{i\in\Lambda} (M \otimes_R N_i) \\ f \otimes 1_N \downarrow & & \downarrow \bigoplus_{i\in\Lambda} (f \otimes 1_{N_i}) \\ M' \otimes_R N & \xrightarrow{\eta_{M'}} & \bigoplus_{i\in\Lambda} (M' \otimes_R N_i) \end{array}$$

Proof. Look at $x \in M$, $y = (y_i)_{i\in\Lambda} \in N$ Define a map:
Ψ: $M \times N \to \bigoplus_{i\in\Lambda} (M \otimes_R N_i)$, $(x,y) \mapsto (x \otimes y_i)_{i\in\Lambda}$. One checks that Ψ is well-defined, $\mathbb{Z}$-bilinear and R-balanced. Consequently, there exists a group morphism: η_M: $M \otimes_R N \to \bigoplus_{i\in\Lambda} (M \otimes_R N_i)$, $x \otimes y \mapsto (x \otimes y_i)_{i\in\Lambda}$.
All the claims in the proposition are now easily checked. □

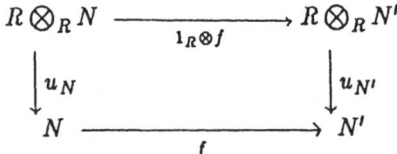

2.11.4. Corollary. If we are given a ring morphism φ: R $\rightarrow$ S then S may be viewed as a left S-, right R-bimodule if we put s.r = s. φ(r) for s $\in$ R, r $\in$ S. If L is a free R-module then S $\bigotimes_R$ L is a free S-module.

2.11.5. Theorem. (Adjunction Property) Consider rings R and S together with modules ${}_RM$, ${}_SU_R$, ${}_SN$. There exists a group isomorphism
ϕ_{MUN}: Hom_R (M,Hom_S(U,N)) $\rightarrow$ Hom_S (U $\bigotimes_R$ M, N), such that
ϕ_{MUN}(f)(y $\otimes$x) = f(x)(y) for f $\in$ Hom_R (M,Hom_S (U,N)), y $\in$ U, x $\in$ M. Given other modules ${}_RM'$, ${}_SU'_R$, ${}_SN'$ and morphisms f: ${}_RM \rightarrow {}_RM'$, g: ${}_RU_S \rightarrow {}_RU_{S'}$, h: ${}_SN \rightarrow {}_SN'$ then we obtain a commutative diagram of groups:

$$\begin{array}{ccc} \mathrm{Hom}_R(M,\mathrm{Hom}_S(U,N)) & \xrightarrow[\phi_{MUN}]{} & \mathrm{Hom}_S(U\bigotimes_R M,N) \\ \Big\downarrow{\scriptstyle \mathrm{Hom}(f,\mathrm{Hom}(g,h))} & & \Big\downarrow{\scriptstyle \mathrm{Hom}(g\otimes f,h)} \\ \mathrm{Hom}_R(M',\mathrm{Hom}_S(U',N')) & \xrightarrow[\phi_{M'U'N'}]{} & \mathrm{Hom}_S(U'\bigotimes_R M',N') \end{array}$$

Proof. Let f$\in$ Hom_R (M, Hom_S(U,N)). The map Ψ_f: U $\times$ M $\rightarrow$ N, (y,x) $\rightarrow$ f(x)(y), y $\in$ U, x $\in$ M, is $\mathbb{Z}$ -bilinear and R-balanced. Thus there exists a homomorphism Φ_f: U $\bigotimes_R$ M $\rightarrow$ N, (y $\otimes$ x) $\mapsto$ f(x)(y). Determine
ϕ_{MUN}: Hom_R (M, Hom_S(U,N)) $\rightarrow$ Hom_S (U $\bigotimes_R$ M, N) by ϕ_{MUN}(f)(y $\otimes$ x) = ϕ_f(y $\otimes$ x) = f(x)(y). It is clear how to check that ϕ_{MUN} is a group isomorphism and that the diagram is commutative. □

2.11.6. Theorem. (Right exactness of $\otimes$). Consider a right R-module M and an exact sequence of left R-modules $0 \longrightarrow N' \xrightarrow[f]{} N \xrightarrow[g]{} N'' \longrightarrow 0$. Then the sequence of abelian groups:

(*) $M \bigotimes_R N' \xrightarrow[1_M\otimes f]{} M \bigotimes_R N \xrightarrow[1_M\otimes g]{} M \otimes_R N'' \longrightarrow 0$

is exact.
Similarly, for any left R-module N and an exact sequence of right R-modules $0 \longrightarrow M' \xrightarrow[e]{} M \xrightarrow[h]{} M'' \longrightarrow 0$, the sequence

(**) $M' \bigotimes_R N \xrightarrow[e\otimes 1_N]{} M \bigotimes_R N \xrightarrow[h\otimes 1_N]{} M'' \bigotimes_R N \longrightarrow 0$

is exact.

Proof. From (1_M $\otimes$ g) $\circ$ (1_M $\otimes$ f) = 1_M $\circ$ (g $\circ$ f) = 1_M $\otimes$ 0 = 0 we derive that X = Im (1_M $\otimes$ f) $\subset$ Ker (1_M $\otimes$ g). If (x,y") $\in$ M $\times$ N" then there exists a y $\in$

N such that g(y) = y". Define a well-defined, $\mathbb{Z}$-bilinear and R-balanced map Ψ: $M \times N" \to M \otimes_R N/X$ by putting $\Psi(x,y") = (x \otimes y)^\wedge$. So there exists a group homomorphism u: $M \otimes_R N" \to M \otimes_R N/X$ such that $u(x \otimes y") = (x \otimes y)^\wedge$. However, $u \circ (1_M \otimes g)$ is nothing but the canonical surjection π: $M \otimes_R N \to M \otimes_R N/$. If $z \in \text{Ker}\,(1_M \otimes g)$ then $\pi(z) = u((1_M \otimes g)(z)) = u(0) = 0$, hence $\text{Ker}\,(1_M \otimes g) = \text{Im}\,(1_M \otimes f)$ and consequently (*) is exact. Exactness of (**) follows in a similar way. □

2.11.7. Example ($\otimes$ **is not left exact).** Consider the eact sequence (with obvious maps): $0 \longrightarrow \mathbb{Z} \underset{i}{\longrightarrow} Q \longrightarrow Q/Z \longrightarrow o$

Putting $M = \mathbb{Z}/n\,\mathbb{Z}$ for some $n \neq 0$ in Theorem 2.11.6. yields an exact sequence: $\mathbb{Z} \otimes_{\mathbb{Z}} (\mathbb{Z}/n\,\mathbb{Z}) \to Q \otimes_{\mathbb{Z}} (\mathbb{Z}/n\mathbb{Z}) \to (Q/\mathbb{Z}) \otimes_{\mathbb{Z}} (\mathbb{Z}/n\,\mathbb{Z}) \to 0$.

We have $Q \otimes_{\mathbb{Z}} (\mathbb{Z}/n\mathbb{Z}) = 0$ because any $r \otimes x$ may e written as $\frac{1}{n} r \otimes n\,r = 0$. On the other hand, $\mathbb{Z} \otimes_{\mathbb{Z}} (\mathbb{Z}/n\mathbb{Z}) \cong \mathbb{Z}/n\mathbb{Z} \neq 0$, thus $\mathbb{Z} \otimes_{\mathbb{Z}} (\mathbb{Z}/n\mathbb{Z}) \to Q \otimes_{\mathbb{Z}} (\mathbb{Z}/n\mathbb{Z})$ is certainly not injective.

2.11.8. Definition A left R-module N is *flat* if a monomorphism f: $M'_R \to M_R$ leads to a monomorphism $f \otimes 1_N$: $M' \otimes_R N \to M \otimes_R N$. Clearly ${}_RR$ is a flat R-module. By Proposition 2.11.3. any direct sum of flat modules is flat; as an example we note that any free R-module is a flat module.

2.11.9. Proposition. Let N be a left flat R-module. Then N is injective if and only if the right R-module $\text{Hom}_{\mathbb{Z}}(N,Q/\mathbb{Z})$ is injective.

Proof. Consider an exact sequence of right R-modules:

$0 \longrightarrow M'_R \underset{f}{\longrightarrow} M_R \underset{g}{\longrightarrow} M''_R \longrightarrow 0$

From Theorem 2.11.5. we conclude that the following diagram is commutative and its rows are exact:

$$\begin{array}{ccccccc} 0 \longrightarrow & \text{Hom}_R(M'',N^*) & \longrightarrow & \text{Hom}_R(M,N^*) & \longrightarrow & \text{Hom}_R(M',N^*) \\ & \downarrow\cong & & \downarrow\cong & & \downarrow\cong \\ 0 \longrightarrow & \text{Hom}_R(M''\otimes_R N,Q/\mathbb{Z}) & \longrightarrow & \text{Hom}_R(M\otimes_R N,Q/\mathbb{Z}) & \longrightarrow & \text{Hom}_R(M'\otimes_R N,Q/\mathbb{Z}) \end{array}$$

If N is a flat R-module then the above diagram implies that $\text{Hom}_R(-,N^*)$ is exact (takes exact sequences to exact sequences) and thus N^* is injective. For the converse one uses that $\mathbb{Q}/\mathbb{Z}$ is an injective cogenerator for the class of abelian groups. □

2.11.10. Corollary. Let $0 \to N' \to N \to N" \to 0$ be an exact sequence of left R-modules. Suppose that N" is a flat R-module, then N' is flat if and only if N is flat.

Proof. Follows from the proposition by applying $\mathrm{Hom}_{\mathbb{Z}}$ (-,$\mathbb{Q}/\mathbb{Z}$) to the given sequence. □

Let M be a left R-module and consider a right ideal I of R. Put $\nu_{I,M}$: $I \bigotimes_R M \to M$, where $\nu_{I,M} = v_M \circ (i \otimes 1_M)$, i: $I \to R$ the canonical inclusion, v_M: $R \bigotimes_R M \to M$ the canonical isomorphism. Then $\nu_{I,M}(\sum_i a_i \otimes x_i) = \sum_i a_i x_i$, $x_i \in M$, $a_i \in R$, and $\mathrm{Im}\, \nu_{I,M} = IM$.

2.11.11. Proposition. The following assertions are equivalent:

1. M is a flat R-module.
2. If I is finitely generated then $\nu_{I,M}$ is a monomorphism.

Proof. 1⇒2. Obvious.

2⇒1. Let J be an arbitrary ideal of R and $z \in \mathrm{Ker}\, \nu_{J,M}$, say $z = \sum_{i=1}^n a_i \otimes x_i$, $a_i \in R$, $x_i \in M$, then $\nu_{J,M}(z) = \sum_{i=1}^n a_i x_i = 0$. Put I equal to $a_1 R + ... + a_n R$ is J and write $z' = \sum_{i=1}^n a_i \otimes x_i \in I \bigotimes_R M$. Then $\nu_{J,M}(z) = \nu_{I,M}(z') = 0$. Injectivity of $\nu_{I,M}$ implies x' = 0 and hence $z = 0$. Injectivity of $\nu_{J,M}$ follows and then $i \otimes 1_M$ is also injective. Pick a cogenerator C for the class of right $\mathbb{Z}$ -modules and write $M^* = \mathrm{Hom}_{\mathbb{Z}}$ (M,C). The proof of Proposition 2.11.9. yields an exact sequence: $\mathrm{Hom}_R\,(R, M^*) \to \mathrm{Hom}_R\,(J, M^*) \to 0$, i.e. M^* is an injective module by Baer's theorem but then M is flat in view of Proposition 2.11.9. □

2.11.12. Corollary. Let K be a submodule of a flat left R-module M. Then M/K is flat if and only if $IK = IM \cap K$ for each (finitely generated) left ideal I of R.

Proof. Let i_K: K → M be the canonical inclusion. The exact sequence

$$0 \longrightarrow K \underset{i_K}{\longrightarrow} M \underset{\pi}{\longrightarrow} M/K \longrightarrow 0$$

yields a commutative diagram of abelian groups with exact rows:

$$\begin{array}{ccccccccc} & & I \otimes_R K & \underset{1_I \otimes i_K}{\longrightarrow} & I \otimes_R M & \underset{1_I \otimes \pi}{\longrightarrow} & I \otimes_R M/K & \longrightarrow & 0 \\ & & \downarrow \alpha & & \downarrow \nu_{I,M} & & \downarrow \nu_{I,M/K} & & \\ 0 & \longrightarrow & K \bigcap IM & \longrightarrow & IM & \longrightarrow & I.(M/K) & \longrightarrow & 0 \end{array}$$

By the proposition, $\nu_{I,M}$ is an isomorphism. Applying the Snake Lemma it follows (or by elementary diagram chasing) that $\nu_{I,M/K}$ is a monomorphism if and only if α is an epimorphism. Now the corollary follows indeed from the proposition because Im α = IK. □

2.11.13. Proposition. A left R-module M is flat exactly then when for each relation $\sum_{i=1}^n b_i x_i = 0$ with $b_i \in$ R, $x_i \in$ M, there exist elements $u_1, \ldots, u_m \in$ M and $a_{ij} \in$ R, i = 1,...,n; j = 1,...,m, such that $\sum_{i=1}^n b_i a_{ij} = 0$ and $x_i = \sum_{j=1}^m a_{ij} u_j$.

Proof. Suppose first that M is flat and consider $\sum_{i=1}^n b_i x_i = 0$ as above. Put I $= \sum_{i=1}^n b_i R$ and let L be a free R-module with basis $\{e_1, \ldots, e_n\}$. Mapping e_i to b_i defines a map f: L → I and an exact sequence:

$$0 \longrightarrow K \underset{i_K}{\longrightarrow} L \underset{f}{\longrightarrow} I \longrightarrow 0$$

Then $\nu_{I,M}(\sum_{i=1}^n f(e_i) \otimes x_i) = \sum_{i=1}^n b_i x_i = 0$ and thus $\sum_{i=1}^n b_i \otimes x_i = 0$ in I $\otimes_R$ M. The flatness of M leads to an exact sequence:

$$0 \longrightarrow K \otimes_R M \underset{i_K \otimes 1_M}{\longrightarrow} L \otimes_R M \underset{f \otimes 1_M}{\longrightarrow} I \otimes_R M \longrightarrow 0$$

We retain: $\sum_{i=1}^n (e_i \otimes x_i) \in$ Ker $(f \otimes 1_M)$ = Im $(i_K \otimes 1_M)$. So there exist u ∈ M, $z_j \in$ K for j = 1,...,m, such that $\sum_{i=1}^n e_i \otimes x_i = \sum_{j=1}^m (z_j \otimes u_j)$. From $z_j \in$ L we conclude: $z_j = \sum_{i=1}^n e_i a_{ij}$, $a_{ij} \in$ R and $\sum_{i=1}^n b_i a_{ij} = \sum_{i=1}^n f(e_i) a_{ij} = f(z_j) = 0$ for all j = 1,...,m. On the other hand: $\sum_{i=1}^n e_i \otimes x_i = \sum_{j=1}^m z_j \otimes u_j = \sum_{j=1}^m (\sum_{i=1}^n e_i a_{ij}) \otimes u_j = \sum_{i=1}^n (e_i \otimes (\sum_{j=1}^m a_{ij} u_i))$. By Proposition 2.11.1. we obtain: L $\otimes_R$ M = $\bigoplus_{i=1}^n$ Im $(i_{e_i R} \otimes 1_M)$, so $x_i = \sum_{j=1}^m a_{ij} u_j$. Conversely, let I be a right ideal of R and look at $\sum_{i=1}^n b_i \otimes x_i \in$ Ker $\nu_{I,M}$. Then $\sum_{i=1}^n b_i x_i = 0$ and hence $\sum_{i=1}^n b_i a_{ij} = 0$ with $x_i = \sum_{j=1}^m a_{ij} u_j$.
In I $\otimes_R$ M we thus obtain: $\sum_{i=1}^n b_i \otimes x_i = \sum_{i=1}^n (b_i \otimes (\sum_{j=1}^m a_{ij} u_j)) = \sum_{j=1}^n ((\sum_{i=1}^n b_i a_{ij}) \otimes u_j) = 0$, and therefore Ker $\nu_{I,M} = 0$ so that we may apply Proposition 2.11.11. □

2.11.14. Proposition. If M is a finitely presented flat left R-module then M is a projective module.

Proof. Recall that M is finitely presented if there is an exact sequence

$$L' \longrightarrow L \underset{f}{\longrightarrow} M \longrightarrow 0$$

where L and L' are finitely generated free R-modules. Write K = Ker f, then we obtain an exact sequence:

$$(*) \qquad 0 \longrightarrow K \underset{i}{\longrightarrow} L \underset{f}{\longrightarrow} M \longrightarrow 0$$

where L is free and K,L,M are finitely generated R-modules.
Consider $\{u_1, ..., u_m\} \subset$ K; we aim to establish a homomorphism φ: L → K such that $\varphi(u_i) = u_i$ and we do this by induction on the number of elements in $\{u_1, ..., u_m\}$. If u ∈ K, then u = $\sum_{i=1}^n b_i e_i$ where e_1 , ... , e_n is a basis of L. Then, 0 = f(u) = $\sum_{i=1}^n b_i f(e_i)$. Applying Proposition 2.11.13. we find x_1 , ... , x_k ∈ M and a_{ij} ∈ R, i = 1,...,n, j = 1,...,k, such that $\sum_{i=1}^n b_i a_{ij} = 0$ and f(e_i) = $\sum_{j=1}^n a_{ij} x_j$. Pick t_j ∈ L so that f(t_j) = x_j, j = 1,...,k. Then f(e_i) = $\sum_{j=1}^k a_{ij} f(t_j)$ and thus e_i = $\sum_{j=1}^k a_{ij} t_j$ ∈ K, i = 1,...,n.
We define θ: L → K by $\theta(e_i) = e_i$ - $\sum_{j=1}^k a_{ij} t_j$, c = 1,...,n (L is free). Then θ(u) = $\sum_{i=1}^n b_i \theta(e_i) = \sum_{i=1}^n b_i e_i$ - $\sum_{i=1}^n \sum_{j=1}^k b_i a_{ij} t_j = \sum_{i=1}^n b_i e_i$ = u.
This proves the case m = 1 and now we assume that the assertion holds for all n < m. So we find φ_m: L→ K such that $\varphi_m(u_m) = u_m$. Denote $v_i = u_i$ - $\varphi_m(u_i)$, i = 1,...,m-1.
The induction hypothesis yields the existence of an R-linear φ': L → K such that φ'(v_i) = v_i, i = 1,...,m-1. It follows that φ'(u_i) = u_i - $\varphi_n(u_i)$+ φ'($\varphi_n(u_i)$), i = 1,...,m-1. Next we may define φ: L → K by φ = φ' + φ_n - φ' ∘ φ_n. Obviously, $\varphi(u_i) = u_i$ for all i = 1,...,m. Finally we may take the u_1 , ... , u_m to be a set of generators for K over R and then we deduce that φ ∘ i = 1_K, i.e the sequence (∗) is split.
It follows that M is a direct summand of L, hence a projective R-module, as stated. □

§2.12 Normalizing Extensions of a Ring.

Let R be a subring of S (with common identity element).If there exists a_1,..., a_n ∈ S such that $Ra_i = a_i$R and S = $\sum_{i=1}^n$ Ra_i then S is said to be a *normalizing extension of R.* If moreover $xa_i = a_ix$ holds for all $\in$ R then S is said to be a *centralizing extension of R.*

Consider an S-module M and write ${}_RM$ for the R-module obtained by restriction of scalars to R. If N is R-submodule of M, put $(N : a_i) = \{m \in M, a_i\ m \in N\}$. The normalizing condition entails that $(N : a_i)$ is an R-submodule M. Define a map f: $M/(N : a_i) \to M/N$, $f_i(\overline{x}) = \overline{a_ix}$; clearly, f_i is an injective morphism of abelian groups and if a_i commutes with all $x \in R$ then f_i is R-linear.

2.12.1. Lemma. If $X \in L_R(M/(N : a_i))$ then $f_i(X) \in L_R(M/N)$ and $X \to f_i(X)$ is a monomorphism from the lattice $L_R(M/(N : a_i))$ to $L_R(M/N)$.

Proof. From $a_iR = Ra_i$ it follows that $f_i(X)$ is an R-submodule of M/N and by the injectivity of f_i the statement follows. □

Write $(0 : a_i) = \{m \in M, a_im = 0\}$; then $(0 : a_i)$ is an R-submodule M. Note that for any R-submodule N of M we also have that each a_iN is an R-submodule.

2.12.2. Lemma. The lattices $L_R(a_iN)$ and $L_R(N/N \cap (0 : a_i)$ are isomorphic.

Proof. For $A \in L_R(a_iN)$ let B be $\{m \in N, a_im \in A\}$; again from $a_iR = Ra_i$ it follows that B is an R-submodule of N. Moreover $N \cap (0 : a_i) \subset B \subset N$ and also $A = a_iB$. Hence we may define a map φ: $L_R(a_iN) \to L_R(N/N \cap (0 : a_i))$ by $\varphi(A) = B/N \cap (0 : a_i)$ and one checks that φ is an isomorphism. □

2.12.3. Lemma. For any R-submodule N of the S-module M the set $N^* = \bigcap_{i=1}^n (N : a_i)$ is the largest S-submodule of M which is contained in N.

Proof. Pick $m \in N^*$, i.e. $a_im \in N$ for $i = 1,\dots,n$. If $s \in S$ then $a_js = \sum_{i=1}^n r_ia_i$ for some $r_i \in R$, $i = 1,\dots,n$. Then $a_j(sm) = \sum r_i\,(a_im) \in N$ or $sm \in (N : a_j)$ for all $j = 1,\dots,n$. Consequently N^* is an S-submodule of M.
Writing $1 = \sum_{i=1}^n \mu_ia_i$ with $\mu_i \in R$ yields $m = \sum_{i=1}^n \mu_ia_im$ hence $m \in N$ and thus

$N^* \subset$ N. Next, consider an S-submodule P contained in N and pick p $\in$ P. Then a_im $\in$ P yields a_im $\in$ N and m $\in$ (N : a_i) or m $\in$ N^* follows, i.e. P $\subset$ N^*. □

2.12.4. Lemma. If N is essential as an R-submodule of M then N^* is essential in M as an R-submodule.

Proof. Let P $\neq$ 0 be an R-submodule of M. If a_iP = 0 then P $\subset$ (N : a_i). If a_iP $\neq$ 0 then N $\bigcap$ a_iP $\neq$ 0 yields P $\bigcap$ (N : a_i) $\neq$ 0 and hence (N : a_i) is an essential R-submodule of M for i = 1,...,n. Obviously N^* is then also an essential R-submodule of M since n is finite. □

2.12.5. Lemma. The S-module M cotains an R-submodule N maximal with respect to the property $N^* = 0$.

Proof. Look at a sequence of R-submodules of M, $(N_i)_{i\in I}$ say, such that $N_i^* = 0$ for all i $\in$ I, and put N = $\bigcup_{i\in I} N_i$. If $N^* \neq 0$, pick an x $\neq$ 0 in N^*. Since Sx $\subset$ $N^* \subset$ N it follows that a_kx $\in$ N for all k = 1,...,n.For each k there is an index i(k) $\in$ I such that a_kx $\in$ $N_{i(k)}$, since n is finite we obtain that there is an i $\in$ I such that x $\in$ $(N_i : a_k)$ for all k = 1,...,n. Then x $\in$ N_i^* leads to a contradiction since $N_i^* = 0$. □

2.12.6. Proposition. Let ${}_SM$ have finite Goldie dimension, say u.dim$_S$M = m, and let N be an R-submodule maximal with the property $N^* = 0$, then:

1. u.dim$_R$ (M/N) $\leq$ m.
2. u.dim$_R$ (M) $\leq$ mn.

Proof. 1. Consider R-submodules N_i in M such that each N_i, i = 1,...,t, strictly contains N and such that the finite family $(N_i/N)_{i=1,...,t}$ is direct in M/N. Since N $\subsetneq$ N_k we have $N_k^* \neq 0$. If t > m then the family $(N_k^*)_{k=1,...,t}$ cannot be direct, therefore there exists an index l such that $(\sum_{k\neq l} N_k^*) \bigcap N_l^* \neq 0$. Thus $((\sum_{k\neq l} N_k) \bigcap N_l)^* \neq 0$ and also N $\subsetneq$ $(\sum_{k\neq l} N_k) \bigcap N_l$, but this yields a contradiction because $(N_i/N)_{i=1,...,t}$ is supposed to be direct.
2. From 0 = N^* = $\bigcap_{i=1}^n$ (N : a_i) we retain that there exists a monomorphism of R-modules: 0 $\to$ M $\to$ $\bigoplus_{i=1}^n$ M/(N : a_i). Lemma 2.12.1. implies that

u.dim$_R$ (M/(N : a_i)) $\leq$ u.dim(M/N) $\leq$ m and accordingly u.dim$_R$M $\leq$ m.n. □

2.12.7. Theorem (Eakin [1], Formanek and Jategaonkar [1]). Let M be a left S-module. Then M is a Noetherian S-module if and only if M is a Noetherian R-module.

Proof. If M is a Noetherian R-module then trivially it is Noetherian as an S-module too. Conversely, let M be a Noetherian S-module but not Noetherian as an R-module. We may consider an S-submodule M' of M which is maximal with the property that M/M' is not a Noetherian R-module (such an M exists since M is Noetherian as an S-module and M' = 0 yields an M/M' which is not Noetherian as an R-module). Up to replacing M by M/M' we may assume that M is not Noetherian as an R-module but such that for all nonzero S-submodules N of M, M/N is a Noetherian R-module.

Now consider an R-submodule P of M maximal with respect to the property $P^* = 0$ (cf. Lemma 2.12.5.). Let there be an ascending chain of R-submodules $(N_i/P)_{i\in I}$ in M/P, $0 \neq N_1/P \subset N_2/P \subset ... \subset N_k/P \subset ...$. From

P $\subsetneq$ N_k it follows that $N_k^* \neq 0$ for all k = 1,...,n. In particular $N_1^* \neq 0$ and by assumption M/N_1^* is a Noetherian R-module (as N_1^* is an S-submodule of M). From $0 \neq N_1^* \subset N_1 \subset N_2 \subset ... \subset N_k \subset ...$, it follows that $N_k = N_{k+1} = ...$, for some k.

We have proved that M/P is a Noetherian R-module. But we have a monomorphism $0 \to M \to \bigoplus_{i=1}^n M/(P : a_i)$, where by Lemma 2.12.1. each M/(P : a_i) is a Noetherian R-module. Hence M is a Noetherian R-module, a contradiction. □

2.12.8. Corollary. If S is a normalizing extension of R then S is left Noetherian if and only if R is left Noetherian.

Proof. If S is left Noetherian then Theorem 2.12.7. implies that S is a left Noetherian R-module and therefore ${}_RR$ is also a Noetherian module. The converse is easy. □

2.12.9. Lemma. Let $S = \sum_{i=1}^n Ra_i$ be a normalizing extension of R. If M is a simple S-module then M is a semisimple R-module of finite length. If S is a centralizing extension then M is an isotypic semisimple R-module.

Proof. Take $x \neq 0$ in M, then $M = Sx = \sum_{i=1}^{n} Ra_ix$ implies that M is a finitely generated R-module. Let $N \neq M$ be a maximal R-submodule of M. From $N^* \subset N \subset M$, $0 = N^*$ follows. This leads to the monomorphism:
$0 \to M \to \bigoplus_{i=1}^{n} M/(N : a_i)$. By Lemma 2.12.. we may conclude that $M/(N : a_i) = 0$ or a simple R-module, consequently M is a semisimple R-module of fiite length and moreover $l(M) \leq n$ holds.
In the centralizing case we actually obtain $M/(N : a_i) = 0$ or $M/(N : a_i) \cong M/N$ as R-modules, hence in this case M even isotypic. □

2.12.10. Corollary. Let S be a normalizing extension of R, then S is left Artinian if and only if R is left Artinian.

Proof. If ${}_SS$ is Artinian then it has a composition sequence and in view of Lemma 2.12.9. it follows that ${}_RS$ is an R-module of finite length. The latter obviously implies that ${}_RR$ has finite length i.e. R is left Artinian.
The converse is easy enough. □

2.12.11. Corollary. Let S be a normalizing extension of R and let M be a left S-module, then M is a semiartinian S-module if and only if M is a semiartinian R-module.

Proof. Let

$$(*) \qquad 0 = s_0(M) \subset s_1(M) \subset ... \subset s_\alpha(M) \subset ...$$

be the Loewy series for ${}_SM$, which is supposed to be semiartinian. For some ordinal α, $M = s_\alpha(M)$. Since $s_{\alpha+1}(M)/s_\alpha(M)$ is a semisimple S-module it follows that $s_{\alpha+1}(M)/s_\alpha(M)$ is a semisimple R-module (cf. Lemma 2.12.9.). By transfinite induction we obtain that ${}_RM$ is semiartinian.
On the other hand if M is a semiartinian R-module but not semiartinian as an S-module, let $\lambda(M)$ be the length of the Loewy series (*) and put $M' = s_{\lambda(M)}(M)$. By our assumptions $M' \neq M$ but $M" = M/M'$ does not contain a nontrivial simple module. The semiartinian property of ${}_RM''$ allows us to replace M by M" i the sequel of the proof. Since ${}_RM$ is semiartininan it contains a simple R-module X; put $Y = SX = \sum_{i=1}^{n} a_iX$. In view of Lemma 2.12.2., a_iX is either zero or simple, so I is a simple R-module of finite length. Consequently Y is a Noetherian and Artinian S-module i.e. Y has finite length as an S-module. The latter entails that Y must

contain a simple S-module, so M must contain a simple S-module too, contradiction. □

2.12.12. Corollary. If S is a normalizing extension of R then S is left semiartinian if and only if R is left semiartinian.

2.12.13. Corollary. Let S be a normalizing extension of R, then
$J(R) \subset J(S) \cap R$.

Proof. Consider a simple S-module M and $a \in J(R)$. Since M is a semisimple R-module (cf. Lemma 2.12.9.) we have $aM = 0$ or $a \in J(S)$ as claimed. □

§2.13 Graded Rings and Modules.

We consider a multiplicatively written group G with identity element e and G is not necessarily abelian.

A ring R is said to be *graded of type G* if there exists a family of additive subgroups $\{R_\sigma \mid \sigma \in G\}$ of R such that $R = \bigoplus_{\sigma\in G} R_\sigma$ as an additive group and $R_\sigma R_\tau \subset R_{\sigma\tau}$ for every $\sigma, \tau \in G$, where $R_\sigma R_\tau$ stands for the set of finite sums of products $r_\sigma r'_\tau$ where $r_\sigma \in R_\sigma$, $r'_\tau \in R_\tau$.
The set of *homogeneous elements* of R is $h(R) = \bigcup_{\tau\in G} R_\tau$ and we put $\deg r = \sigma$ when $r \in R_\sigma$. From the definition it is clear that each nonzero r of R may be written as $\sum_{\sigma\in G} r_\sigma = r$ with $r_\sigma \in R_\sigma$ and only finitely many of the r_σ are nonzero. Clearly this decomposition is unique, so we refer to it as the *homogeneous decomposition* of r and we say that r_σ is the *homogeneous component of r of degree σ.*

2.13.1. Proposition. Let $R = \bigoplus_{\sigma\in G} R_\sigma$ be a graded ring of type G, then R_e is a subring of R (note: $1 \in R_e$) and each R_σ is an R_e-bimodule.

Proof. The statement is obvious if we establish that $1 \in R_e$. Let $1 = r_1 + ... + r_n$ be the homogeneous decomposition of 1 with $\deg r_i = \sigma_i \in G$. For an arbitrary

$x_\tau \in R_\tau$, some $\tau \in$ G, $1.x_\tau = r_1 x_\tau + ... + r_n x_\tau$ gives rise to two different homogeneous decompositions of the element x_τ unless $r_i x_\tau = 0$ for all i such that $\sigma_i \neq$ e. The latter holds for all x_τ and all $\tau \in$ G, hence $r_i = 0$ for all i such that $\sigma_i \neq$ e or $1 = r_e \in R_e$ follows. □

Some easy examples of graded rings appear frequently in several areas of commutative algebra (projective geometry), representation theory, rings of differential operators etc... .

Examples. 1. Any ring may be considered as a graded ring of type G for any group G by defining R_e = R, $R_\sigma = 0$ if $\sigma \neq$ e.

2. Let A be a ring and G a group. The group ring AG is the free A-module with basis G and multiplication defined by $(\sum_{\sigma\in G} a_\sigma \sigma)(\sum_{\tau\in G} b_\tau \tau) = \sum_{\theta\in G} c_\theta \theta$, where $c_\theta = \sum_{\sigma\tau=\theta} a_\sigma b_\tau$. The group ring AG may be viewed as a graded ring of type G with gradation defined by $AG_\sigma = A\sigma = \{a\sigma, a \in A\}$, for $\sigma \in$ G.

3. Let A be a ring and let φ: A → A be an injective ring homomorphism. Consider the ring of polynomials, A[X], in one indeterminate X. Any polynomial may be written as $\sum_{i=0}^{u} a_i X^i$ with $a_i \in$ A, i ∈ ℕ . Now we define a new multiplication on the abelian group of A[X] by putting: X.a = φ(a)X, for all a ∈ A. The ring thus obtained will be denoted by A[X,φ] and it is is called the *skew polynomial ring* determined by φ. The ring A[X,φ] is a graded ring of type ℤ with: $A[X,\varphi]_m = 0$ if m < 0, $A[X,\varphi]_n = AX^n$ if n ≥ 0.

4. Let A be a ring and G a group. Let Z(A) = {c ∈ A, ca = ac for all a ∈ A} be the center of A. Consider the free A-module with basis G as in 2. but now define multiplication by the rule: $(\sum_{\sigma\in G} a_\sigma \sigma)(\sum_{\tau\in G} b_\tau \tau) = \sum_{\theta\in G} c_\theta \theta$ where $c_\theta = \sum_{\sigma\tau=\theta} a_\sigma b_\tau f(\sigma,\tau)$ where f: G × G → UZ(A), $(\sigma,\tau) \mapsto f(\sigma,\tau)$, is a 2-cocycle (with trivial action) into the units UZ(A) of Z(A), i.e. f satisfies: $f(\sigma,\tau) f(\sigma\tau,\gamma) = f(\tau,\gamma) f(\sigma,\tau\gamma)$ for all $\sigma,\tau,\gamma \in$ G. It is easy to verify that this operation defines a ring AG^f, called the *twisted group ring with respect to f* (note that the cocycle condition explicited above is exactly stating the associativity of AG^f). Again AG^f may be viewed as a graded ring of type G by putting $(AG^f)_\sigma = A\sigma = \{a\sigma, a \in A\}$.

5. Let A be a ring and G a group. Suppose there is given a group morphism ϕ: G → Aut A where Aut A is the group of ring automorphisms g: A → A. Write $\phi(\sigma) = \alpha_\sigma$ for $\sigma \in$ G. We let f: G × G → UZ(A) be a 2-cocycle satisfying: $f(\sigma,\tau) f(\sigma\tau,\gamma) = f(\tau,\gamma)^{\alpha_\sigma} f(\sigma,\tau\gamma)$, and we define multiplication on the free A-module with basis G as follows: $(\sum_{\sigma\in G} a_\sigma \sigma)(\sum_{\tau\in G} b_\tau \tau) = \sum_{\theta\in G} c_\theta \theta$, where $c_\theta = \sum_{\sigma\tau=\theta} a_\sigma b_\tau^{\alpha_\sigma}$

$f(\sigma,\tau)$. Again we obtain a ring structure on that free A-module. The ring obtained is denoted by $A *_{(\alpha,f)} G$ or simply $A * G$ if there is no ambiguity and it is called the *crossed product* of A and G with respect to α and f. As in foregoing examples, $A * G$ may be viewed as a graded ring of type G if we put $(A * G)_\sigma = A\sigma =$
$\{a\sigma, a \in A\}$.

In the sequel of this section $R = \bigoplus_{\sigma\in G} R_\sigma$ is a graded ring of type G.
A left R-module M is said to be a *graded left R-module* if there exists a family of additive subgroups, M_σ, $\sigma \in G$, of M such that $M = \bigoplus_{\tau\in G} M_\tau$ and $R_\sigma M_\tau \subset M_{\sigma\tau}$ for all $\sigma,\tau \in G$. The set $h(M) = \bigcup_{\sigma\in G} M_\sigma$ is the set of *homogeneous elements* of M. From $R_e M_\sigma \subset M_\sigma$ it follows that M_σ is an R_e-module for all $\sigma \in G$.
Any $m \neq 0$ in M_σ is said to be *homogeneous of degree* σ, written $\deg(m) = \sigma$. Any nonzero $m \in M$ has a unique epression as $\sum_{\sigma\in G} m_\sigma$ where $m_\sigma \in M_\sigma$ are zero up to a finite number of $\sigma \in G$. The m_σ are the *homogeneous components*. A submodule N of M is called a *graded (left) submodule* if $N = \bigoplus_{\sigma\in G} (N \bigcap M_\sigma)$ or equivalently, if for any $x \in N$ the homogeneous components of x in M are contained in N. If N is a graded submodule of M then the factor module M/N is graded in the following way: $(M/N)_\sigma = (N + M_\sigma) / N$.

Examples. 1. If M is a graded R-module then every submodule generated by a set of homogeneous elements of M is also a graded submodule.

2. Let M be a graded R-module, N any R-submodule of M. By $(N)_g$ we denote the R-submodule of M generated by $N \bigcap h(M)$. Clearly, $(N)_g$ is the largest graded submodule contained in N.

If M is a graded R-module then $L^g_R(M)$ (or more simply $L^g(M)$) will be the set of graded R-submodules of M.

2.13.2. Proposition. $L^g_R(M)$ is a sublattice of $L_R(M)$. More precisely, $L^g_R(M)$ is a complete modular, upper-continuous and compactly generated lattice.

Proof. If $(N_i)_{i\in I}$ is a family of elements of $L^g_R(M)$ then $\bigcap_{i\in I} N_i$ and $\sum_{i\in I} N_i$ are graded R-submodules of M. □

By $L^g(R)$ we denote the set of graded ideals of R.

2.13.3. Corollary. $L^g(R)$ is a modular an complete sublattice of L(R).

Observations. 1. If R is a graded ring and M is a graded R-module then $L^g(M) \neq L(M)$ in general. Indeed, consider a ring A with trivial $\mathbb{Z}$-gradation On the A-module N = M × M we may introduce a nontrivial gradation by putting: $N_0 = M \times \{0\}$, $N_1 = \{0\} \times M$ and $N_i = 0$ for all $i \neq 0,1$. Obviously $L^g(N)$ consists of all submodules M' × M" of M × M where M' and M" are arbitrary submodules of M. On the other hand $\Delta_M = \{(m,m), m \in M\}$ is a submodule of M which is not graded, $\Delta_M \in L_A(N)$, $\Delta_M \notin L^g_A(N)$.

2. It is already obvious from 1 that there may exist many different gradations on an R-module M whereas on the other hand each R-module may be viewed as a graded R-module. Specific structure results for graded rings may thus only be expected when the gradation(s) considered have restrictive properties.

3. The graded R-module M is finitely generated if and only if M may be generated by a finite number of homogeneous elements, in other words, L(M) is compact if and only if $L^g(M)$ is compact.

Let $R = \bigoplus_{\sigma \in G} R_\sigma$ be a graded ring of type G and $M = \bigoplus_{\sigma \in G} M_\sigma$ a graded R-module. If the lattice $L^g_R(M)$ is Noetherian resp. Artinian, resp. of finite length then M is said to be *gr-Noetherian*, res. *gr-Artinian*, resp. of *finite gr-length.*

We say that the graded R-module is *gr-simple* if 0 and M are the only graded submodules of M. Obviously, M is gr-simple if and only if $M \neq 0$ and M is an atom in $L^g_R(M)$. The proof of the following proposition is similar to the proof of Proposition 2.3.1., i.e. using the results of Section 1.2.

2.13.4. Proposition. For a graded R-module M the following statements are equivalent:

1. M is gr-Noetherian, resp. gr-Artinian.
2. Any ascending chain, resp. descending chain, of graded submodules of M is stationary.
3. Any graded submodule of M, resp. for any family $(N_i)_{i \in I}$ of graded submodules

of M, is finitely generated, resp. there exists a finite part J ⊂ I such that $\bigcap_{i\in I} N_i = \bigcap_{j\in J} N_j$.

It is obvious that a graded R-module M is gr-simple if and only if for all x ∈ h(M), x ≠ 0, we have Rx = M. Exactly as for ungraded modules in Section 2.3. one shows also that M has finite gr-length if and only if M is gr-Noetherian and gr-Artinian, if and only if M has a composition series with gr-simple factor modules.

Consider graded R-modules M = $\bigoplus_{\sigma\in G}$. An R-linear map f: M → N is said to be a *gr-homomorphism* if f(M_σ) ⊂ N_σ for all σ ∈ G. Write Hom^g_R(M,N) for the set of gr-homomorphisms of M to N; this set is a subgroup of Hom_R(M,N) (there is difference of notation here when compared to the notation used in C. Năstăsescu, F. Van Oystaeyen [3], the reason for this is that here we emphasize the intrinsic properties of graded modules e.g. by looking at the lattice of graded submodules, while on the other hand we keyed down the role of categorical considerations in this theory).
If f ∈ Hom^g_R(M,N) then it is clear that both Ker f and Im f are graded submodules. If f ∈ Hom^g_R(M,N) is bijective then it is called a *gr-isomorphism*.The reader may convince himself that f ∈ Hom^g_R(M,N) is a gr-isomorphism if and only there exists a g ∈ Hom^g_R(M,N) such that g ∘ f = 1_M and f ∘ g = 1_N. We say that M and N are *gr-isomorphic*, and we denote this by M $\underset{gr}{\cong}$ N (sometimes M ≅ N if no confusion arises from this), if and only if there exists a gr-isomorphism f: M → N. The isomorphism theorems, cf. 2.1., also hold for graded modules and gr-isomorphisms

§2.14 Graded Rings and Modules of Type ℤ Internal Homogenisation.

In this section R is a graded ring of type ℤ and M is always a graded (left) R-module.
We say that M is *left limited* (resp. *right limited*) if M_i = 0 for all i < n_0 for some n_0 ∈ ℤ (resp. M_i = 0 for all i ≥ n_0). If M_i = 0 for all i< 0 then M is said to be *positively graded.*

2.14.1. Lemma. Let R = $\bigoplus_{n\in\mathbb{Z}} R_n$ be a left limited graded ring and let M be a graded R-module If M is finitely generated then M is left limited.

Proof. Let $x_1, \ldots, x_m \in h(M)$ generate M and suppose deg $x_1 \leq \ldots \leq$ deg x_m. Since R is left limited there exists an $n_0 \in \mathbb{Z}$ such that $R_i = 0$ for all $i < n_0$. Take $j < n_0 +$ deg x_1 then for any $x \in M_j$ we have $x = r_1x_1 + \ldots + r_mx_m$. The homogeneous part of lowest degree appearing in the right hand member is equal to or higher than $n_0 +$ deg x_1 so $x \in M_j$ must be zero i.e. $M_j = 0$ for all $j < n_0 +$ deg x_1. □

For a graded R-module M we write $M^+ = \bigoplus_{i \geq 0} M_i$, $M^- = \bigoplus_{i \leq 0} M_i$. It is obvious that R^- and R^+ are subrings of R and that M^+ is a graded R^+-module, M^- is a graded R^--module.

If N is a graded R-submodule of M then: $(M/N)^+ = M^+/N^+$ and $(M/N)^- = M^-/N^-$.

Now consider an arbitrary R-submodule X of M. An $x \neq 0$ inX may be decomposed as $x_1 + \ldots + x_n$, where deg $x_1 <$ deg $x_2 < \ldots <$ deg x_n, in a unique way. Write $X^\sim$, resp. $X_\sim$, for the R-submodule of M generated by the elements x_n, resp. x_1, when x varies through X (note that n depends on x each time).

2.14.2. Lemma. With notations and conventions as before:

1. $X^\sim$ and $X_\sim$ are graded R-submodules of M.
2. $X = X^\sim$ if and only if X is a graded R-submodule of M.
3. If $X \subset Y$ are R-submodules of M then $X^\sim \subset Y^\sim$ and $X_\sim \subset Y_\sim$.
4. $X = 0$ if and only if $X^\sim = 0$ if and only if $X_\sim = 0$.

2.14.3. Proposition. Let $X \subset Y$ be R-submodules of the graded R-submodule M, then there is equivalence between the assertions:

1. $X = Y$.
2. $X^\sim = Y^\sim$ and $X \cap M^- = Y \cap M^-$.
3. $X_\sim = Y_\sim$ and $X \cap M^+ = Y \cap M^+$.

Proof. Let us show 2⇒1., the implication 3⇒1. may be established in formally the same way. Pick $y \in Y$, say $y = y_1 + \ldots + y_m$ with deg $y_1 < .. <$ deg $y_m = t$. We show that $y \in X$ by induction on t. If $t \leq 0$ then $y \in M^-$ and then $y \in Y \cap M^- \subset X$. Suppose $t > 0$. Then $y_m \in Y^\sim = X^\sim$ or there exists an $x \in X$ such that $x = x_1 + \ldots + x_r + y_m$ with deg $x_1 < \ldots <$ deg $x_2 < t$. Since y - x $\in$ Y and y - x has homogeneous component of highest degree of degree lower that t, the induction hypothesis yields y - x ∈ X and thus y ∈ X is a direct consequence. □

2.14.4. Corollary. If M is left limited, resp. right limited, then X = Y if and only if $X^{\sim} = Y^{\sim}$, resp. $X_{\sim} = Y_{\sim}$.

2.14.5. Corollary. If X ⊂ M is an R-submodule of the left limited (right) graded R-module M such that $X^{\sim}$ ($X_{\sim}$) is generated by r homogeneous elements, then X may be generated by e elements.

Proof. Let $x_1, \ldots, x_r \in h(M)$ generate $X^{\sim}$. Select $y_1, \ldots, y_r \in X$ such that x_i appears as the homogeneous component of highest degree in the decomposition of y_i in M. Let Y be the R-submodule of M generated by $y_1, \ldots, y_r$. Obviously Y ⊂ X and $Y^{\sim} = X^{\sim}$. Corollary 2.14.5. then finishes the proof. □

Homogenisation can also be defined "externally" in a polynomial extension M[X] of M (over R[X]); actually this may be done for arbitrary grading groups utilizing M[G] over RG but we do not go into these matters here because the applications in dimension theory deal either with the case G = $\mathbb{Z}$ or the case where G is finite.

§2.15 Noetherian Modules over Graded Rings of Type $\mathbb{Z}$. Applications.

Let $R = \bigoplus_{n\in\mathbb{Z}} R_n$ be a graded ring of type $\mathbb{Z}$ and let $M = \bigoplus_{n\in\mathbb{Z}} M_n$ be a graded (left) R-module.

2.15.1. Proposition. If M has left (or right) limited gradation then M is gr-Noetherian (gr-Artinian) if and only if M is Noetherian (Artinian).

Proof. Assume that M is gr-Noetherian and let $X_1 \subset X_2 \subset \ldots \subset X_n \subset \ldots$ be an ascending chain of submodules of M. Associated to this chain we have another ascending chain, this time of graded R-modules: $X_1^{\sim} \subset ... \subset X_n^{\sim} \subset ...$. By the assumptions $X_n^{\sim} = X_{n+1}^{\sim} = ...$ for some $n \in \mathbb{N}$, but then Corollary 2.14.4. entails that $X_n = X_{n+1} = ...$. Thus M is Noetherian. The similar statement in the Artinian case can be proved in the same manner. □

2.15.2. Proposition. If M is a gr-Noetherian graded R-module of type $\mathbb{Z}$ then we have

1. For each $i \in \mathbb{Z}$, M_i is a Noetherian R_0-module
2. M^+ is a Noetherian R^+-module.
3. M^- is a Noetherian R^--module.

Proof. 1. Consider an R_0-submodule N_i of M_i. Then RN_i is a graded R-submodule of M, so by assumption RN_i may be generated by finitely many $x_1 , ... , x_k \in h(M)$ and in fact, we may assume $x_1 , ... , x_k \in N_i$. If $y \in N_i$ then $y = \sum_{i=1}^{n} r_i x_i$ for some $r_i , .. , r_k \in R$. Up to replacing the r_i by some homogeneous component in their decompositions we may assume that $r_i \in h(R)$ for i = 1,...,k. In that case deg $r_i = 0$ for i = 1,...,k and it follows that N_i is a finitely generated R_0-module; so M_i is a Noetherian R_0-module.

2. In view of Proposition 2.14.1. it suffices to show that M^+ is gr-Noetherian over the positively graded ring R^+. Let $N = \bigoplus_{i \geq 0} N_i$ be a graded R^+-submodule of M^+. The extended R-module RN is a graded R-submodule of M and so we may assume that $x_1 , ... , x_k \in h(M)$ generate RN and even that $x_1 , ... , x_k \in N$. Put t = max (deg x_1 , ... , deg x_k) and pick $y \in h(N)$ with deg $y \geq t$. There exist $r_1 , ... , r_k \in h(R)$ such that $y = \sum_{i=1}^{k} r_i x_i$. Since deg $y \geq t$ it follows that deg $r_i \geq 0$ for all i = 1,...,k and thus $r_1 , ... , r_k \in R^+$. From 1. we derive that $M_0 \oplus ... \oplus M_{t-1}$ is a Noetherian R_0-module, hence $N_0 \oplus ... \oplus N_{t-1}$ is a finitely generated R_0-submodule. Clearly, $\{x_1 , ... , x_k , y_1 , ... , y_s\}$ will generate N over R^+ whenever $y_1 , ... , y_s$ generate $N_0 \oplus ... \oplus N_{t-1}$ over R_0, hence M^+ is gr-Noetherian as a graded R^+-module.

3. Similar to the proof of 2. □

2.15.3. Remark. For any R_0-submodule N_i of M_i we have that $RN_i \cap M_i = N_i$. From this one easily deduces that each M_i is an Artinian R_o-module when M is gr-Artinian.

2.15.4. Theorem. Let M be a graded R-module of type $\mathbb{Z}$. The following assertions are equivalent:

1. M is gr-Noetherian.
2. M is Noetherian.

Proof. (cf. C. Năstăsescu [6]) 2⇒1 is obvious.

1⇒2. If $X_1 \subset X_2 \subset ... \subset X_n \subset ...$ is an ascending chain of R-submodules of M then there exists $n_0 \in \mathbb{N}$ (cf. Proposition 2.15.2.) such that $M^- \bigcap X_i = M^- \bigcap X_{i+1}$ and $X_i^{\sim} = X_{i+1}^{\sim}$ for all i $\geq n_0$. By Proposition 2.14.3. it follows that $X_i = X_{i+1}$ for all i $\geq n_0$. □

2.15.5. Remark. If M is gr-Artinian then it need not be Artinian. The obvious example is $k[X, X^{-1}]$ where k is a field and X is an indeterminate; in this ring every homogeneous element is invertible (i.e. it is a gr-field) hence it is certainly gr-Artinian but not Artinian.

We now mention some results that may be viewed as applications of the foregoing.

2.15.6. Lemma. Let R be a positively graded ring and let M be a graded R-module. Suppose there is an element a $\in R_1$ such that: $R_0 a = aR_0$ and $M_i = a^i M_0$ for all i ≥ 0. If M_0 is a Noetherian R_0-module than M is a Noetherian R-module.

Proof. Consider a graded R-submodule N of M. For i ≥ 0 put $P_i = \{x \in M_0, a^i x \in N_i\}$. That P_i is an R_0-submodule of M_0 is clear, moreover $P_o = N_o$ and $P_i \subset P_{i+1}$ for all i ≥ 0. Furthermore, $a^i P_i = N_i$ for all i ≥ 0. The Noetherian property of the R_0-module M_0 yields that $P_n = P_{n+1} = ...$ for some n $\in \mathbb{N}$.
Choose a system of generators $\{z_i^{(1)}, ..., z_i^{(r_i)}\}$ for P_i over R_0. It is not hard to verify that $\bigcup_{i=0}^n \{a_i z_i^{(1)}, ..., a^i z_i^{(r_i)}\}$ generates N as an R-module. It follows that M is gr-Noetherian and it suffices to apply Theorem 2.15.4. to obtain that M is a Noetherian R-module as claimed. □

Let A be an arbitrary ring, M any left A-module. By M[X] we mean the module of polynomials with coefficients in M and indeterminate X, i.e. $M[X] = \{m_0 + m_1X + \ldots + m_rX^r,\ m_i \in M\}$. Clearly M[X] is an A[X]-module and it is actually a (positively) graded A[X]-module with gradation defined by $M[X]_i = \{mX^i,\ m \in M\}$ for all $i \geq 0$.

2.15.7. Theorem. (Hilbert's Basis Theorem). If M is a Noetherian A-module then M[X] is a Noetherian A[X]-module.

Proof. The element X satisfies the conditions of the lemma. □

2.15.8. Remark. If φ is an automorphism of A then $A[X,\varphi]$ (cf. 2.13.) is a graded ring satisfying the conditions of the lemma. Hence if A is left Noetherian then $A[X,\varphi]$ is left Noetherian.

Consider rings A and B together with A-B-bimodules ${}_AM_B, {}_AN_B$. Look at the ring $T = \begin{pmatrix} A & M \\ N & B \end{pmatrix}$ consisting of all matrices $\begin{pmatrix} a & m \\ n & b \end{pmatrix}$ with $a \in A$, $m \in M$, $n \in N$, $b \in B$, with addition and multiplication defined as follows:

$$\begin{pmatrix} r & m \\ n & s \end{pmatrix}\begin{pmatrix} a & m' \\ n' & b \end{pmatrix} = \begin{pmatrix} ra & rm' + mb \\ na + sn' & sb \end{pmatrix}$$

and with the usual addition of matrices.
We may define a $\mathbb{Z}$-gradation on T by putting: $T_{-1} = \begin{pmatrix} 0 & 0 \\ N & 0 \end{pmatrix}$, $T_0 = \begin{pmatrix} A & 0 \\ 0 & B \end{pmatrix}$, $T_1 = \begin{pmatrix} 0 & M \\ 0 & 0 \end{pmatrix}$ and $T_i = 0$ for all $|i| \geq 2$.

2.15.9. Lemma. The graded left ideals of T are of the form: $\begin{pmatrix} I & M' \\ N' & J \end{pmatrix}$ where I and J are left ideals in A and B resp., and where M' is an A-submodule of M, N' a B-submodule of N, such that $M J \subset M'$, $N I \subset N'$.

Proof. Straightforward. □

2.15.10. Theorem. The ring T is left Noetherian, resp. Artinian, if and only if: A and B are left Noetherian, resp. Artinian, M is a finitely generated A-module, N is a finitely generated B-module.

Proof. If T is Noetherian, resp. Artinian then Proposition 2.15.2. and the remark following it, entail that T_0 is Noetherian, resp. Artinian, and T_{-1} and T_1 are left Noetherian, resp. Artinian, T_0-modules. The desired properties of A, B, M, N follow easily.
For the converse it suffices to apply Lemma 2.15.9. and Proposition 2.15.1. □

2.15.11. Remark. As a consequence of the foregoing theorem we point out that the ring T = $\begin{pmatrix} \mathbb{Q} & \mathbb{Q} \\ 0 & \mathbb{Z} \end{pmatrix}$ is left Noetherian but not right Noetherian, whereas on the other hand T' = $\begin{pmatrix} \mathbb{R} & \mathbb{R} \\ 0 & \mathbb{Q} \end{pmatrix}$ is left Artinian but not right Artinian.

§2.16 Strongly Graded Rings and Clifford Systems for Finite Groups.

If R = $\bigoplus_{\sigma\in G} R_\sigma$ is a graded ring of type G such that $R_\sigma R_\tau = R_{\sigma\tau}$ for all $\sigma,\tau \in$ G then R is a *strongly graded ring* (or a *generalized crossed product*). In the study of graded rings it is natural to look at epimorphic images of a graded ring even if the kernel of the map need not be a graded ideal. These factor rings of graded rings are not graded themselves, nevertheless there are many properties (and methods) of graded rings that prevail when considering such factor rings. Therefore it is worthwile to introduce the following definitions
A ring R is said to be a *G-system* if there exists a family of additive subgroups $(R_\sigma)_{\sigma\in G}$ of R satisfying: R = $\sum_{\sigma\in G} R_\sigma$, $R_\sigma\, R_\tau \subset R_{\sigma\tau}$ for all $\sigma,\tau \in$ G. A G-system such that $R_\sigma\, R_\tau = R_{\sigma\tau}$ holds for all $\sigma,\tau\ \in$ G is said to be a *Clifford system*.
More detail concerning the notions introduced above and their applications may be found in the references given in the Bibliographical Comments at the end of this Chapter; this section is based in large upon the results of F. Van Oystaeyen [5], P. Grzeszczuk [1] and C. Năstăsescu [2].

If R = $\sum_{\sigma\in G} R_\sigma$ is a G-system then the multiplication of R defines maps of R_e-bimodules $R_\sigma \otimes_{R_e} R_\tau \to R_{\sigma\tau}$ that can be used to define a multiplicaton on the R_e-bimodule $\bigoplus_{\sigma\in G} R_\sigma$ = R'. It follows that R' is graded of type G and hence we obtain that each G-system is an epimorphic image of a graded ring of type G (and

vice-versa). One easily checks that each Clifford system is an epimorphic image of a strongly graded ring.
Note that we implicitely used that $1 \in R_e$ for a G-system R, but this is not so obvious here, actually it is not true for infinite G. Indeed, let n,m be integers and let A be the ring (!) of all rational numbers of the form $\frac{2n}{2m+1}$. Define Ψ: A[X] $\to$ Q by f(X) $\mapsto$ f($\frac{1}{2}$). Since A[X] is $\mathbb{Z}$ -graded and since Ψ is epimorphic it follows that Q is a $\mathbb{Z}$ -system with unit $1 \notin A = Q_0$. For finite groups however we have:

2.16.1. Lemma. For a G-system $R = \sum_{g\in G} R_g$, $1 \in R_e$.

Proof. Consider a non-empty $H \subset G - \{e\}$. By induction on $n = |G - H|$ we will show that there exists an $x_H \in R_e$ such that $1 - x_H \in \sum_{h\in H} R_h$. If n = 1 then $H = G - \{e\}$ and the claim obviously holds.
Suppose $n \geq 2$ and consider $g \in G - H$, $g \neq e$. Put $K = Hg^{-1} \cup \{g^{-1}\}$, $L = H \cup \{g\}$. Then $|K| = |L| = |H| + 1$. The induction hypothesis provides x_K, $x_L \in R_e$ such that

$$1 - x_K \in \sum_{k\in K}^{R_k} = R_{g^{-1}} + \sum_{h\in H} R_{hg} \qquad (*)$$

$$1 - x_L \in \sum_{l\in L} R_l = R_g + \sum_{h\in H} R_h \qquad (**)$$

From (*) we derive that $(1 - x_K) R_g \subset R_e + \sum_{h\in H} R_h$ and thus
$(1 - x_K)(1 - x_L) \in (1 - x_K) R_g + \sum_{h\in H} (1 - x_K) R_H \subset R_e + \subset_{h\in H} R_h$.
This yields that $1 - x_H \in \sum_{h\in H} R_h$ for some $x_H \in R_e$.
Putting H = {g}, g ≠ e in the relation (*) provides an $x \in R_e$ such that $1 - x \in R_g$. Let g have exponent m in G, then $(1 - x)^m \in R_{g^m} = R_e$ or $1 - y \in R_e$ for some $y \in R_e$, i.e. $1 \in R_e$. □

2.16.2. Corollary. For a G-system R the following properties hold:

1. If I is a left ideal of R_e then RI = R if and only if $I = R_e$.
2. An $x \in R_e$ is right (left) invertible in R if and only if x is right (left) invertible in R_e.
3. $J(R) \cap R_e \subset J(R_e)$.

Proof. 1. If $R = RI = \sum_{g\in G} R_g I$ then it follows from $R_g I R_h I = (R_g I R_h)I \subset R_{gh} I$ that $RI = \sum_{g\in G} R_g I$ is a G-system with unit 1. By the lemma $1 \in R_e I = I$ or $I = R_e$.
2. Let $x \in R_e$ be right invertible in R. Consider $xR_e = I$. Since $IR = R$ we have $I = R_e$ or $xx' = 1$ for some $x' \in R_e$.
3. It follows from 2 that $J(R) \bigcap R_e$ is a quasi-regular ideal of R, hence $J(R) \bigcap R_e \subset J(R_e)$.

2.16.3. Lemma. Let M be a left module over the G-system R and let $0 \neq M_1 \subset M_2 \subset \ldots \subset M$ be a chain of R_e-submodules of ${}_{R_e}M$ such that $\bigcup_n M_n$ is essential in ${}_{R_e}M$. Then there is an $m \geq 1$ such that M_m contains an R-submodule (nonzero) of M.

Proof. We aim to show by induction on $k = 1,2,\ldots, |G|$ that there exists $H_k \subset G$ and a nonzero $m_k \in M$ such that: $e \in H_k$, $|H_K| = k$, for some $s(k) \geq 1$, $\sum_{h\in H_k} R_h m_k \subset M_{s(k)}$ is nonzero.
For $k = 1$ put $H_1 = \{e\}$, and take $m_1 \neq 0$ arbitrary in M_1, puting $s(1) = 1$. Now let $|G| > k \geq 1$ and suppose that m_k and H_k are such that the above conditions hold. Take $g \neq H_k$. If $R_g m_k = 0$ then $m_{k+1} = m_k$, $H_{k+1} = H_k \cup \{g\}$ satisfy the same conditions. If $R_g m_k \neq 0$ then there exists an $r_g \in R_g$ such that $r_g m_k \in M_t$ for some $t \geq 1$ since $\bigcup_n M_n$ is an essential R_e-submodule of M. Put $m_{k+1} = r_g m_k$, $H_{k+1} = H_k g^{-1} \cup \{e\}$ and $s(k+1) = \max\{s(k), t\}$. Clearly $|H_{k+1}| = k+1$ and from $R_{hg^{-1}} r_g \subset R_h$ we now derive:
$0 \neq \sum_{h\in H_k+1} R_h m_{k+1} = R_e m_{k+1} + \sum_{h\in H_k} R_{hg^{-1}} r_g m_k \subset M_t + \sum_{h\in H} R_h m_k \subset M_t + M_{s(k)} = M_{s(k+1)}$.
Finally, the third condition with $k = |G|$ leads to the existence of a nonzero R-submodule of M in M_m for some $m = s(|G|)$. □

2.16.4. Theorem. Let R be a G-system for a finite group G. If M is a simple left R-module then ${}_{R_e}M = M_1 \bigoplus \ldots \bigoplus M_k$ is a direct sum of $k \leq |G|$ simple R_e-modules.

Proof. Let N be an R_e-submodule of M and let K be maximal with respect to $N \bigcap K = 0$. Then $N \bigoplus K$ is an essential R_e-submodule of ${}_{R_e}M$. By the lemma we obtain $N \bigoplus K = M$. Thus every nonzero R_e-submodule of M contains a simple R_e-module. Consider $e \in H \subset G$ such that for some $m \in M$, $\sum_{h\in H} R_h m \neq 0$ and for all $h \in$

H either $R_h m = 0$ or $R_h m$ is a simple R_e-module; moreover that we have taken an H with these properties and such that |H| is maximal. We claim: H = G. If $g \in G - H$ then $R_g m \neq 0$. Let $R_e m'$ be a simple R_e-submodule of $R_g m$. For each $h \in H$, $R_{hg^{-1}} m' \subset R_h m$. Thus if $h' \in Hg^{-1} \cup \{e\}$ then $R_{h'} m' = 0$ or $R_{h'} m'$ is a simple R_e-module. This contradicts the maximality of H. Consequently M is a direct sum of at most |G| simple R_e-modules. □

2.16.5. Corollary. If R is a G-system then:

1. For any left R-module M, $J({}_{R_e}M) \subset J({}_RM)$
2. For any left R-module M, Soc $({}_RM) \subset$ Soc $({}_{R_e}M)$
3. $J(R_e) = J(R) \cap R_e$.

2.16.6. Theorem. Let M be a left R-module where R is a G-system, then M is a Noetherian R-module if and only if M is a Noetherian R_e-module.

Proof. Suppose M is a Noetherian R-module but not Noetherian as an R_e-module. We may reduce the problem to the case where M is such that for each nonzero R-submodule N of M, ${}_{R_e}(M/N)$ is Noetherian. Let $X_1 \subsetneq X_2 \subsetneq \dots \subset_{R_e} M$ be a strictly ascending chain of R_e-submodules and let Y be maximal with respect to $(\bigcup_n X_n) \cap Y = 0$. We obtain a strictly ascending chain of R_e-submodules $X_1 \oplus Y \subsetneq X_2 \oplus Y \subsetneq \dots$, of M, such that $\bigcup_n (X_n \oplus Y) = (\bigcup_n X_n) \oplus Y$ is an essential R_e-submodule of M. In view of Lemma 2.16.3. there is an $m \geq 1$ such that $X_m \oplus Y$ contains an R-submodule $N \neq 0$ of M. In M/N we obtain the following strictly ascending chain:

$$X_{m+1} \oplus Y/N \subsetneq X_{m+2} \oplus Y/N \subsetneq \dots \subset M/N,$$

and this contradicts the assumption on M/N. □

2.16.7. Corollary. If the G-system R is left Noetherian then R_e is left Noetherian too.

2.16.8. Proposition. If R is a G-system, resp. graded of type G, then the following statements are equivalent:

1. R is a Clifford system, resp. R is strongly graded by G.

2. For all $\sigma \in G$, $R_\sigma R_{\sigma^{-1}} = R_e$.

3. For any set of generators of G, $(\sigma_i)_{i\in I}$ say, $R_{\sigma_i} R_{\sigma_i^{-1}} = R_e$ and $R_{\sigma_i^{-1}} R_{\sigma_i} = R_e$ for all $i \in I$.

4. In case R is graded then every graded R-module M satisfies $R_\sigma M_\tau = M_{\sigma\tau}$ for all $\sigma, \tau \in G$.

Proof. We phrase the proof for a graded ring R, the proof for a G-system is formally similar. The implications 1⇒3⇒2 and 4⇒1 are completely obvious.

2⇒1. For $\sigma, \tau \in G$ we have $R_{\sigma\tau} = R_e\, R_{\sigma\tau} = R_\sigma\, R_{\sigma^{-1}}\, R_{\sigma\tau} \subset R_\sigma\, R_\tau$, thus $R_\sigma\, R_\tau = R_{\sigma\tau}$.

1⇒4. For $\sigma, \tau \in G$ we have $M_{\sigma\tau} = R_e\, M_{\sigma\tau} = R_\sigma\, (R_{\sigma^{-1}}\, M_{\sigma\tau}) \subset R_\sigma M_\tau \subset M_{\sigma\tau}$. □

2.16.9. Corollary. If R is strongly graded by G and M is a graded R-module then $RM_e = M$. In particular if M_e is (Noetherian) Artinian, as an R_e-module then M is (gr-Noetherian) gr-Artinian.

2.16.10. Corollary. Let R be strongly graded by G, then R_σ is a finitely generated projective left (or right) R_e-module.

Proof. Since $1 \in R_{\sigma^{-1}}\, R_\sigma$, $1 = \sum_i u^{(i)}_{\sigma^{-1}}\, v^{(i)}_\sigma$ with $u^{(i)}_{\sigma^{-1}} \in R_{\sigma^{-1}}$, $v^{(i)}_\sigma \in R_\sigma$ for any $\sigma \in G$. Take $r_\sigma \in R_\sigma$, then we obtain: $r_\sigma = r_\sigma.1 = \sum_{i=1}^n (r_\sigma\, u^{(i)}_{\sigma^{-1}})\, v^{(i)}_\sigma$ and thus R_σ is generated by $v^{(1)}_\sigma, \dots, v^{(k)}_\sigma$ as a left R_e-module. A similar proof works for the right hand case. In order to derive projectivity let us define φ_i: $R_\sigma \to R_e$ $x \mapsto x\, u^{(i)}_{\sigma^{-1}}$ for $i = 1, \dots, k$, φ: $R_\sigma \to R_e^k$, $x \mapsto (\varphi_i(x), i = 1, \dots, k)$, Ψ: $R_e^k \to R_\sigma$, $(y_1, \dots, y_k) \mapsto \sum_i y_i\, v^{(i)}_\sigma$.

For $x \in R_\sigma$, $\Psi \circ \varphi(x) = \sum_{i=1}^k \varphi_i(x)\, v^{(i)}_\sigma = \sum_{i=1}^k x\, u^{(i)}_{\sigma^{-1}}\, v^{(i)}_\sigma = x$. Consequently $\Psi \circ \varphi = 1_{R_\sigma}$ and therefore R_σ is isomorphic to a direct summand of the free R_e-module R_e^k, i.e. R_σ is a projective left (similarly: right) R_e-module. □

2.16.11. Corollary. Let R be a Clifford system then R_σ is a finitely generated projective left (and right) R_e-module.

Proof. If R = $\sum_{\sigma\in G} R_\sigma$ let R' = $\bigoplus_{\sigma\in G} R_\sigma$ be the strongly graded ring (cf. second paragraph in the introduction to this section) such that R' → R → 0 is the canonical ring epimorphism. Apply Corolary 2.16.10. to R' and then the statement is established because $R'_\sigma = R_\sigma$ for all $\sigma \in$ G. □

2.16.12. Remark. Propositon 2.16.8. and all its corollaries hold for arbitrary G. For the sequel of this section the assumption that G is finite is essential.

Let R be a Clifford system for a finite group G and let M be an arbitrary R-module, N any R_e-submodule of M. Put $N^* = \bigcap_{\sigma\in G} R_\sigma$N.For $\tau \in$ G we have $R_\tau N^* \subset R_\tau R_\sigma$N = $R_{\tau\sigma}$N. so it follows that $R_\tau N^* \subset \bigcap_{\sigma\in G} R_{\tau\sigma}$N = N^*, i.e. N^* is an R-submodule of M contained in N. In fact, it is clear that N^* is the largest R-submodule of M contained in N.

2.16.13. Lemma. If N is an essential left R_e-submodule of M then N^* is an essential left R_e-submodule of M too.

Proof. Let X ⊂ M be a nonzero R_e-submodule of M. Since $R_\sigma R_{\sigma^{-1}} = R_e$ it follows that $R_{\sigma^{-1}}$X $\neq 0$ for all $\sigma \in$ G, hence N $\bigcap R_{\sigma^{-1}}$X $\neq 0$ and $R_\sigma)N \bigcap R_{\sigma^{-1}}X)$ $\neq 0$. But the latter yields $R_\sigma N \bigcap$ X $\neq 0$ or R_σN is an essential left R_e-module of M for all $\sigma \in$ G. Finiteness of G implies that N^* is an essential left R_e-submodule of M. □

2.16.14. Lemma. M contains a left R_e-submodule N which is maximal with respect to the property $N^* = 0$.

Proof. Similar to the proof of Lemma 2.12.5., taking into account that each R_σ, hence R, is a finitely generated R_e-module in view of Corolary 2.16.11. □

2.16.15. Lemma. Let L be an R_e-submodule of M, then:

1. $L_{R_e}(M/R_\sigma L) \cong L_{R_e}(M/L)$, for all $\sigma \in$ G.

2. $L_{R_e}(R_\sigma L) \cong L_{R_e}(L)$, for all $\sigma \in G$.

Proof. 1. If X/L *in* L_e (M/L) then $X/L \to R_\sigma X/R_\sigma L$ is a lattice isomorphism.
2. Obvious since R_σ is an invertible R_e-bimodule. □

2.16.16. Corollary. 1. If R has finite Goldie dimension then R_e has finite Goldie dimension too. If ${}_RM$ has finite u.dim then similarly for ${}_{R_e}M$.
2. R is left Artinian if and only if R_e is left Artinian.

2.16.17. Theorem (Generalized Clifford Theorem). Let R be strongly graded of type G and M a simple R-module. There exists a simple R_e-submodule W such that $M = \sum_{\sigma \in G} R_\sigma W$; in particular ${}_{R_e}M$ is a semisimple R_e-module of finite length.

Proof. Pick $x \neq 0$ in M. Then $M = Rx = \sum_{\sigma \in G} R_\sigma x$ and hence M is a finitely generated left R_e-module. So we may consider a maximal R_e-submodule K of ${}_{R_e}M$. Since K^* is an R-submodule contained in K and since M is simple we have $K^* = 0$. By Lemma 2.16.15., $R_\sigma K$ is a maximal R_e-submodule of M for each $\sigma \in G$. From the exact sequence $0 \to M \to \bigoplus_{\sigma \in G} M/R_\sigma K$, it follows that M is a semisimple R_e-module. Let W be a simple R_e-submodule of M then $M = RW = \sum_{\sigma \in G} R_\sigma W$. □

2.16.18. Remark. For a Clifford system R a result similar to the foregoing holds, cf. C. Năstăsescu and F. Van Oystaeyen [3].

2.16.19. Corollary. If the Clifford system R is semisimple then R_e is semisimple. The module M is semiartinian if and only if M is semiartinian as an R_e-module.

Proof. The first statement follows directly from the foregoing, the second statement may be proved by running along the lines of proof of Corollary 2.12.11. □

Let R be a Clifford system for a finite group G and let M and N be R-modules. Let us fix, for every $\sigma \in G$, a decomposition of 1 in $R_\sigma R_{\sigma^{-1}} = R_e$, say:

$1 = \sum_{i=1}^{n_\sigma} u_\sigma^{(i)} v_{\sigma^{-1}}^{(i)}$, $u_\sigma^{(i)} \in R_\sigma$, $v_{\sigma^{-1}} \in R_{\sigma^{-1}}$. To a given f $\in$ Hom_{R_e}(M,N) we associate a function $\tilde{f}$: M $\to$ N, m $\mapsto$ $\sum_{\sigma\in G} \sum_{i=1}^{n_\sigma} u_\sigma^{(i)} f(v_{\sigma^{-1}}^{(i)} m)$.

2.16.20. Lemma. With notations as above, $\tilde{f}$ is R-linear.

Proof. Consider $r_\tau \in R_\tau$ and calculate for m $\in$ M:

$$\tilde{f}\,(r_\tau m) = \sum_{\sigma\in G} \sum_{i=1}^{n_\sigma} u_\sigma^{(i)} f\,(v_{\sigma^{-1}}^{(i)} r_\tau m)$$

$$= \sum_{\sigma\in G} \sum_{i=1}^{n_\sigma} u_\sigma^{(i)} f(v_{\sigma^{-1}}^{(i)} r_\tau (\textstyle\sum_{j=1}^{n_\tau - 1_\sigma} u_{\tau^{-1}\sigma}^{(j)} v_{\sigma^{-1}\tau}^{(j)})\, m)$$

$$= \sum_{\sigma\in G} \sum_{i=1}^{n_\sigma} \sum_{j=1}^{n_\tau - 1_\sigma} u_\sigma^{(i)} v_{\sigma^{-1}}^{(i)} r_\tau u_\tau - 1_\sigma f\,(u_{\sigma^{-1}\tau}^{(j)} m)$$

$$= \sum_{\sigma\in G} \sum_{j=1}^{n_\tau - 1_\sigma} r_\tau u_{\tau^{-1}\sigma}^{(j)} f\,(v_{\sigma^{-1}\tau}^{(j)} m)$$

$$= r_\tau \sum_{j\in G} \sum_{j=1}^{n_\gamma} u_\gamma^{(j)} f\,(v_{\gamma^{-1}}^{(j)} m) = r_\tau\ \mathrm{f(m)}.$$

□

2.16.21. Theorem (Essential Version of Maschke's theorem). Let R be a Clifford system for the finite group G. Let V and W be R-modules such that W is a direct summand of V as an R_e-module.

1. There exists a left R-submodule U of V such that V $\oplus$ U is an essential left R_e-submodule of V.
2. Suppose R has no |G|-torsion and suppose moreover that V = |G| V, then there exists a left R-submodule U of V such that V = W $\oplus$ U.

Proof. 1. Consider an R_e-module W' such that V = W $\oplus$ W'. Since W is a left R-module, R_σW = W for all $\sigma \in$ G, thus it is evident that $(W \oplus W')^* = W \oplus (W')^*$ By Lemma 2.16.13. it follows that $W \oplus (W')^*$ is an essential left R_e-submodule of V. The statement 1. now follows from the fact that $(W')^*$ is a left R-module.
2. Let f: V $\to$ W be the canonical left R_e-linear epimorphism such that f | W = 1_W. By the lemma, $\tilde{f}$: V $\to$ W is left R-linear. For m $\in$ W we also have, $v_{\sigma^{-1}}^{(i)}$ m $\in$ W since W is a left R-module. Now we calculate: $\tilde{f}$(m) = $\sum_{\sigma\in G} \sum_{i=1}^{n_\sigma} u_\sigma^{(i)} f(v_{\sigma^{-1}}^{(i)} m)$ = $\sum_{\sigma\in G} \sum_{i=1}^{n_\sigma} u_\sigma^{(i)} v_{\sigma^{-1}}^{(i)} m$ = |G| m.
The fact that V = W $\oplus$ W' whilst |G| V = V entails that |G| W = W too. Replacing $\tilde{f}$ by $\tilde{f}$/|G| defined by ($\tilde{f}$/|G|)(m) = $\tilde{f}$(m)/|G| we finally obtain a left R-linear splitting map for W $\to$ V.

□

2.16.22. Remark. If we have that $|G|^{-1} \in R$ then the conditions on $|G|$ in the theorem are fulfilled; then we may write $\tilde{f}.|G|^{-1}$ for $\tilde{f}/|G|$. The decompositions of 1 $\in R_e$ fixed in the proof are not unique, therefore $\tilde{f}$ cannot a priori be considered as a canonical map. Up to the prescription of some element "of trace 1" it is canonical, but we do not go in to this matter here.

2.16.23. Corollary. 1. In the situation of the theorem it is obvious that there is a left R-submodule U_1 of V such that $W \bigoplus U_1$ is an essential left R-submodule of V.

2. If R and V are as above and L is a left R-submodule of V then L is essential as a left R-submodule if and only if it is essential as a left R_e-submodule.

3. Assume $|G|^{-1} \in R$. If V is semisimple as a (left) R_e-module then V is a semisimple R-module. If R_e is a semisimple ring then R is also a semisimple ring (Maschke's theorem).

Proof. 1. and 3. are direct consequences of the theorem.

2. If L were not essential as a left R_e-submodule of V then for some left R-submodule L' of V we have that $L \bigoplus L'$ is essential as a left R_e-submodule of V. This would contradict the fact that L is an essential left R-submodule of V. □

Combining the foregoing with Theorem 2.16.17. we obtain:

2.16.24. Corollary. If R is a Clifford system for G such that $|G|^{-1} \in R$ then an R-module M is a finitely generated semisimple R-module only if it is a finitely generated semisimple R_e-module.

If R is a Clifford system then $\sigma \mapsto [R_\sigma]$ (where [ ] denotes the isomorphism class of R_e-bimodules) defines a group morphism Φ: $G \to \text{Pic}(R_e)$.

Recall that $\text{Pic}(R_e)$ is the group of isomorphism classes of invertible R_e-bimodules, where the group law is induced by the tensor product $\bigotimes_{R_e}$. An *invertible* R_e-bimodule is an R_e-bimodule X such that there exists another R_e-bimodule Y such that $X \bigotimes_{R_e} Y \cong Y \bigotimes_{R_e} X \cong R_e$. The Clifford system R is said to be *outer* if the morphism ϕ: $G \to \text{Pic}(R_e)$ is a monomorphism.

2.16.25. Theorem (F. Van Oystaeyen [5]). Let R be an outer Clifford system such that R_e is simple then R is simple.

Proof. First assume that R is strongly graded; then we may talk about the length of a homogeneous decomposition i.e. the number of nonzero homogeneous components of an element. Let I be a nonzero ideal of R and pick x $\neq 0$ in I such that its decomposition has minimal length. Up to multiplying by some element of R_τ for some $\tau \in G$ we may assume that $x_e \neq 0$. Since R_e is simple: $1 = \sum_{i=1}^{m} \lambda_i x_e \mu_i$ with $\lambda_i, \mu_i \in R_e$. Up to replacing x by $\sum_{i=1}^{m} \lambda_i x \mu_i \in$ I we may assume that $x_e = 1$. If x $= x_e = 1$ then there is nothing to prove, so assume x $\neq 1$, i.e. some $x_\tau \neq 0$ for some $\tau \neq$ e in G. For every r $\in R_e$, rx - xr $\in$ I and the length of its decompositions is less than the length of x, therefore rx - xr = 0 for all r $\in R_e$. Consider J = $R_{\tau^{-1}} X_\tau$. Since x_τ commutes with R_e, J is an ideal of R_e and as $R_{\tau^{-1}} x_\tau \neq 0$ (otherwise $R_\tau R_{\tau^{-1}} x_\tau = 0$, contradiction) it follows that $R_{\tau^{-1}} x_\tau = R_e$. So there must exist $y_{\tau^{-1}} \in R_{\tau^{-1}}$ such that $y_{\tau^{-1}} x_\tau = 1$.
By symmetry, $x_\tau R_{\tau^{-1}} = R_e$ yields $x_\tau z_{\tau^{-1}} = 1$ for some $z_{\tau^{-1}} \in R_{\tau^{-1}}$. Clearly $z_{\tau^{-1}} = y_{\tau^{-1}}$ follows, thus $x_\tau y_{\tau^{-1}} = y_{\tau^{-1}} x_\tau = 1$. From $y_{\tau^{-1}} R_\tau \subset R_e$ we derive that $R_\tau = R_e x_\tau$. The latter states that $[R_\tau] = [R_e] = 1$ in Pic(R_e), contradicting the assumption that the system is outer. If R is only a Clifford system let R' be the strongly graded ring $\bigoplus_{\sigma \in G} R_\sigma$ constructed earlier. By what we proved so far, R' will be a simple ring; since R is an epimorphic image of R' it follows then that R = R', hence R is also simple. □

2.16.26. Remark. The foregoing extends a similar property in the theory of fixed rings for finite group actions in which case G is supposed to act by outer automorphisms, cf. S. Montgomery [1].

We refer to Van Oystaeyen [5] for the following:

2.16.27. Theorem. Let A be a Clifford system such that Z(A_e) $\subset$ Z(A) ("quasi-inner") and assume that $|\ G\ |^{-1} \in$ A. If A_e is an Azumaya algebra over Z(A_e) then A is an Azumaya algebra over Z(A). (for Azumaya algebras, cf. Van Oystaeyen [5]).

Let us provide some concrete constructions of strongly graded rings. First note that example 2. and 4. are strongly graded, example 5. is also strongly graded (see

Examples at the beginning of Section 2.13.).

Let $R \subset S$ be rings. A subbimodule ${}_RA_R$ of ${}_RS_R$ is said to be *invertible* if there exists a subbimodule ${}_RB_R$ of ${}_RS_R$ such that AB = BA = R. Let Inv (S,R) be the set of all invertible R-submodules of ${}_RS_R$, then Inv (S,R) is a group with identity element ${}_RR_R$. Let G be a group and φ: G $\to$ Inv (S,R) a group morphism. Write $\varphi(\sigma) = A_\sigma$ for all $\sigma \in$ G. Define $\check{R}(\varphi) = \bigoplus_{\sigma\in G} A_\sigma$, where multiplication is defined by the R_e-bimodule isomorphisms $A_\sigma \bigotimes_{R_e} A_\sigma \to A_{\sigma\tau}$, $x \otimes y \mapsto xy$. It is clear that $\check{R}(\varphi)$ is a strongly graded ring of type G; we call $\check{R}(\varphi)$ the *generalized Rees ring* associated to φ. The structure of these rings and its invariants like Picard, class and Brauer groups have been studied in some detail in F. Van Oystaeyen [6], F. Van Oystaeyen, A. Verschoren [1] etc... .
As a particular case of the general construction we single out the case G = $\mathbb{Z}$, $A \in$ Inv (S,R) and φ_A: $\mathbb{Z} \to$ Inv (S,R), $n \mapsto A^n$. The ring $\check{R}(\varphi_A) = \bigoplus_{n\in\mathbb{Z}} A^n$ is strongly graded of type $\mathbb{Z}$. Note that $\check{R}(\varphi_A)$ is left Noetherian if R is left Noetherian.

The graded Jacobson radical $J_G(R)$ of a G-graded ring R is defined to be the ideal of R satisfying the following equivalent conditions:

1. $J_g(R)$ is the intersection of all maximal graded right ideals of R.
2. $J_g(R)$ is the largest graded ideal I of R such that $I \bigcap R_e$ is a quasi-regular ideal of R_e.

2.16.28. Theorem. For every ring R graded by a finite group G, $J_g(R) \subset J(R)$.

Proof. Clearly $I = R\, J(R_e)\, R$ is a graded ideal of R. By Corolary 2.16.2. (3), $I \subset J(R)$ and $J(R_e) \subset I \bigcap R_e \subset J(R) \bigcap R_e = J(R_e)$. Thus $I \bigcap R_e = J(R_e)$ and $I \subset J_g(R)$ (look at condtion 2° above).
Consider $S = J_g(R)/I$. This is a G-graded ring (possibly without unit) such that $S_e = 0$, hence $S^{|G|} = 0$. This shows that $J_g(R)^{|G|} \subset J(R)$ and $J_g(R) \subset J(R)$ follows. □

Some properties of prime and maximal ideals of graded rings, G-systems and Clifford systems have been included in the set of exercises at the end of the chapter.

§2.17 Invariants of a Finite Group Action.

Let G be a finite group of automorphisms of a ring R. We write $R^G = \{r \in R, g(r) = r$ for all $g \in G\}$ for the ring of G-invariants in R (note: instead of g(r) we also write r^g for notational convenience).
Throughout this section we assume that $| G |^{-1} \in R$. Recall example 5. after Proposition 2.13.1., i.e. the crossed product rings $R \underset{\alpha,f}{*} G$ where α is group morphism α: $G \to \mathrm{Aut}\, R$, f a factorsystem for G. We look at the particular case where f is trivial (f = 1) and we call $R \underset{\alpha}{*}$ the *skew group ring* with respect to α. If α is monomorphic then we simply write $R * G$. Obviously $R * G$ is strongly graded over $(R * G)_e =$ R.e, e the identity of G.
A left (right, two-sided) ideal I of R is G-invariant if $g(I) \subset I$ for all $g \in G$. For $r \in R$ we may define the *trace* of r to be the element: $t_G(r) = \frac{1}{n} \sum_{g \in G} g(r)$, $n = | G |$. The map t_G: $R \to R_G$, $r \to t_G(r)$ is a morphism of R^G-bimodules such that $t_G(R^G) = R^G$. We may define a left $R * G$ -module structure on R by putting: $(rg)x = rg(x)$ for all $x \in R$.

2.17.1. Lemma. 1. The left $R * G$ -submodules of R are exactly the left G-invariant ideals.
2. The map φ: $R^G \to \mathrm{End}_{R*G}({}_{R*G}R)$, $\varphi(a)(x) = xa$ for all $a \in R^G$ and $x \in R$, is a ring isomorphism.
3. The map Ψ: $R^G \to f(R * G)f$, $a \mapsto af = faf$, where $f = \frac{1}{n} \sum_{g \in G} g$, is an isomorphism of rings.

Proof. 1. is obvious and 2. may be proved by straightforward verification.
3. Put $f = \frac{1}{n} \sum_{g \in G} g$, then $gf = fg$ for all $g \in G$, hence $f^2 = f$. If $r \in R$ then we calculate: $fRf = \frac{1}{n} (\sum_{g \in G} g(r)\, f) = t_G(r)f$. The latter implies that
$f\,(R * G)\, f = t_G\,(R)\, f = R^G.f$ □

Let M be a left R-module and let M^* be the left R∗G-module
$M^* = \bigoplus_{g \in G} gM = \{\sum_{g \in G} gm_g,\ m_g \in M\}$ with scalar multiplication: $rg.(hm) = gh\,(h^{-1}(g^{-1}(r))m)$, for all $r \in R$, $m \in M$. We may identify M with 1.M in M^*. Define τ: $M^* \to M$, $\sum_{g \in G} gm_g \mapsto \sum_{g \in G} m_g$. It is clear that τ is R^G-linear, hence if $X \in L_{R*G}(M^*)$ then $\tau(X) \in L_{R^G}(M)$ and τ defines an isotone map (again denoted by τ): $L_{R*G}(M^*) \to L_{R^G}(M)$. Let $Y \in L_{R^G}(M)$ then $fY \subset M^*$ and therefore

(R $*$ G)(fY) is an R $*$ G -submodule of M^*. Define σ: $L_{R^G}(\mathrm{M}) \to L_{R^G}(M^*)$ by $\sigma(\mathrm{Y}) = (\mathrm{R} * \mathrm{G})(\mathrm{fY})$. With these notations we have:

2.17.2. Lemma. The map σ is strictly ascending and $\tau \circ \sigma = 1_{L_{R^G}(M)}$. Moreover τ commutes with internal direct sums, i.e. $\tau\ (X_1 \oplus X_2) = \tau(X_1) \oplus \tau(X_2)$ for $X_1, X_2 \in L_{R*G}(M^*)$.

Proof. If $m^* = \sum_{g\in G} g\ m_g \in M^*$ then $\mathrm{f}m^* = (\frac{1}{n} \sum_{h\in G} h) \sum_{g\in G} g\ m_g = \frac{1}{n} \sum_{g\in G} (\sum_{h\in G} hg)\ m_g = \sum_g \frac{1}{n} (\sum_h hg)\ m_g = \sum_g \mathrm{f}\ m_g = \mathrm{f}\ \tau\ (m^*)$.
Take Y $\in L_{R^G}(\mathrm{M})$. From $\sigma(\mathrm{Y}) = (\mathrm{R} * \mathrm{G})(\mathrm{fY})$ it follows that for y $\in$ Y and ag in R $*$ G we have $(\mathrm{ag})(\mathrm{fy}) = (\mathrm{ag})(\frac{1}{n} \sum_{h\in G} \mathrm{hy}) = \frac{1}{n} \sum_{h\in G} \mathrm{gh}\ (h^{-1} \circ g^{-1})(\mathrm{a})\mathrm{y}$, and consequently: $\tau\ ((\mathrm{ag})(\mathrm{fy})) = \frac{1}{n} \tau\ (\sum_{h\in G} \mathrm{gh}\ (h^{-1} \circ g^{-1})(\mathrm{a})\mathrm{y}) = \frac{1}{n} \sum_{h\in G} h^{-1}(g^{-1}(a))\mathrm{y}$ $= t_G(g^{-1}(a))y$. Since $t_G(g^{-1}(a)) \in R^G$ we obtain that $\tau\sigma(\mathrm{Y}) \subset \mathrm{Y}$. On the other hand, for y $\in$ Y we have $\frac{1}{n}$ fy $\in$ fY $\subset \tau(\mathrm{Y})$ and $\tau\ (\frac{1}{n}\ \mathrm{fy}) = \frac{1}{n}\ (\mathrm{ny}) = \mathrm{y}$, i.e.
$\mathrm{y} \subset \tau\ (\sigma(\mathrm{Y}))$.
The equality of $\tau \circ \sigma$ and $1_{L_{R^G}(M)}$ yields that σ is strictly ascending. Furthermore, additivity of τ entails that $\tau\ (X_1 + X_2) = \tau(X_1) + \tau(X_2)$. If $X_1 \bigcap X_2 = 0$ then $\mathrm{f}X_1 \bigcap \mathrm{f}X_2 = 0$. Suppose $x_1 \in X_1$, $x_2 \in X_2$ are such that $\tau(x_1) = \tau(x_2)$, then $\mathrm{f}\tau(x_1) = \mathrm{f}\tau(x_2)$, so $\mathrm{f}x_1 = \mathrm{f}x_2$ but then $\mathrm{f}x_1 = 0$ because $\mathrm{f}X_1 \bigcap \mathrm{f}X_2 = 0$. Consequently $\tau(X_1) \bigcap \tau(X_2) = 0$. □

2.17.3. Theorem. Let G be a finite group of automorphisms of the ring R such that $|\ \mathrm{G}\ |^{-1} \in \mathrm{R}$. For a left R-module M we have:

1. M is a Noetherian R-module if and only if M is a Noetherian R^G-module.
2. If M has finite length then ${}_{R^G}M$ has finite length and $l_{R^G}(\mathrm{M}) \leq |\ \mathrm{G}\ |.l_R(\mathrm{M})$
3. $s_R(\mathrm{M}) \subset s_{R^G}(\mathrm{M})$.
4. $J_{R^G}(\mathrm{M}) \subset J_R(\mathrm{M})$.

Proof. 1. It suffices to prove that ${}_{R^G}M$ is Noetherian if M is. For g $\in$ G we look at the automorphism g^{-1}: R $\to$ R and the transformed module $(g_*^{-1})(\mathrm{M})$. In $M^* = \bigoplus_{g\in G}$ gM we see that gM $\cong$ $(g^{-1})_*(\mathrm{M})$ as R-modules. Since M is Noetherian and G is finite it follows that gM, $(g^{-1})_*$M and M^* are Noetherian R-modules. By Theorem 2.16.6., M^* is a Noetherian R $*$ G -module but then Lemma 2.17.2. implies that M is a Noetherian R^G-module.

2. If M has finite length then gM has finite length and $l_R(M) = l_R(gM)$. Hence $l_R(M^*) = |G|\, l_R(M)$. Since $l_{R*G}(M^*) \leq l_R(M^*)$ we may conclude from Lemma 2.17.2. that: $l_{R^G}(M) \leq l_{R*G}(M^*) \leq |G|\, l_R(M)$.

3. If M is a semisimple R-module then M^* is a semisimple R-module; by Corollary 2.16.23 (3) we know that M^* is then also a semisimple R $*$ G - module. Consider an R^G-submodule Y of M. Then $\sigma(Y)$ is an R $*$ G-submodule of M^*, hence there exists an X' $\in L_{R*G}(M)$ such that $\sigma(Y) \oplus X' = M^*$. By Lemma 2.17.2. again, it follows that $M = \tau(M^*) = \tau(\sigma(Y)) \oplus \tau(X') = Y \oplus \tau(X')$ and this states that Y is an R^G-direct summand of M. The assertion follows.

4. Look at a maximal R submodule $N \subsetneq M$. From 3. it follows that M/N is a semisimple R^G-module of finite length and thus $J_{R^G}(M) \subset N$ or $J_{R^G}(M) \subset J_R(M)$ follows. □

2.17.4. Corollary. With assumptions on R as in the theorem:

1. If R is left Noetherian, resp. Artinian, then R^G is left Noetherian, resp. Artinian and R is finitely generated as a left R_G-module.
$J(R) \bigcap R^G = J(R^G)$.

Proof. 1. Obvious from the theorem.
2. Clearly $I = J(R) \bigcap R^G$ is a nontrivial ideal of R^G. Pick a $\in$ I, then 1-a is invertible in R, so there is a b $\in$ R such that (1-a)b = b(1-a) = 1. Applying t_G we obtain: n = $t_G((1\text{-}a)b) = (1\text{-}a)t_G(b) = t_G(b(1\text{-}a)) = t_G(b)(1\text{-}a)$. It follows that 1-a is invertible in R^G and thus $J(R) \bigcap R^G \subset J(R^G)$. Conversely if a' $\in J(R^G)$ and M is any simple R-module then the theorem entails that M is a semisimple R^G-module, hence a'M = 0 or a' $\in$ J(R). Consequently $J(R) \bigcap R^G = J(R^G)$. □

2.17.5. Remarks. 1. Let R be the ring $\begin{pmatrix} \mathbb{Q} & \mathbb{Q} \\ 0 & \mathbb{R} \end{pmatrix}$ which is left Artinian but neither right Artinain nor right Noetherian. Let $\sigma \in \mathrm{Aut}(R)$ be defined by $\begin{pmatrix} a & b \\ 0 & c \end{pmatrix} \mapsto \begin{pmatrix} a & -b \\ 0 & c \end{pmatrix}$, and let G be the group $\{1,\sigma\}$. It is easily seen that $R^G = \begin{pmatrix} \mathbb{Q} & 0 \\ 0 & \mathbb{R} \end{pmatrix}$ and $R^G = \mathbb{R} \oplus \mathbb{Q}$ is obviously right Artinian. Since 2 is a unit in R we have obtained a counter-example to the converse of the first statement in Corolary 2.17.4.
2. In [1], S. Montgomery pointed out an example of a Noetherian integral domain of characteristic zero (where 2 is not invertible) and $G = \{1,\varphi\}$ such that R is not a finitely generated R^G-module and R^G is not Noetherian.

§2.18 Exercises

(26) Let R = $R_1 \times R_2$ be a direct product of rings and consider an R_i-module M_i, i = 1,2. The abelian group M = $M_1 \times M_2$ is an R-module with the obvious (component-wise) scalar multiplication.
Prove that L $({}_RM)$ = L $({}_{R_1}M) \times$ L $({}_{R_2}M)$.

(27) Prove that for an infinite family of rings $(R_i)_{i\in I}$: $L(\Pi_{i\in I} R_i) \neq \Pi_{i\in I} L(R_i)$.

(28) Consider the $\mathbb{Z}$-module M = $\mathbb{Z} \times \mathbb{Z}$. Show that the lattice L(M) is not distributive.

(29) Let M be an R-module and N $\neq$ M a submodule of M. Show that N is an (infinite) intersection of irreducible submodules.

(30) A left ideal I is a direct summand of ${}_RR$ if and only if there exists an idempotent f $\in$ R such that I = Rf.

(31) Let R be a commutative ring and I a finitely generated ideal such that I^2 = I. Establish an idempotent f in R such that I = Rf.

(32) An R-module M has finite coirreducible dimension if and only if every submodule of M is an essential extension of a finitely generated submodule.

(33) $\mathbb{Z}^{\mathbb{N}}$ is an essential submodule in $Q^{\mathbb{N}}$ but $\mathbb{Z}^{\mathbb{N}}$ is not essential in $Q^{\mathbb{N}}$.

(34) Consider an essential R-submodule N of M, then Z(M/N) = M/N.

(35) Let M be an R-module. By recurrence we define an ascending chain of submodules: Z_1(M) = Z(M), ..., $Z_n(M)/Z_{n-1}(M) = Z(M/Z_{n-1}(M))$. Prove that $Z_2(M) = Z_3(M) = \ldots$

(36) Let R be a ring satisfying the descending chain condition for principal left ideals. If a $\in$ R, then a is invertible if and only if $l_R(a) = 0$.

(37) A commutative ring is Noetherian if and only if each prime ideal is finitely generated.

(38) If a commutative ring extension B of A such that B = $A[b_1,...,b_n]$ is Noetherian then A is Noetherian too.

(39) Let M be an R-module of finite length and take f $\in$ $\mathrm{End}_R(M)$. Show that f is a monomorphism if and only if f is an epimorphism if and only if f is an automorphism.

(40) If M is an indecomposable R-module of finite length then the endomorphism ring $\mathrm{End}_R(M)$ is local.

(41) Suppose that R is a semiprime ring and consider the lattice, L(R), of ideals of R. Prove the following statements:

a. If $A \in L(R)$ then $l_R(A) = r_R(A) = \mathrm{Ann}(A)$ ($l_R = Ann^l_R$, $r_R = Ann^l_R$)
b. Ann(A) is the unique pseudocomplement of A in L(R).
c. $\mathrm{Ann}(A) = \bigcap \{P \in \mathrm{Spec}(R), P \not\supset A\}$.
d. A is essential in L(R) if and only if Ann(A) = 0.
e. L(R) has finite Goldie dimension if and only if 0 is a finite intersection of prime ideals.
f. If L(R) has finite Goldie dimension and $A \in L(R)$ then A is essential in L(R) if and only if A is not contained in a minimal prime ideal.

(42) Let R be a Dedekind domain, M an R-module. Show that the lattice $L^\circ_R(M)$ (the dual of $L_R(M)$) has a basis. Hint: it is enough to show that a nonzero R-module U contains an R-submodule U', uniform in $L^\circ_R(U)$. The injective envelope E(U) is a direct sum of copies of K (the field of fractions of R) and copies of the P-primary components of K/R for certain nonzero prime ideals P of R. Reduce the problem to certain R-submodules either in R or in some P-primary component of K/R; then use that L_R (P-primary component) is linearly ordered.

(43) The Goldie dimension of $L^\circ(R)$ is equal to the cardinality of the set of all maximal ideals of K.

(44) Recall that R is said to be a *Von Neumann regular ring* if for all $a \in R$ there is a $b \in R$ such that $a = aba$. Prove that the following statements are equivalent:

1. R is Von Neumann regular.
2. Every finitely generated left ideal is a direct summand.
3. Every finitely generated right ideal is a direct summand.
4. Every principal left (or right) ideal is a direct summand.

(45) R is semisimple if and only if R is left Noetherian and Von Neumann regular.

(46) If R is Von Neumann regular then J(R) = 0, Z(R) = 0. Prove that the center of R is again Von Neumann regular.

(47) A direct product of Von Neumann regular rings is again a Von Neumann regular ring.

(48) If R is a Von Neumann regular without nilpotent elements then it is called a *strongly regular* ring. Prove that the following statements are equivalent:

a. R is strongly regular.
b. R is Von Neumann regular and its idempotent elements are central.
c. R is Von Neumann regular and all left ideals are ideals.
d. For each a ∈ K there is a b ∈ R such that $a = a^2b$.

(49) If R is a strongly regular ring then the principal left ideals form a Boolean algebra isomorphic with the Boole algebra of the idempotents of R.

(50) For a semisimple R-module M, $\text{End}_R(M)$ is a Von Neumann regular ring.

(51) A direct product of (noncommutative) fields is a strongly regular ring.

(52) Any strongly regular ring is isomorphic to a subring of a direct product of fields.

(53) An R-module M is *finitely presented* if there exists an exact sequence $R^m \to R^n \to M \to 0$. If we have an exact sequence $0 \to K \to L \to M \to 0$ where M is finitely presented and L is finitely generated then K is finitely generated.

(54) If R is a left Noetherian ring then every finitely generated left R-module is also finitely presented.

(55) A finite direct sum of finitely presented R-modules is a finitely presented R-module.

(56) A ring R is called a *left coherent* ring if each finitely generated ideal is finitely presented. Prove that R is left coherent if and only if for all a ∈ R, $Ann^l_R(a)$ is finitely generated and two finitely generated left ideals have a finitely generated intersection.

(57) A Von Neumann regular ring as well as a left Noetherian ring is left coherent.

(58) If R is left coherent then the set of finitely generated left ideals forms a sublattice of $L_R({}_RR)$.

(59) Let K be a left coherent ring and let I be an ideal of R which is finitely generated as a left ideal, then R/I is a left coherent ring.

(60) Let K be a field of charK = p ≠ 0. Let G be a finite group p | | G |. Let L be the lattice of K-subspaces of K^G then r(L) = 0 (r(L) is the radical of L). For g ∈ G define φ_g: L → L, N ↦ gN. Then $r(L^G) \neq 0$, so $r(L^G) \neq r(L)$ is possible.

(61) A ring R is a Baer ring if each left annihilator ideal is generated by an idempotent element. Show that R is a Baer ring if and only if each right annihilator ideal is generated by an idempotent element.

(62) If R is a Baer ring then the set of principal left ideals of the form Rf, where f is an idempotent element, is a complete lattice isomorphic to the dual lattice of the

principal ideals of the form eR where e is an idempotent element.

(63) Let R be a Von Neumann regular ring. Prove that R is a Baer ring only if the lattice of left (or right) principal ideals is complete.

(64) Let I be a minimal left ideal of a ring R then $I^2 = 0$ or else I = Re where e is an idempotent.

(65) Let R be a semiprime ring If e in R is an idempotent element such that eRe is a field then Re (resp. eR) is a minimal left (resp. right) ideal of R.

(66) Let e and f be idempotent elements of R. Then Re $\cong$ Rf if and only if there exists a,b $\in$ R such that e = ab and f = ba.

(67) Prove that the center of a left semiartinian ring is a semiartinian ring.

(68) If R is a left semiartinian ring of finite Loewy length then R is right semiartinian of finite Loewy length.

(69) Prove that a commutative semiprime ring which is semiartinian is also a Von Neumann regular ring.

(70) If R is Von Neumann regular then R is left semiartinian if and only if it is right semiartinian.

(71) Consider a finitely generated projective left R-module P. Prove that $J(\mathrm{End}_R P) = \mathrm{Hom}_R(P, J(P))$, $\mathrm{End}_R(P)/J(\mathrm{End}_R(P)) \cong \mathrm{End}_R(P/J(P))$.

(72) Prove that a ring R is left Noetherian if and only if each direct sum of injective modules is injective, if and only if each injective module is a direct sum of indecomposable injective modules.

(73) (Utumi, Johnson, Wong) Let Q be an injective left R-module and put $T = \mathrm{End}_R(Q)$, then:

a. $J(T) = \{f \in T, \mathrm{Ker}\, f$ is essential in $Q\}$.
b. T/J(T) is a Von Neumann regular ring
c. If ${}_RQ$ is nonsingular, i.e. Z(Q) = 0, then T is regular too.

(74) Let Q be an injective R-module and put $T = \mathrm{End}_R(Q)$. If I is a finitely generated right ideal of T then $I = \{f \in T, \mathrm{Ker}\, f \supset Q(I) = \bigcap_{f \in I} \mathrm{Ker}\, f\}$.

(75) If Q is an injective Artinian left R-module then $T = \mathrm{End}_R(Q)$ is right Noetherian.

(76) A ring is *semiprime* if R/J(R) is semisimple and J(R) is a nilpotent ideal. Prove that the following statements are equivalent:

a. R is a semiprimary left Noetherian ring.

b. R is semiprimary and each quotient ring of R has finite Goldie dimension.

(77) If Q is an injective Noetherian left R-module then the ring $End_R(Q)$ is semiprimary.

(78) (Johnson, Wong) Consider an injective left R-module Q and $T = End_R(Q)$. Consider the contramodule ${}_TQ$. Show that for every finitely generated left T-submodule K of ${}_TQ$ we have $r_T\ (l_R(K)) = K$.

(79) A ring is said to be *quasi-Frobenius* (QF) if it is left Artinian and ${}_RR$ is injective. Prove that the following statements are equivalent:

a. R is a QF-ring.

b. ${}_RR$ is left injective and Noetherian.

c. ${}_RR$ is injective and R is right Noetherian.

(80) If R is a QF-ring then φ: $L({}_RR) \to L(R_R)^\circ$, $I \mapsto Ann^r_R\ (I) = \{a \in R, Ia = 0\}$, is an isomorphism of lattices.

(81) Let R be a domain (no zerodivisors) such that every finitely generated left ideal is principal. Prove that R is a left Ore domain.

(82) If R is a left Ore domain then R[X] is a left Ore domain.

(83) Let D be a left Ore domain with field of fractions K, then ${}_DK$ is the injective hull of ${}_DD$.

(84) Let S be a multiplicatively closed set in R. If P is a maximal ideal with the property $P \bigcap S = \phi$ then P is a prime ideal of R (assuming $1 \in S$, $0 \notin S$).

(85) Let R be a commutative ring and M an R-module. For $p \in Spec(R)$ we let M_p, resp. R_p, be the module of fractions $S_p^{-1}M$, resp. the ring of fractions $S_p^{-1}R$, where $S_p = R - p$.

a. The sequence $0 \to M' \to M \to M'' \to 0$ is exact if and only if
$0 \to M'_p \to M_p \to M''_p \to 0$ is exact for all $p \in Spec(R)$.

b. $M = 0$ if and only if $M_p = 0$ for all $p \in Spec(R)$.

c. For all $p \in Spec(R)$, the ring R_p is local.

(86) For a commutative local ring, L(R) is distributive if and only if L(R) is totally ordered.

(87) For a commutative ring, L(R) is distributive if and only if $L(R_p)$ is distributive for every $p \in Spec(R)$. For a commutative Von Neumann regular ring L(R) is distributive.

(88) Let k be a commutative field and let k(y) be the field of rational functions. Look at the abelian group k(y)[X] but with multiplication defined by $X.f(y) = f(y^2)X$, for all $f \in k(y)$. Prove that k(y).[X] defined above is a left Ore domain but not a right Ore domain.

(89) A left R-module M is *divisible* if for all $s \in S_{Reg}$ and for all $y \in M$, there is an $x \in M$ such that $sx = y$. Prove that each injective module is divisible. If R is a domain such that left ideals are principal then a left R-module Q is injective if and only if Q is divisible.

(90) Consider $R = \mathbb{Z}[X]$ with field of fractions $K = \mathbb{Q}(X)$. Show that K/R is a divisible R-module which is not injective.

(91) (Bit-David, Robson, [1]) Let S be a normalizing extension of R and let I be a prime ideal of S, then there exist prime ideals $P_1, ..., P_m$ ($m \leq n$) of R such that $I \bigcap R = \bigcap_{i=1}^{m} P_i$.

a. If $P \in \text{Spec}(R)$ then there exists a $Q \in \text{Spec}(R)$ such that P is a prime ideal which is minimal over $Q \bigcap R$.
b. Let $P_1 \subsetneq P_2$ be prime ideals of R and let Q_1 be a prime ideal of S such that P_1 is minimal over $R \bigcap Q_1$. Then there exists a prime ideal Q_2 of S such that $Q_1 \subsetneq Q_2$ and P_2 is minimal over $Q_2 \bigcap R$.
c. Prove that $\text{rad}(R) = R \bigcap \text{rad}(S)$.

(92) Let R be a positively graded ring and put $R_+ = R_1 \bigoplus R_2 \bigoplus ...$. If M is a nonzero graded R-module with left limited gradation then $R_+M \neq M$.

(93) Let R be a positively graded domain. Then R is a left and right principal ring if and only if $R = R_0[X,\varphi]$ for some automorphism $\varphi: R_0 \rightarrow R_0$.

(94) If $R = \bigoplus_{g\in G} R_g$ is graded by an ordered group. If $P \in \text{Spec}(R)$ then P_g is a prime ideal (recall: $P_g = \bigoplus_{h\in G} (P \bigcap R_h)$). Prove that rad(R) is a graded ideal.

(95) (Bergman [1]) Let R be graded of type $\mathbb{Z}/n\mathbb{Z}$ and let $a = a_0 + a_1 + ... + a_{n-1} \in J(R)$, then $na_i \in J(R)$ for all i = 0,...,n-1.

(96) (Bergman [1]) Let R be of type $\mathbb{Z}$, then J(R) is a graded ideal. Prove that J(A[X]) = I[X] for some nilideal I of A, A an arbitrary ring, X an indeterminate.

(97) (Năstăsescu, Van Oystaeyen) If M is a graded R-module then s(M) is a graded submodule of M.

(98) If M is a graded R-module of type $\mathbb{Z}$ then the singular radical Z(M) is a graded R-submodule of M.

(99) (Năstăsescu, Van Oystaeyen, [2]) A group G is polycyclic-by-finite if there exists

a chain of subgroups $\{e\} = G_0 \subset G_1 \subset ... \subset G_n = G$ where each G_{i-1} is normal in G_i and G_i/G_{i-1} is finite or cyclic for all i = 1,...,n. If R is strongly graded by a polycyclic-by-finite group and R_e is left Noetherian then R is left Noetherian.

(100) Let A be a left Noetherian ring and G a polycyclic-by-finite group then AG is left Noetherian.

(101) Let R be strongly graded by a finite group G such that $| G |^{-1} \in R$. Then R is Von Neumann regular if and only if R_e is Von Neumann regular. If M is an R-module then $Z_R(M) = Z_{R_e}(M)$. If M is a graded R-module then $Z_R(M)$ is a graded submodule of M.

(102) (Ulbrich [1]) Let R be strongly graded by a finite group G and let M be an R-module, N a graded R-module.

a. The map φ: $\mathrm{Hom}_R(M,N) \to \mathrm{Hom}_R(M,N_\sigma)$, $f \mapsto \pi_\sigma \circ f$ where π_σ: $N \to N_\sigma$ is the canonical map, is an isomorphism.
b. The map Ψ: $\mathrm{Hom}_R(N,M) \to \mathrm{Hom}_{R_e}(N_\sigma,M)$, $g \mapsto g \circ i_\sigma$ where i_σ: $N_\sigma \to N$ is the canonical map, is an isomorphism.
c. ${}_R R$ is injective if and only if ${}_{R_e}R_e$ is injective.
d. R is a QF-ring if and only if R_e is a QF-ring.

(103) Let R be strongly graded by a finite group G. Prove:

a. If $P \in \mathrm{Spec}(R)$ then there is a $Q \in \mathrm{Spec}(R_e)$ such that $P \bigcap R_e = \bigcap_{\sigma \in G} Q^\sigma$ where $Q^\sigma = R_{\sigma^{-1}} Q R_\sigma$.
b. If $R_{\sigma 1}$ A R_σ = A for all $\sigma \in G$ then the ideal A of R_e is said to be G-invariant. If $Q \in \mathrm{Spec}(R_e)$ then there is a $P \in \mathrm{Spec}(R)$ such that $Q = P \bigcap R_e$ if and only if Q is G-invariant.
c. Establish that $\mathrm{rad}(R_e) = R_e \bigcap \mathrm{rad}(R)$.
d. If R_e is semiprime and without n-torsion where n = |G| then R is semiprime.

(104) Let R be strongly graded by a finite group G, |G| = n. The following properties hold (check!):

a. If R is a semiprime left Goldie ring then R_e is a semiprime left Goldie ring. In this case the classical ring of fractions $Q_{cl}(R)$ is strongly graded of type G and $Q_{cl}(R)_e = Q_{cl}(R_e)$.

b. If R_e is a semiprime left Goldie ring and suppose that R_e is without n-torsion, then R is a semiprime left Goldie ring.

(105) (Bergman, Isaacs [1], Fisher and Osterburg [1]) Let R be a ring, G a finite group of automorphisms of R, say |G| = n. Assume that R is semiprime and without n-torsion. Prove the following assertions:

a. If I $\neq 0$ is a left (resp. right) G-invariant) ideal of R then $I^G \neq 0$ and tr I $\neq 0$.
b. R^G is a semiprime ring.
c. R is a left Goldie ring if and only if R^G is a left Goldie ring

(106) Let G be a finite group of automorphisms of R such that multiplication by |G| is bijective on R then R is semilocal if and only if R^G is semilocal.

(107) (F. Van Oystaeyen) Let R be a Clifford syatem for a finite group G. For each maximal ideal Ω of R, $\Omega \bigcap R_e$ is a finite intersection of maximal ideals of R_e.

(108) (P. Grzeszczuk [1]) Prove (107) for a G-system. For every G-system, B(R_e) $\subset$ B(R), where B(-) denotes the Brown-McCoy radical.

(109) Any invertible module is a finitely generated left and right flat R-module.

Bibliographical Comments to Chapter 2.

The first part of this chapter deals with the basic notions and fundamental results concerning finiteness conditions on modules. The reader may check the obvious links between the first sections of Chapter 1 and those of Chapter 2.

We have followed the classical approach to the theory of rings and modules: Noetherian rings, Artinian rings, semisimple and simple rings, rings of fractions, etc... . Goldie's theorems [1958–60] take a central place in the theory of left Noetherian rings; the notion of Goldie dimension evolves naturally from this theory.

In Section 2.9. we study sufficient conditions for an Artinian module to be Noetherian and we follow in large the paper by C. Năstăsescu [7]. After a short survey of some properties of injective and projective modules we present a selection of special classes of rings, playing an important part in contemporary ring theory e.g. normalizing extensions, graded rings, rings of invariants under group actions.

The main result of section 2.12. is Theorem 2.12.7. first proved in this form by Formanek and Jategaonkar in [1] but generalizing a result of Eakin published in 1968 in [1].

For a detailed account on the theory of graded rings we refer to C. Năstăsescu, F. Van Oystaeyen [1],[3].

For Clifford systems, resp. G–systems, we refer to F. Van Oystaeyen [5], P. Greszczuk [1] and C. Năstăsescu [2]. In Section 2.17. we study the fixed ring R^G for G a group of automorphisms of R. The lemma 2.17.2. is basic here, our presentation of it is an adaption of Lemma 3.1. in [1] by Lorenz and Passman. In the latter reference we also find the principal result for this section, i.e. Theorem 2.17.3.

A more complete treatment of the properties of rings of invariants for finite group actions may be found in Montgomery's book [1]. Some typical references for the contents of Chapter 2. are:

T. Albu, C. Năstăsescu [1]; F. Anderson, K. Fuller [1]; G. Bergman [1; G. Bergman, I. Isaacs [1]; J. Bit David [1]; E. Dade [1]; P. Eakin [1]; J. Fisher, J. Osterburg [1]; E. Formanek, A. Jategaonkar [1]; J. Bit David, J.C. Robson [1]; M. Lorenz, D. Passman [1],[2]; S. Montgomery [1]; J. Fisher, S. Montgomery [1]; C. Năstăsescu [2],[5],[6],[7]; C. Năstăsescu, F. Van Oystaeyen [1],[2],[3]; D. Passman [1],[2]; F. Van Oystaeyen [3],[4],[5]; B. Stenström [2]; K.H. Ulbrich [1]; M. Cohen and S. Montgomery [1]; P. Grezsczuk [1]; M. Van den Bergh [1].

Chapter 3

Krull Dimension and Gabriel Dimension of an Ordered Set.

§3.1 Definitions and Basic Properties.

For a partially ordered set $(L,\leq)$ we let $\Gamma(L)$ be the set $\{(a,b), a \leq b, a,b \in L\}$. By transfinite recursion we may define on $\Gamma(L)$ a filtration in the following way:

$\Gamma_{-1}(L) = \{(a,b) \in \Gamma(L), a = b\}$

$\Gamma_0(L) = \{(a,b) \in \Gamma(L), [a,b] \text{ is Artinian}\}$

$\Gamma_\alpha(L) = \{(a,b) \in \Gamma(L)$, for all $b \geq b_1 \geq ... \geq b_n \geq ... \geq a$, there is an $n \in \mathbb{N}$ such that $(b_{i+1}, b_i) \in \bigcup_{\beta<\alpha} \Gamma_\beta(L)$ for each $i \geq n\}$, where we assume that for each ordinal β, $\beta < \alpha$, $\Gamma_\beta(L)$ is already defined. In this way we obtain an ascending chain $\Gamma_{-1}(L) \subset \Gamma_0(L) \subset ... \subset \Gamma_\alpha(L) \subset ...$. Since L is a set it follows that $\Gamma_\varsigma(L) = \Gamma_{\varsigma+1}(L) = ...$. If $\Gamma(L) = \Gamma_\alpha(L)$ for some ordinal α then we say that *the Krull dimension of L* is defined.

3.1.1. Lemma. Let L' be a subset of the partially ordered set $(L,\leq)$.

1. For each ordinal α, $\Gamma_\alpha(L) \bigcap \Gamma(L') \subset \Gamma_\alpha(L')$.
2. If L' is convex then $\Gamma_\alpha(L) \bigcap \Gamma(L') = \Gamma_\alpha(L')$ for all α, where a subset of a partially ordered set is said to be *convex* if for all $x,y \in L'$ such that $x \leq y$ and $z \in [x,y]$, then $z \in L'$.
3. If the Krull dimension of L is defined then so is the Krull dimension of L'.

Proof. 1. Consider a $\leq$ b in L. If [a,b] is Artinian in L then it is Artinian in L', hence $\Gamma_0(L) \bigcap \Gamma(L') \subset \Gamma_0(L')$. Assume that the claim holds for all ordinals β, $\beta < \alpha$, and let (a,b) $\in \Gamma_\alpha(L) \bigcap L'$. Consider a descending chain in L', b $\geq b_1 \geq \ldots \geq b_n \geq \ldots \geq$ a. Since (a,b) $\in \Gamma_\alpha(L)$, there exists an n $\in \mathbb{N}$ such that $(b_{i+1}, b_i) \in \bigcup_{\beta<\alpha} \Gamma(L)$ for all i $\geq$ n. By the induction hypothesis: (b_{i+1}, b_i) is in $\bigcup_{\beta<\alpha} \Gamma_\beta(L) \bigcap \Gamma(L') \subset \bigcup_{\beta<\alpha} \Gamma_\beta(L')$ for all i $\geq$ n. Thus (a,b) $\in \Gamma_\alpha(L')$.

2. Follows in a similar way and the final assertion 3. follows from 1. □

To a partially ordered set (L,$\leq$) we may associate a partially ordered set $(\overline{L}, \leq)$ defined as follows. If L has a least element then we put $\overline{L}$ = L; if not then we pick $a_0 \notin$ L and put $\overline{L} = \{a_0\} \bigcup L$ with $a_0 < x$ for all x $\in$ L. By transfinite recursion one easily establishes.

3.1.2. Lemma. If $\Gamma_\alpha(L) = \Gamma(L)$ for some ordinal α then $\Gamma_{\alpha+1}(\overline{L}) = \Gamma(\overline{L})$.

For example, $\mathbb{Z}$ does not have a least element, so we have $\Gamma_0(\mathbb{Z}) = \Gamma(\mathbb{Z})$ and $\Gamma_1(\overline{\mathbb{Z}}) = \Gamma(\overline{\mathbb{Z}})$.

3.1.3. Remark. If (L,$\leq$) has a least element and we construct $\overline{L} = \{a_0\} \bigcup L$ with $a_0 < x$ for all x $\in$ L (for some $a_0 \notin$ L) then one easily checks that $\Gamma_\alpha(L) = \Gamma(L)$ implies $\Gamma_\alpha(\overline{L}) = \Gamma(\overline{L})$.

If the Krull dimension of (L,$\leq$) is defined then we write Kdim L = α if α is the least ordinal for which $\Gamma_\alpha(\overline{L}) = \Gamma(\overline{L})$, we call this ordinal the *Krull dimension of L.* The Krull dimension of the dual ordered set L° is called the *Krull codimension of L*; it is denoted by co-Kdim L = Kdim L°.

It is clear that Kdim L = 0 if and only if L is Artinain and co-Kdim L = 0 if and only if L is Noetherian. For example, Kdim $\mathbb{N}$ = 0, Kdim $\mathbb{Z}$ = 1.

3.1.4. Proposition. (Gabriel, Rentschler, [1]). Let φ: L $\rightarrow$ M be a strictly increasing map between partially ordered sets L and M. If M has Krull dimension so does L and moreover, Kdim L $\leq$ Kdim M.

Proof. By Remark 3.1.3. we may assume that both L and M have a least element. The map φ induces a map Ψ: $\Gamma(L) \to \Gamma(M)$, $\Psi(a,b) = (\varphi(a), \varphi(b))$. For each ordinal α, $\Psi^{-1}(\Gamma_\alpha(M)) \subset \Gamma_\alpha(L)$. Indeed, if $(a,b) \in \Psi^{-1}(\Gamma(M))$ then $(\varphi(a), \varphi(b)) \in \Gamma_0(M)$ and hence $[\varphi(a), \varphi(b)]$ is Artinian. Since φ is strictly increasing it follows that $[a,b]$ is Artinian and hence $(a,b) \in \Gamma_0(L)$.
Starting from $\Psi-1$ $(\Gamma_0(M)) \subset \Gamma_0(L)$ the general statement follows by transfinite recursion. From the latter the statement of the proposition follows. □

3.1.5. Proposition (Lemonnier, [4]). Let $(L,\leq)$ be an ordered set and consider ordinals α and β. The following assertions hold:

1. We have $(a,b) \in \Gamma_\alpha(L)$ if and only if $[a,b]$ has Krull dimension and Kdim $[a,b] \leq \alpha$.
2. Kdim L $\leq \beta$ if and only if in each decreasing sequence $(a_n)_{n\geq 1}$ in L we have Kdim $[a_{n+1}, a_n] < \beta$ when n is large enough.
3. L has Krull dimension if and only if for each increasing sequence $(a_n)_{n\geq 1}$ in L, $[a_{n_1}, a_n]$ has Krull dimension for n sufficiently large.
4. If Kdim L $= \beta$ then for each $\alpha < \beta$ there exists a decreasing sequence $(a_n)_{n\geq 1}$ such that Kdim $[a_{n+1}, a_n] \geq \alpha$ for each n.
5. If $(a,b) \in \Gamma(L)$ - $\Gamma_{\alpha+1}(L)$ then there exists a $c \in L$ such that $a < c < b$, $(a,c) \notin \Gamma_{\alpha+1}(L)$ and $(c,b) \in \Gamma_\alpha(L)$.

Proof. 1. If $(a,b) \in \Gamma_\alpha(L)$ then $\Gamma([a,b]) \subset \Gamma_\alpha(L)$ and in view of Lemma 3.1.1. we obtain that $\Gamma_\alpha([a,b]) = \Gamma([a,b])$, i.e. Kdim $[a,b] \leq \alpha$. Conversely, since $[a,b]$ is a convex set it follows that $\Gamma_\alpha([a,b]) \subset \Gamma_\alpha(L)$ and hence $\Gamma([a,b]) \subset \Gamma_\alpha(L)$, so $(a,b) \in \Gamma_\alpha(L)$.
2. Since $\Gamma_\beta(\overline{L}) = \Gamma(\overline{L})$ there exists $n_0 \in \mathbb{N}$ such that (a_{n+1}, a_n) is in $\bigcup_{\alpha<\beta} \Gamma_\alpha(\overline{L})$ for all $n \geq n_0$. Therefore there exists an $\alpha_n < \beta$ such that $(a_{n+1}, a_n) \in \Gamma_{\alpha_n}(\overline{L})$. From 1. it follows that Kdim $[a_{n+1}, a_n] \leq \alpha_n < \beta$ for all $n \geq n_0$. Consequently (the converse of the foregoing being obvious) the statement in 2. follows.
3. If L has Krull dimension then the implication $\Rightarrow$ follows from 2. Conversely, suppose that there is an ordinal γ such that $\Gamma_\gamma(L) = \Gamma_{\gamma+1}(L) = \dots$. Consider a,b $\in$ L with $a \leq b$ and let $b \geq b_1 \geq \dots \geq b_n \geq \dots \geq a$ be a decreasing sequence of elements of L. There is an $n_0 \in \mathbb{N}$ such that for each $n \geq n_0$,
Kdim $[b_{n+1}, b] \leq \alpha(n)$. By Lemma 3.1.1. we obtain:
$(b_{n+1}, b_n) \in \Gamma_{\alpha(n)}([b_{n+1}, b_n]) \subset \Gamma_{\alpha(n)}(L) \subset \Gamma_\gamma(L)$. Hence, $(a,b) \in \Gamma_{\gamma+1}(L) = \Gamma_\gamma(L)$ and so $\Gamma(L) = \Gamma_\gamma(L)$ i.e. L has Krull dimension.
4. Direct from 2.
5. If $(a,b) \notin \Gamma_{\alpha+1}(L)$ then there exists a decreasing sequence $b \geq b_1 \geq \dots \geq b_n \geq$

... $\geq$ a such that $(b_{i+1}, b_i) \notin \Gamma_\alpha(L)$ for infinitely many values of i, say $i_1 < i_2 <$ ... . Take c = b_{i_2}. □

3.1.6. Proposition (Gabriel, Rentschler [1]). Let L and M be non-empty ordered sets having Krull dimension. The direct product L × M ordered by (a,b) $\leq$ (a',b') if and only if a $\leq$ a' and b $\leq$ b' has Krull dimension, Kdim L × M equal to sup (Kdim L,Kdim M).

Proof. It is clear that we may reduce the problem to the case where L and M both have a least element. Consider the bijection:

φ: $\Gamma(L \times M) \rightarrow \Gamma(L) \times \Gamma(M)$, ((x,y),(x',y')) $\mapsto$ ((x,x'),(y,y')). By transfinite recursion we establish that $\varphi(\Gamma_\alpha(L \times M)) = \Gamma_\alpha(L) \times \Gamma_\alpha(M)$, for each ordinal α. The statement is now clear. □

3.1.7. Theorem (Lemonnier [4]). The ordered set L has Krull dimension if and only if L has Krull codimension.

Proof. Let co-Kdim L be defined, i.e. $\Gamma(L^\circ) = \Gamma_\alpha(L^\circ)$ for some ordinal α. We prove by transfinite recursion on α that L also has Krull dimension.

Assume the contrary. Then there is an ordinal β such that $\Gamma_\beta(L) = \Gamma_{\beta+1}(L) = ...$ and $\Gamma_\beta(L) \neq \Gamma(L)$. Take (a,b) $\in \Gamma(L) - \Gamma_\beta(L)$.By Proposition 3.1.5.(5), there exists a $b_1 \in L$ such that a $< b_1 <$ b and $(b_1,b) \notin \Gamma_\beta(L)$. Again by Proposition 3.1.5.(5) there exists a $b_2 \in L$ such that $b_1 < b_2 <$ b, $(b_2,b) \notin \Gamma_\beta(L)$ and $(b_1, b_2) \in \Gamma_\beta(L)$. Continuing this way, we obtain a strictly increasing sequence

$a < b_1 < b_2 < ... < b_n < ... < b$, such that (b_n, b_{n+1}) is not in $\Gamma_\beta(L)$ for all

$n \geq 1$.

Since $\Gamma(L^\circ) = \Gamma_\alpha(L^\circ)$, there is an $n_0 \in \mathbb{N}$ such that $(b_n, b_{n+1}) \in \Gamma_{\alpha(n)}(L^\circ)$ for all $n \geq n_0$, where $\alpha(n) < \alpha$. The induction hypothesis entails that for all $n \geq n_0$, $[b_n, b_{n+1}]$ has Krull dimension, hence $(b_n, b_{n+1}) \in \Gamma_\beta(L)$ because of Proposition 3.1.5.(1), contradiction. □

3.1.8. Corollary. A Noetherian ordered set has Krull dimension.

Define the real diadic numbers $\mathbb{D} = \{m2-n, m \in \mathbb{Z}, n \in \mathbb{N}\}$.

3.1.9. Lemma. The sets $\mathbb{D}$, $\mathbb{Q}$, $\mathbb{R}$ with their usual orderings do not have Krull dimension.

Proof. If suffices to establish the claim for $\mathbb{D}$. Assume the contrary. Then $L = \{x \in \mathbb{D}, 0 \leq x < 1\}$ also has Krull dimension, say Kdim $L = \alpha$. By Proposition 3.1.5.(2) there exists $n_0 \in \mathbb{N}$ such that in the decreasing sequence, $\frac{1}{2} > \frac{1}{4} > \ldots > \frac{1}{2^n} > \ldots$ we have Kdim $[2^{-(n+1)}, 2^{-n}] < \alpha$ for $n \geq n_0$.
Put $M = \{x \in \mathbb{D}, 2^{-(n+1)} \leq x < 2^{-n}\}$. Clearly: Kdim $(M) < \alpha$. The function φ: $L \to M$, $x \mapsto (x+1)\, 2^{-(n+1)}$ is bijective and it preserves the order, hence Kdim $M =$ Kdim $L = \alpha$, yields a contradiction. □

3.1.10. Theorem (Lemonnier,[4]). The ordered set $(L,\leq)$ does not have Krull dimension if and only if L contains a subset isomorphic to $\mathbb{D}$.

Proof. One implication is obvious in view of the Lemma. Suppose L has Krull dimension. By Proposition 3.1.5(5) , we may select an $e \in L$ such that $L' = \{x \in L, x \leq e\}$ and $L'' = \{x \in L, x \geq e\}$ do not have Krull dimension. Again using Proposition 3.1.5(5) we construct a decreasing (resp. increasing) sequence in L' (resp. in L"), $(x_t)_{t\leq 0}$ (resp. $(x_t)_{t\geq 0}$) such that $x_0 = e$ and $[x_t, x_{t+1}]$ does not have Krull dimension for all $t \in \mathbb{Z}$.
Let $[a,b] \subset L$ be such that $[a,b]$ does not have a Krull dimension. We will construct a strictly increasing φ: $\mathbb{D} \bigcap [0,1] \to [a,b]$ such that $\varphi(0) = a$, $\varphi(1) = b$ and for $\alpha, \beta \in \mathbb{D} \bigcap [0,1]$ with $\alpha < \beta$ the interval $[\varphi(\alpha), \varphi(\beta)]$ does not have Krull dimension. Put $A_n = \{m2^{-n}, 0 \leq m \leq 2^n\}$. Clearly $\mathbb{D} \bigcap [0,1] = \bigcup_{n\geq 0} A_n$. Assume that we have already found a strictly increasing φ_n: $A_n \to [a,b]$ with the properties announced for φ. We show that φ_n extends to a strictly increasing φ_{n+1}: $A_{n+1} \to [a,b]$ with the same properties.
Obviously, $C_n = A_{n+1} - A_n = \{k\, 2^{-n}, 1 \leq k < 2^n, k \text{ an odd integer}\}$. If $k\, 2_n \in C_n$ then there exist $\alpha, \beta \in A_n$ with $\alpha < \beta$ such that $k2^{-n}$ is the middle of $[\alpha, \beta]$. Since $[\varphi_n(\alpha), \varphi_n(\beta)]$ does not have Krull dimension we deduce the existence of $u \in L$, $\varphi_n(\alpha) < u < \varphi(\beta)$ such that both $[\varphi_n(\alpha),u]$ and $[u,\varphi_n(\beta)]$ do not have Krull dimension.
Put $\varphi_{n+1}(k2^{-n}) = u$. Define φ by extending the φ_n $(n \geq 1)$. Let $t \in \mathbb{Z}$. Proceeding as indicated above we obtain a strictly increasing

Ψ_t: $\mathbb{D} \cap [t,t+1] \to [x_t, x_{t+1}]$ such that $\Psi_t(t) = x_t$, $\Psi_t(t+1) = x_{t+1}$ and we then define Ψ: $\mathbb{D} \to L$ by extending the functions Ψ_t, $t \in \mathbb{Z}$. □

3.1.11. Theorem (Gabriel, Rentschler [1]). Let $(E,\leq)$ be an ordered set with a least and a largest element. Let $S_c(E)$ be the set of series $(e_i)_{i\geq 1}$ in E which are eventually constant. On $S_c(E)$ we may define an ordering: $(e_i)_{i\geq 1} \leq (f_i)_{i\geq 1}$ if and only if $e_i \leq f_i$ for all $i \geq 1$. Let C(E) be the subset of $S_c(E)$ consisting of the increasing series. If E has Krull dimension, say Kdim E = α, then Kdim $S_c(E)$ = Kdim C(E) = $\alpha + 1$.

Proof. Let $(S_n)_{n\geq 1}$ be a decreasing series of elements of $S_c(E)$, say $S_n = (s_n^k)_{k\geq 1}$. Put $a_n = \lim_{k\to\infty} s_n^k$. There is an $r(n) \in \mathbb{N}$ such that $s_n^k = a_n$ and $s_{n+1}^k = a_{n+1}$ for all $k \geq r(n)$. It is easy to verify that there is an increasing function: φ: $[s_{n+1}, s_n] \to [s_{n+1}^1, s_n^1] \times [s_{n+1}^2, s_n^2] \times \ldots \times [s_{n+1}^{r(n)}, s_n^{r(n)}] \times S_c([a_n, a_{n+1}])$. Using induction on α we obtain: Kdim $[s_{n+1}, s_n] \leq \alpha$ and thus Kdim $S_c(E) \leq \alpha + 1$ (Proposition 3.1.5.). Now let a and b be the least and the largest element of E, respectively. Define $s_n = (s_n^k)_{k\geq 1}$, $s_n^k = a$ for all $1 \leq k \leq n$ and $s_n^k = b$ for $k > n$. Clearly, $s_n \in C(E)$ for all $n \geq 1$, $s_n \geq s_{n+1}$, and Kdim $[s_{n+1}, s_n]$ = Kdim [a,b] = Kdim E = α.
Since $C(E) \subset S_c(E)$ it follows that Kdim C(E) exists and Kdim C(E) $\geq 1 + \alpha$. On the other hand, Kdim C(E) $\leq \alpha + 1$.
Thus Kdim C(E) = Kdim $S_c(E) = \alpha + 1$. □

§3.2 The Krull Dimension of a Modular Lattice

3.2.1. Proposition. Let L be a modular lattice with 0 and 1. If $a \in L$ then Kdim L = sup (Kdim [0,a],Kdim [a,1]).

Proof. Define φ: $L \to [0,a] \times [a,1]$, $x \mapsto (a \wedge x, a \vee x)$. If $x \leq y$ then $\varphi(x) \leq \varphi(y)$ and moreover, if $\varphi(n) = \varphi(y)$ then $a \wedge x = a \wedge y$ and $a \vee x = a \vee y$. Modularity of L then entails: $y = y \wedge (a \vee y) = y \wedge (x \vee a) = x \vee (y \wedge a) = x \vee (a \wedge x) = x$.
Consequently φ is strictly increasing. Application of Proposition 3.1.4. and 3.1.6. finishes the proof. □

3.2.2. Proposition. Let L be a modular and upper continuous lattice. If L has Krull dimension then L has finite uniform dimension.

Proof. Suppose the statement is false, then we may assume that L is such that Kdim L $= \alpha$ and α being the least ordinal such that there exists a lattice of Kdim exactly α which contains an infinite direct union $\bigvee_i a_i$, $0 < a_i \leq 1$. For $k \geq 0$ we put $I_k = \{i2^k, i = 1,2,...\}$ and $b_k = \bigvee_{i \in I_k} a_i$.
It is clear that $b_0 > b_1 > ... > b_k > b_{k+1} > ...$. Since $\Gamma(L) = \Gamma_\alpha(L)$, there is an $n \in \mathbb{N}$ such that $(b_{k+1}, b_k) \in \Gamma_{\alpha(k)}(L)$ with $\alpha(k) < \alpha$ for all $k \geq n$. Consequently, Kdim $[b_{k+1}, b_k] \leq \alpha(k) < \alpha$.
The minimality assumption on α entails that $[b_{k+1}, b_k]$ has finite uniform dimension for all $k \geq n$. On the other hand we have $b_k = b_{k+1} \vee (\bigvee_{i \in I_k - I_{k+1}} a_i)$. It is clear that $J = I_k - I_{k+1}$ is finite and $b_k = \bigvee_{j \in J} (b_{k+1} \vee a_j)$. Using the results of Section 1.5. we obtain: $(a_j \vee b_{k+1}) \wedge (\bigvee_{k \in J-j} b_{k+1} \vee a_k) = (a_j \vee b_{k+1}) \wedge (b_{k+1} \vee (\bigvee_{k \in J-j} a_k)) = b_{k+1} \vee ((a_j \vee b_{k+1}) \wedge \bigvee_{k \in J-j} a_k) = b_{k+1}$.
This implies that $\bigvee_{j \in J} (b_{k+1} \vee a_i)$ is direct in the lattice $[b_{k+1}, b_k]$ and since J is infinite this yields a contradiction. □

3.2.3. Proposition. Let L be a modular upper continuous lattice and assume that L has Krull dimension. Put α equal to sup{1 + Kdim [a,1], where a is an essential element of L}. Then, Kdim L $\leq \alpha$.

Proof. Suppose Kdim L $> \alpha$. There is an infinite decreasing sequence $a_1 > a_2 > ... > a_n ...$ such that Kdim $[a_{i+1}, a_i] \geq \alpha$ for all $i \geq 1$. By Proposition 3.2.2. there is an integer n such that dim $([0,a_n])$ = dim $([0,a_{n+1}])$.
Consider a pseudocomplement b of a_n. By Lemma 1.7.3., $b \vee a_n$ is essential in L and the foregoing equality gives rise to: dim $([0,b \vee a_n])$ = dim $([0,b \vee a_{n+1}])$. Hence $b \vee a_{n+1}$ is essential in $b \vee a_n$ (Section 1.5.). The lattices $[a_{n+1}, a_n]$ and $[b \vee a_{n+1}, b \vee a_n]$ are isomorphic under the map sending $x \in [a_{n+1}, a_n]$ to $b \vee x$.
We obtain: Kdim $[a_{n+1}, a_n]$ = Kdim $[b \vee a_{n+1}, b \vee a_n] \leq$ Kdim $[b \vee a_{n+1}, 1]$, and thus: Kdim $[a_{n+1}, a_n] + 1 \leq$ Kdim $[b \vee a_{n+1}, 1] + 1 \leq \alpha$.
The latter contradicts Kdim $[a_{n+1}, a_n] \geq \alpha$. □

Let L be a lattice with 0 and 1. We say that L is *α-critical* for the ordinal α if Kdim L $= \alpha$ and Kdim $[a,1] < \alpha$ for all $a \neq 0$ in L. We say that L is *critical* if there exists an ordinal α for which L is α-critical. An *$a \in L$ is α-critical* if [0,a] is

α-critical. It is clear that a $\in$ L is 0-critical if and only if a is an atom of L.

3.2.4. Proposition. Let L be a modular lattice with 0 and 1. If L has Krull dimension then it contains a critical element.

Proof. Put Kdim L $= \alpha$; the proof is by transfinite recursion on α. If $\alpha = 0$ then L is Artinian and hence L contains an atom. We assume that the claim is verified for any modular lattice M with Kdim M $< \alpha$. The induction hypothesis allows to assume that Kdim L = Kdim [0,a] for all a $\neq$ 0.
If L is not α-critical then there is an $a_1 \neq 0$ such that Kdim $[a_1,1] = \alpha$. If a_1 is not α-critical, there is an $a_2 < a_1$, $a_2 \neq 0$ such that Kdim $[a_2, a_1] = \alpha$. So we eventually arrive at the following decreasing sequence: $a_1 > a_2 > ... \, a_n > ...$, such that Kdim $[a_{n+1}, a_n] = \alpha$ for all n $\geq$ 1, but that is a contradiction (cf. Proposition 3.1.5 (2)). Thus L is α-critical or 1 is an α-critical element of L. □

3.2.5. Proposition. Let L be a modular lattice with 0 and 1. If L is α-critical then each nonzero a in L is α-critical.

Proof. Since Kdim [a,1] $< \alpha$ we may use Proposition 3.2.1. to derive that Kdim [0,a] $= \alpha$. Consider b $\neq$ 0, b $\leq$ a. Since Kdim [b,1] $< \alpha$ whilst [b,a] $\subset$ [b,1] it follows that Kdim [b,a] $< \alpha$ and therefore a is α-critical. □

3.2.6. Corollary. If L is a modular lattice with 0 and 1 which is α-critical then it is co-ireducible.

Proof. If L were not co-irreducible then there are a,b $\in$ L - {0} such that a $\wedge$ b = 0. We have: Kdim [a,1] $< \alpha$, Kdim [b,1] $< \alpha$ and [0,a] = [b $\wedge$ a,a] $\cong$ [b,b $\wedge$ a]. By Proposition 3.2.5., Kdim [b,b $\vee$ a] $< \alpha$, hence Kdim [0,a] $< \alpha$. Proposition 3.2.1. leads to Kdim L $< \alpha$, a contradiction. □

3.2.7. Corollary. Let L be a modular upper continuous lattice with 0 and 1. If L has Krull dimension, then there exists a direct union of critical elements which is essential in L.

Proof. Let $\{a_i\}_{i\in J}$ be an independent family of critical elements of L which is maximal as such. By Proposition 3.2.2., J is finite, by Proposition 3.2.4. the join $\bigvee_{i\in J} a_i$ is essential in L. □

3.2.8. Lemma. Let L be a modular and upper-continuous lattice with 0 and 1. Consider an increasing sequence in L, $0 = a_0 < a_1 < a_2 < \ldots < a_n < \ldots$, together with a decreasing sequence in L, $1 = b_0 > b_1 > b_2 > \ldots$. If for all $i \geq 0$ we have that: (*) $a_{i+1} \wedge b_i \not\leq a_i \vee b_{i+1}$, then L cannot have Krull dimension.

Proof. For $b = \bigvee_{i\geq 1} (a_i \wedge b_i)$, (*) holds for the lattice [b,1] with respect to the sequences $b = b \vee a_0 < b \vee a_1 < b \vee a_2 < \ldots$, and $1 = b_0 \vee b > b_1 \vee b > b_2 \vee b > \ldots$. Indeed, if $(b_i \vee b) \wedge (a_{i+1} \vee b) \leq a_i \vee b_{i+1} \vee b$ then $b_i \wedge a_{i+1} \leq a_i \vee b_{i+1} \vee b$ and since $b \leq a_i \vee b_{i+1}$ this yields $b_i \cap a_{i+1} \leq a_i \vee b_{i+1}$, a contradiction. Secondly, let us show that $b = (b \vee a_n) \wedge (b \vee b_n)$. That b is dominated by the right hand member is obvious. Conversely, $(b \vee a_n) \wedge (b \vee b_n) = b \vee ((b \vee a_n) \wedge b_n)$ by modularity of L, but then we obtain (using modularity of L again):
$(a_n \vee b) \wedge b_n = a_n \vee \bigvee_{i=1}^{n} (a_i \wedge b_i) \vee \bigvee_{i=n+1}^{\infty} (a_i \wedge b_i) \wedge b_n =$
$((a_n \vee \bigvee_{i=1}^{n} (a_i \wedge b_i)) \vee b_n) \vee \bigvee_{i\geq n+1} (a_i \wedge b_i)$.
From the inequalities: $\bigvee_{i\geq n+1} (a_i \cap b_i) \leq b$ and $\bigvee_{i=1}^{n} (a_i \wedge b_i) \leq a_n$ we then derive the following: $(a_n \vee b) \wedge b_n \leq (a_n \wedge b_n) \vee b \leq b$, which leads to the equality $b = (b \vee a_n) \wedge (b \vee b_n)$.
Now finally we assume that L has Krull dimension. Then we may select $n \in \mathbb{N}$ such that the uniform dimension dim $[b, b_n \vee b]$ is minimal (in the lattice [b,1]) From the inequality: $(a_{n+1} \vee b) \wedge (b_n \vee b) \not\leq a_n \vee b_{n+1} \vee b$ we deduce that b is not equal to $(a_{n+1} \vee b) \wedge (b_n \vee b)$ and hence:
dim $[b, b_n \wedge b] \geq$ dim $[b, b_{n+1} \vee b]$ + dim $[b, (a_{n+1} \vee b) \wedge (b_n \wedge b)] \geq$
1 + dim $[b, b_{n+1} \vee b]$ = 1 + dim $[b, b \vee b_n] >$ dim $[b, b_n \vee b]$,
which is a contradiction. □

3.2.9. Theorem. Consider a modular upper continuous lattice L with Krull dimension. Let ϵ be a limit ordinal such that there exists an increasing sequence of elements of L, say $\{a_\lambda, \lambda < \epsilon\}$, with $1 = \bigvee_{\lambda<\epsilon} a_\lambda$. If α is an ordinal such that Kdim $[0, a_\lambda] \leq \alpha$ for all $\lambda < \epsilon$, then Kdim L $\leq \alpha$.

Proof. If the theorem were false then there would exist a sequence $1 = b_0 > b_1 > b_2 > \ldots$, such that Kdim $[b_{i+1}, b_i] \geq \alpha$ for all $i \geq 1$. Since $b_1 \neq 1$ there must exist a $\lambda_1 < \epsilon$ such that $a_{\lambda_1} \not\leq b_1$. Put $c_0 = 0$, $c_1 = a_{\lambda_1}$, hence $c_1 \wedge b_0 \not\leq c_0 \vee b_1$. Suppose we have already defined $c_0 < c_1 < c_2 < \ldots < c_n$ in $\{a_\lambda, \lambda < \epsilon\}$ such that $b_1 \wedge c_{i+1} \not\leq b_{i+1} \vee c_i$, $0 \leq i \leq n-1$. If $b_n \leq c_n \vee b_{n+j}$ for all $j \geq 1$, then $b_{n+j-1} \leq c_n \vee b_{n+j}$ and $b_{n+j-1} = b_{n+j-1} \wedge (c_n \vee b_{n+j})$ $= b_{n+j} \vee (c_n \wedge b_{n+j-1})$, and also $[b_{n+j}, b_{n+j-1}] \cong [b_{n+j} \wedge c_n, b_{n+j-1} \wedge c_n]$ follows. In the sequence $c_n \wedge b_n > c_n \wedge b_{n+1} > c_n \wedge b_{n+2} > \ldots$, we have Kdim $[b_{n+j} \wedge c_n, b_{n+j-1} \wedge c_n] \geq \alpha$. The latter implies that Kdim $[0, c_n] > \alpha$, a contradiction. □

3.2.10. Corollary. Let L be a modular upper continuous lattice with Krull dimension such that $1 = \bigvee_{i \in J} a_i$, then Kdim L $= sup_{i \in J}$ { Kdim $[0, a_i]\} = \alpha$.

Proof. It is clear that $\alpha \leq$ Kdim L. For the converse, look at an $a \neq 1$ in L. By the hypothesis there is an $x > a$ such that Kdim $[a,x] \leq \alpha$. So we arrive at an increasing sequence $\{a_\lambda, \lambda \leq \epsilon\}$ with $a_\epsilon = 1$, such that for each $\lambda \leq \epsilon$ we have that Kdim $[a_\lambda, a_{\lambda+1}] \leq \alpha$ and $a_\lambda = \bigvee_{\mu<\lambda} a_\mu$ whenever λ is a limit ordinal. If Kdim L $< \alpha$ let ς be the least ordinal such that Kdim $[0, a_\varsigma] \not\leq \alpha$. By Theorem 3.2.9., ς is a nonlimit ordinal, but the fact that Kdim $[0,a] = \sup$ {Kdim $[0, a_{\varsigma-1}]$, Kdim $[a_{\varsigma-1}, a_\varsigma]\}$ leads to the conclusion that Kdim $[0, a_\varsigma] \leq \alpha$, a contradiction. □

3.2.11. Corollary. Let L be a modular and upper continuous lattice with Kdim L $= \alpha$. For each $\beta \leq \alpha$ there exists a maximal element a_β such that Kdim $[0, a_\beta] \leq \beta$.

3.2.12. Corollary. Let L be a modular and upper continuous lattice having Krull dimension, then Kdim L equals γ where $\gamma = \sup$ {α an ordinal, there exists $a,b \in L$, $a < b \leq 1$, such that $[a,b]$ is α-critical}.

Proof. Obviously Kdim L $\geq \gamma$. Transfinite recursion and Proposition 3.2.4. yields that there is n increasing sequence $\{a_\lambda, \lambda \leq \epsilon\}$ with $a_\epsilon = 1$ and for each $\lambda \leq \epsilon$, $a_{\lambda+1}$ being the join in $[a_\lambda, 1]$ of all critical elements in $[a_\lambda, 1]$. If λ is a limit ordinal we put $a_\lambda = \bigvee_{\mu<\lambda} a_\mu$. By Corollary 3.2.10. and transfinite recursion agin, we conclude that Kdim $[0, a_\lambda] \leq \gamma$ for all $\gamma \leq \epsilon$. Consequently Kdim L $\leq \gamma$. □

§3.3 Critical Composition Series of a Lattice

In this section L is a modular lattice with 0 and 1. A *critical composition series of L* is a chain $1 = a_0 > a_1 > {}_2 > ... > a_{n-1} > a_n = 0$ such that $[a_{i+1}, a_i]$ is a critical lattice for all i = 0,1,...,n-1. Moreover, if there exist ordinals $\alpha(1) > \alpha(2) > ... > \alpha(k) \geq 0$ such that the n_1 first intervals in the critical composition series are $\alpha(1)$-critical, the next n_2 intervals are $\alpha(2)$-critical, etc... , then L is said to have a *critical composition series of type* $(\alpha(1), n_1; \alpha(2), n_2; ... ; \alpha(k), n_k)$. If $\alpha(1) = ... = \alpha(k) = \alpha$ then the critical composition series is said to be an *α-critical composition series.*

3.3.1. Lemma. If L has a critical composition series of type $(\alpha(1), n_1; \alpha(2), n_2; ... ; \alpha(k)\, n_k)$ then Kdim L = $\alpha(1)$. If a $\neq$ o in L then Kdim [0,a] $\geq \alpha(k)$.

Proof. The first assertion is evident in view of Proposition 3.2.1.. If a $\neq$ 0 in L then there is an a_i in the critical composition series such that a $\leq a_i$ but a $\not\leq a_{i+1}$, hence $a_{i+1} < a \vee a_{i+1} \leq a_i$. From $[a_{i+1}, a \vee a_{i+1}] \cong [a \wedge a_{i+1}, a]$ and Proposition 3.2.5. we derive: Kdim $[a \wedge a_{i+1}, a]$ = Kdim $[a_{i+1}, a \vee a_{i+1}]$=
Kdim $[a_{i+1}, a_i] \geq \alpha(k)$. Hence Kdim [0,a] $\geq \alpha(k)$. □

3.3.2. Lemma. A Noetherian lattice L has a central composition series.

Proof. Put α = Kdim L, $a_0 = 1$, $\alpha(1) = \alpha$ Let $a_2 \in$ L be maximal with the property Kdim $[a_2,1] = \alpha$. That $[a_2,1]$ is α-critical is clear and Kdim $[0,a_2] \leq \alpha$. If Kdim $[0,a_2] = \alpha$, consider an $a_3 < a_2$ which is maximal with the property Kdim $[a_3, a_2] = \alpha$. Again it is clear that $[a_3, a_2]$ is α-critical and then we proceed as before. The definition of Krull dimension entails that only finitely many elements $1 = a_0 \geq a_1 \geq a_2 \geq ... \geq a_{n_1}$ with Kdim $[a_i, a_{i-1}] = \alpha = \alpha(1)$, i = 1,...,$n_1$, exist. Let Kdim $[0,a_{n_1}] = \alpha(2) < \alpha(1)$. Since L is Noetherian there must exist an element $a_{n_1+1} < a_{n_1}$ which is maximal with the property Kdim $[a_{n_1+1}, a_{n_1}] = \alpha(2)$. Clearly, $[a_{n_1+1}, a_{n_1}]$ is $\alpha(2)$-critical.
Repetition of this proces, using the fact that any set of ordinals is well-ordered, leads to a critical composition series for L. □

3.3.3. Lemma. Let L be a lattice and a ∈ L. Suppose that the lattice [a,1] has a critical composition series of type $(\alpha(1),n_1; \dots ; \alpha(k),n_k)$. If Kdim [0,a] < $\alpha(k)$ then a is the unique maximal element with the property that Kdim [0,a] < $\alpha(k)$.

Proof. Pick b ∈ L such that Kdim [0,b] < $\alpha(k)$; we have: Kdim [a,a ∨ b] ≤ Kdim [a ∧ b,b] ≤ Kdim [0,b] ≤ $\alpha(k)$. By Lemma 3.3.1., [a,a ∨ b] = 0 or a = a ∨ b or b ≤ a. □

3.3.4. Lemma. Any two α-critical composition series for the lattice L have the same length.

Proof. Consider the α-critical composition series: $1 = a_0 > a_1 > \dots > a_n = 0$, $1 = b_0 > b_1 > \dots > b_m = 0$. In view of Theorem 1.3.3. there exist equivalent refinements: $1 = a_0 = a_{11} > a_{12} > \dots > a_1 t_1 = a_1 > \dots > a_{n-1} = a_{n-1} = a_{(n-1)t_{n-1}} > a_{n2} > \dots > a_{nt_n} = a_n = 0$; $1 = b_0 = b_{11} > b_{12} > \dots > b_{1s_1} = b_1 > \dots > b_{m-1} = b_{m-1\ t_{m-1}} > \dots > b_{m\ s_m} = b_m = 0$. Since $[a_{i+1}, a_i]$, i = 1,...,n, and $[b_{j+1}, b_j]$, j = 1,...,m, are α-critical we have that Kdim $[b_j, b_{j\ s_{j-1}}] = \alpha$, j = 1,...,m, and Kdim $[a_i, a_{i\ t_{i-1}}] = \alpha$, i = 1,...,n, while all the other intervals have Krull dimension strictly larger than α. Using Corollary 1.3.2. we may conclude that m = n. □

Note that in the situation of the lemma we may find elements $x_1, \dots, x_n$ such that $a_0 \geq x_1 > a_1 \geq x_2 > a_2 > \dots > a_{n-1} \geq x_n > a_n = 0$ and $y_1,\dots,y_n$ such that $b_0 \geq y_1 > b_1 \geq y_2 > b_2 > \dots > b_{n-1} \geq y_n > b_n = 0$ together with a permutation Π of {1,...,n} such that the intervals $[a_i, x_i]$ and $[b_{\Pi(i)}, y_{\Pi(i)}]$ are projective, for all i = 1,...,n.

3.3.5. Theorem. Suppose that the modular lattice L with 0 and 1 has a critical composition series of type $(\alpha(1),n_1; \dots ; \alpha(k), n_k)$ and another one of type $(\beta(1), m_1; \dots ; \beta(f), m_f)$ then: k = f, $\beta(i) = \alpha(i)$ and $m_i = n_i$ for all i = 1,...,k.

Proof. By Lemma 3.3.2., Kdim L = $\alpha(1) = \beta(1)$. We proceed the proof by transfiite recursion on Kdim L = $\alpha(1)$. The case $\alpha(1) = 0$ folows from Corolary 1.3.4. So assume that $\alpha(1) \geq 1$ and let a be the maximal element in the composition series of type $(\alpha(1), n_1; \ldots ; \alpha(k), n_k)$ with the property: Kdim [0,a] = $\alpha(2)$; let b be the maximal element in the other critical composition series with the property Kdim [0,b] = $\beta(2)$. Lemma 3.3.3. entails that a = b and Kdim [0,a] = $\alpha(2) = \beta(2) < \alpha(1)$. The individual hypothesis then entails k = f and $m_i = n_i$ for i = 2,3,...,k. Finally, [a,1] = [b,1] has an α-critical composition series, so $m_1 = n_1$ by Lemma 3.3.4. □

Note that, as pointed out after Lemma 3.3.4., we may again find elements $x_1,\ldots,x_n$ and $y_1,\ldots,y_n$, such that:

$$a_0 \geq x_1 > a_1 \geq x_2 > a_2 \ldots > a_{k-1} \geq x_k > a_k = 0,$$

$$b_0 \geq y_1 > b_1 \geq y_2 > b_2 .. > b_{k-1} \geq y_k > b_k = 0,$$

together with a permutation of {1,..,k} such that the intervals $[a_i, x_i]$ and $[b_{\Pi(i)}, y_{\Pi(i)}]$ are projective for all i = 1,...,k.

Let L be a lattice with Krull dimension. An a ∈ L is said to be a *basis element* if a is maximal amongst the α-critical elements of L where α = min { Kdim [0,x], x ≠ 0 in L }. A *basis series* is a chain $0 = b_0 < b_1 < \ldots < b_{n_1} < b_n = 1$, where each b_i is a basis element of $[b_{i-1}, 1]$ for all i = 1,...,n.

3.3.6. Lemma. Let L be a lattice having Krull dimension, then the following assertions are equivalent:

1. L is critical.
2. L is co-irreducible and there exists a critical a in L such that Kdim [a,1] < Kdim [0,a].

Proof. 1⇒2. L is co-irreducible in view of Corollary 3.2.6. Pick a ≠ 0 in L. Since a is critical we have Kdim [a,1] < Kdim L and Kdim [0,a] = Kdim L.

2⇒1. By Proposition 3.2.1., Kdim L = Kdim [0,a]. Take b ≠ 0 in L such that Kdim [b,1] = Kdim L. Since L is co-irreducible we have a ∧ b ≠ 0 and from [b, a ∨ b] ≅ [a ∧ b, a] it follows that Kdim [b, a ∨ b] < Kdim [0,a] = Kdim L and also, Kdim [a ∨ b, 1] ≤ Kdim [a,1] < Kdim [0,a] = Kdim L. Using Proposition 3.2.1. once more, we arrive at the contradiction: Kdim [b,1] < Kdim L.

3.3.7. Lemma. Let L be a lattice with Krull dimension and consider a $\in$ L. If a dominates a basiselement b of L such that b $\neq$ 0 then Kdim [0,b] $\leq$ Kdim [b,a].

Proof. Assume Kdim [b,a] < Kdim [0,b]. If there is a nonzero c $\leq$ a such that c $\wedge$ b = 0 then [0,c] = [c $\wedge$ b, c] $\cong$ [b, c$\vee$ b] and so: Kdim [0,c] $\leq$ Kdim [b,a] < Kdim [0,b]. The latter contradicts the fact that b is critical. Consequently a must be an essential extension of b. Since b is co-irreducible, a must be co-irreducible too. Because Kdim [b,a] < Kdim [0,b], Lemma 3.3.6. implies that a is critical. Hence Kdim [0,b] = Kdim [0,a]. Since b $\neq$ a, b cannot be a basis element, contradiction.

□

3.3.8. Proposition. Consider a chain

$1 = a_0 > a_1 > \ldots > a_n = 0$ (*)

in a lattice L having Krull dimension. The following assertions are equivalent:

1. (*) is a basis series of L.
2. (*) is a critical composition series of L.

Proof. 1⇒2. Since $[a_{i+1}, a_i]$ is critical for $0 \leq i \leq n-1$, it remains to be proved that Kdim $[a_{i+1}, a_i] \geq$ Kdim $[a_{i+2}, a_{i+1}]$. But $[a_{i+2}, a_i]$ contains the basis element a_{i+1} so Lemma 3.3.7. yields: Kdim $[a_{i+2}, a_{i+1}] \leq$ Kdim $[a_{i+1}, a_i]$.

2⇒1. Let us establish that a_i is a basis element of $[a_{i+1}, 1]$. Pick x $\neq a_{i+1}$ in $[a_{i+1}, 1]$. Lemma 3.3.1. entails: Kdim $[a_{i+1}, x] \geq$ Kdim $[a_{i+1}, a_i] = \beta$. Consequently, $\beta = \min$ { Kdim $[a_{i+1}, x]$, x $\in [a_{i+1}, 1]$, x $\neq a_{i+1}$ }.
Assume that there is an x $\in [a_{i+1}, 1]$ such that $[a_{i+1}, x]$ is β-critical but with a_i < x. Then $\beta \geq$ Kdim $[a_i, x] \geq$ Kdim $[a_i, a_{i-1}] \geq$ Kdim $[a_{i+1}, a_i] = \beta$, because of Lemma 3.3.1. and therefore Kdim $[a_i, x] = \beta$. But the latter contradicts the assumption that $[a_{i+1}, x]$ is β-critical, thus a_i must be a basis element in the lattice $[a_{i+1}, 1]$.

§3.4 The Gabriel Dimension of a Modular Lattice.

Throughout this section L is a modular upper continuous lattice having 0 and 1.

We define the *Gabriel dimension* of L, denoted by Gdim L, using transfinite recursion. We put Gdim L = 0 if and only if L = {0}. Let α be a nonlimit ordinal and assume that the Gabriel dimension Gdim M = β has already been defined for lattices with $\beta < \alpha$. We say that L is it α-simple if for each a $\neq$ 0 in L we have Gdim [0,a] $\not< \alpha$ and Gdim [a,1] $< \alpha$ We then say that Gdim L = α if Gdim L $\not< \alpha$ but for every a $\neq$ 1 in L there exists a b > 0 such that [a,b] is β-simple for some $\beta \leq \alpha$.

This definition implies that for an α-simple L, Gdim L = α and α is a nonlimit ordinal. Consider a $\in$ L. If Gdim [0,a] = α then we say that α is the *Gabriel dimension of a* and we write Gdim a = α. If [0,a] is α-simple then a is said to be an *α-simple element of L.* It is rather obvious that an a in L is 1-simple if and only if a is an atom in L whereas Gdim L = 1 exactly then when L is semiartinian. By definition it is also clear that Gdim L $\leq \alpha$ if and only if for all a $\neq$ 1 in L, [a,1] contains a β-simple element for some $\beta \leq \alpha$.

3.4.1. Proposition. L has Gabriel dimension if and only if for any a $\in$ L both [0,a] and [a,1] have Gabriel dimension. Moreover Gdim L = sup {Gdim [0,a], Gdim [a,1]}.

Proof. First assume that L has Gabriel dimension α. Then Gdim [a,1] $\leq \alpha$. We claim that Gdim [0,a] $\leq \alpha$. The case $\alpha = 1$ follows from Proposition 1.9.3., so we assume that the proposition holds for all ordinals β, $\beta < \alpha$. If L is α-simple then Gdim [0,a] $\not< \alpha$. Pick b $\neq$ 0 in [0,a]. Then Gdim [b,1] $< \alpha$ and from [b,a] $\subset$ [b,1] we derive that Gdim [b,a] $< \alpha$ by the induction hypothesis. Consequently [0,a] is α-simple. If L were not α-simple, pick b $\neq$ 0 in [0,a] and let c be a pseudocomplement of a in the lattice [b,1]. The lattice [c,1] contains an element d which is β-simple with $\beta \leq \alpha$. Since c < d, d $\wedge$ a $\neq$ b. Put e = d $\wedge$ a, we claim that e is β-simple in the lattice [b,a].

Modularity of L implies that: (a $\vee$ c) $\wedge$ d = c $\vee$ (a $\wedge$ d) = c $\vee$ e $\neq$ c. Since [c, c$\vee$ e] $\subset$ [c,d] and [c,d] being β-simple, it follows that [c, c $\vee$ e] is β-simple. But from [c, c $\vee$ e] $\cong$ [e, $\wedge$ c, e] = [b,c] it follows that [b,e] is β-simple in [b,a]. Thus [0,a] has Gabriel dimension and Gdim [0,a] $\leq$ Gdim L. Consequently, we obtain: sup (Gdim [0,a], Gdim [a,1]) $\leq$ Gdim L.

Secondly, assume that sup (Gdim [0,a], Gdim [a,1]) = α exists and we aim to show that Gdim L exists and moreover Gdim L $\leq \alpha$. Pick b $\in$ L, b $\neq$ 1. If a $\leq$ b then [b,1] contains a β-simple element with $\beta \leq \alpha$. If a $\not\leq$ b then a $\wedge$ b $\neq$ a and [a $\wedge$ b, a] contains a β-simple element with $\beta \leq \alpha$. From [a $\wedge$ b, a] $\cong$ [b, a $\vee$ b] we conclude that [b, a $\vee$ b] contains a β-simple element with $\beta \leq \alpha$, hence [b,1] contains a β-simple element. This establishes the existence of Gdim L and

Gdim L $\leq \alpha$. □

The foregoing proof also yields:

3.4.2. Corollary. 1. If a $\in$ L is α-simple then each b $\neq$ 0, b $\leq$ a is also α-simple. 2. There is a direct family $\{a_i\}_{i\in J}$ such that eacch a_i is α_i-simple and $\bigvee_{i\in J}\ a_i$ is essential in L.

3.4.3. Proposition. Consider $\{a_i\}_{i\in J}$ in L such that $1 = \bigvee_{i\in J}\ a_i$, then Gdim L = $\sup_{i\in J}$ { Gdim $[0,a_i]$} if one of these members is defined.

Proof. First, if Gdim L = α then Gdim $[0,a_i] \leq\ \alpha$ then Gdim $[0,a_i] \leq\ \alpha$ by Proposition 3.4.1., hence sup {Gdim $[0,a_i]$,i} $\leq\ \alpha$. Conversely, assume that α = $\sup_{i\in J}$ {Gdim $[0,a_i]$} does exist and consider an a $\neq$ 1 in L. For some i $\in$ J, $a_i\ \not\leq$ a. Hence a $\wedge\ a_i\ <\ a_i$. Since Gdim $[0,a_i] \leq\ \alpha$, Gdim $[a \wedge a_i,\ a_i] \leq\ \alpha$ and therefore $[a \wedge a_i,\ a_i]$ contains a β-simple element with $\beta\ \leq\ \alpha$.
From $[a \wedge a_i,\ a_i] \cong [a, a \vee a_i]$ it follows that the lattice $[a, a \vee a_i]$ contains a β-simple element and then [a,1] also contains a β-simple element with $\beta\ \leq\ \alpha$. Consequently Gdim L exists and Gdim L $\leq\ \alpha$. □

3.4.4. Corollary. If the lattice L has Gabriel dimension, then Gdim L = sup {α, [a,b] is α-simple for some a < b $\leq$ 1} = Θ.

Proof. By transfinite recursion we find an increasing sequence $\{a_\lambda,\ \lambda\ \leq\ \varepsilon\}$, a_ε = 1 and for each $\lambda\ \leq\ \varepsilon$, $a_{\lambda+1}$ is the join of all β-simple elements $[a_\lambda,1]$ for some ordinal β. If λ is a limit ordinal then $a_\lambda = \bigvee_{\mu<\lambda}\ a_\mu$. Applying Proposition 3.4.3. and transfinite recursion we obtain Gdim $[0,a_\lambda] \leq\ \Theta$ for all $\lambda\ \leq\ \varepsilon$, or Gdim L $\leq\ \Theta$. On the other hand $\Theta\ \leq$ Gdim L in view of Proposition 3.4.1., thus Gdim L = Θ. □

3.4.5. Corollary. Let L be a lattice such that the Gabriel dimension of [a,1] exists for all a $\neq$ o in L, then: Gdim L $\leq 1 + \alpha$ where
α = sup {Gdim [a,1], 0 $\neq$ a $\leq$ 1}.

Proof. If Gdim L $\not\leq\ \alpha$ the Proposition 3.4.1. implies that Gdim [0,a] $\not\leq\ \alpha$ for all a

$\neq 0$. But then L is $(1+\alpha)$-simple by the definition of Gabriel dimension, hence Gdim L $= 1 + \alpha$. □

A chain $0 = a_0 < a_1 < ... < a_n = 1$ in the lattice L is said to be a *Gabriel simple composition series for L* if for all i $\in \{1,...,n\}$, $[a_{i-1}, a_i]$ is α_i-simple and $\alpha_1 \leq \alpha_2 \leq ... \leq \alpha_n$. Clearly, a Noetherian lattice has a Gabriel simple composition series.

3.4.6. Theorem. If L has Gabriel simple composition series $0 = a_0 < a_1 < ... < a_n = 1$ and $0 = b_0 < b_1 < ... < b_m = 1$. Then m = n and moreover, there exist x_i, y_i, i = 1,...,n, such that $a_{i-1} < x_i \leq a_i$, $b_{i-1} < y_i \leq b_i$ together with a permutation Π of $\{1,...,n\}$ such that $[a_{i-1}, x_i]$ and $[b_{\Pi(i)-1}, y_{\Pi(i)}]$ are projective intervals for all i = 1,...,n.

Proof. Similar to the proof of Lemma 3.3.4. □

§3.5 Comparison of Krull and Gabriel Dimension.

In this section L is a modular, upper continuous lattice with 0 and 1.

3.5.1. Theorem. If L has Krull dimension then L has Gabriel dimension and Kdim L $\leq$ Gdim L $\leq$ Kdim L + 1.

Proof. Put α = Kdim L. By transfinite recursion on α we establish that Gdim L $\leq \alpha + 1$. If $\alpha = 0$ then L is Artinian and obviously L is semiartinian i.e. Gdim L $\leq$ 1. Assume that we proved the claim for ordinals $\beta < \alpha$ and pick a $\neq 1$ in L. Then, [a,1] contains a β-critical element with $\beta \leq \alpha$, say b. Since [a,b] is β-critical, Kdim [x,b] $< \beta$ for some x $\in$ [a,b], x $\neq$ a, hence the induction hypothesis entails that Gdim [x,b] $\leq \beta$. But Gdim [a,b] $\leq 1 + \beta \leq 1 + \alpha$ (Corollary 3.4.5.), hence [a,1] contains a μ-simple element with $\mu \leq \alpha + 1$. Consequently Gdim L exists and Gdim L $\leq 1 + \alpha$.

Next we show that Kdim L is not larger than Gdim L = β. Assume $\beta = 0$, i.e. L is semiartinian, and let $a_1 \geq a_2 \geq ... \geq a_n \geq ...$ be a descending chain in L. Put a = $\bigwedge a_i$. From the fact that [a,1] is semiartinian and having finite uniform dimension, it follows that the socle s(L) is the direct join of a finite number of atoms. Consequently, the length of [0,s(L)] is finite and so there is an n $\in$ IN such that $a_n \wedge$ s(L) = $a_{n+1} \wedge$ s(L) = ... From $\bigwedge_{i=1}^{\infty} a_i = 0$ it follows that $a_n \wedge$ s(L) = 0 and a_n = 0 then follows because s(L) is esential in L. Thus L is Artinian and Kdim L = 0. Now suppose that Kdim M $\leq$ Gdim M for any lattice M with Gdim M $< \beta$. If Gdim L = β then there is an ascending chain $\{a_\lambda, \lambda \leq \varepsilon\}$ with $a_\varepsilon = 1$ and for each $\lambda \leq \varepsilon$, $a_{1+\lambda}$ is the join of all μ-simple elements of $[a_\lambda,1]$ for some ordinal $\mu \leq \beta$. If λ is a limit ordinal then $a_\lambda = \bigvee_{\mu<\lambda} a_\mu$. The problem may be reduced to the case where L is β-simple by using Corollary 3.2.10. For a $\neq$ 0 in L we then have that Gdim [a,1] $< \beta$ and by the induction hypothesis Kdim [a,1] $< \beta$. But finally, Kdim L $\leq \beta$ follows from Proposition 3.2.3. □

3.5.2. Theorem. If L is a Noetherian lattice then Gdim L = 1 + Kdim L.

Proof. Since L has Kdim, Theorem 3.5.1. implies that L has Gdim and Gdim L $\leq$ Kdim L + 1. Since L is Noetherian, there is an ascending chain $0 = a_0 < a_1 < ... < a_n = 1$ such that $[a_i.a_{i+1}]$ is α_i-simple for all i = 0,...,n-1. By Proposition 3.4.1., α = Gdim L = $\sup_{0\leq i\leq n-1}\{\alpha_i\}$ and so Gdim L = α must be a nonlimit ordinal, say $\alpha = \beta + 1$.
By transfinite recursion we prove that Kdim L = β. If $\beta = 0$ then $\alpha = 1$ and L is semiartinian but as L is also Noetherian it is Artinian i.e. Kdim L = 0. Consider now a decreasing sequence in L, say $b_1 \geq b_2 \geq ...$, together with a descending chain for each i = 0,...,n-1: $(a_i \vee b_1) \wedge a_{i+1} \geq (a_i \vee b_2) \wedge a_{i+1} \geq ... \geq ...$, in $[a_i, a_{i+1}]$. The α_i-simplicity of $[a_i, a_{i+1}]$ entails the existence of an $n_i \in$ IN such that Gdim $[(a_i \vee b_{k+1}) \wedge a_{i+1}, (a_i \vee b_k) \wedge a_{i+1}] < \alpha_i \leq \alpha$, for all k $\geq n_i$. Put m = max $\{n_i$, i = 0,...,n-1$\}$. By induction on n we show that Gdim $[b_{k+1}, b_k] < \alpha$ for all k $\geq$ m. The case n = 1 needs no coment. By the induction hypothesis we obtain:
Gdim $[(a_{n-2} \vee b_{k+1}) \wedge a_{n-1}, (a_{n-2} \vee b_k) \wedge a_{n-1}] < \alpha$ for all k $\geq$ m. On the other hand we know that Gdim $[a_{n-1} \vee b_{k+1}, a_{n-1} \vee b_k] < \alpha$ for all k $\geq$ m. Up to replacing L by $[a_{n-2},1]$ we may assume that $a_{n-2} = 0$ and, putting a = a_{n-1}, we obtain: Gdim $[b_{k+1} \wedge a, b_k \wedge a] < \alpha$, Gdim $[a \vee b_{k+1}, a \vee b_k] < \alpha$ for all k $\geq$ m. Taking into account the following isomorphisms:
$[b_{k+1} \wedge a, b_k \wedge a] \cong [b_{k+1}, b_k \vee (b_k \wedge a)] = [b_{k+1}, b_k \wedge (b_k \vee a)]$, $[b_k \wedge (b_{k+1} \vee a), b_k] \cong [b_{k+1} \vee a, b_k \wedge a]$, we arrive at: Gdim $[b_{k+1}, b_k \wedge (b_{k+1} \vee a)] < \alpha$, Gdim $[b_k \vee (b_{k+1} \wedge a), b_k] < \alpha$.
Applying Proposition 3.2.1. yields Gdim $[b_{k+1}, b_k] < \alpha$ for all k $\geq$ m and the latter

implies that Gdim $[b_{k+1}, b_k] \leq \beta$ for all $k \geq m$. The induction hypothesis leads to Kdim $[b_{k+1}, b_k] < \beta$ for all $k \geq m$.
By the definition of Krull dimension it is now clear that Kdim L $\leq \beta$ and hence 1 + Kdim L $\leq$ Gdim L $\leq$ Kdim L + 1, i.e. Gdim L = Kdim L + 1 as claimed. □

3.5.3. Note. By looking at the proof of the foregoing theorem we see that Gdim L = 1 + Kdim L remains true whenever Kdim L is a finite ordinal.

§3.6 Exercises.

(110) (Lemonnier [2]) Let E and F be ordered sets and order $E \times F$ lexicographically. If E and F have Krull dimension then so does $E \times F$ and Kdim $E \times F$ = Kdim E + Kdim F.

(111) If L is well ordered then L° has Krull dimension.

(112) By transfinite recursion on α we define ordinals $\omega(\alpha)$ by: $\omega(o) = 1$, $\omega(\alpha+1) = \omega(\alpha) + \omega(\alpha-1) + \dots$, for a limit ordinal $\omega(\alpha)$ is defined to be the ordinal associated to the set $\bigcup_{\beta<\alpha} [0,\omega(\beta)]$. Show that co-Kdim L = α if L = $[0,\omega(\alpha)]$.

(113) Show that the set C(IN) of all increasing sequences in IN does not have a Krull dimension.

(114) Let E and F be ordered sets of finite Krull dimension and let f: E → F be an increasing map. Suppose that there exist $n,r \in$ IN such that Kdim $[a,b] \geq n$ yields Kdim $[f(a),f(b)] \geq r$. Then Kdim E + r $\leq$ sup (n + r, Kdim F + n). In particular, if $a < b$ implies Kdim $[f(a),f(b)] \geq r$ then r + Kdim E $\leq$ Kdim F.

(115) Let L be a modular lattice with 0 and 1 having critical composition series of type $(\alpha(1), n_1; \alpha(2), n_2; \dots ; \alpha(k), n_k)$. For $a \neq 0$ in L, [0,a] has a critical composition series of type $(\beta(1), m_1; \dots ; \beta(h), m_h)$ where for each $i \in \{1,\dots,h\}$ there is a $j \in \{1,\dots,k\}$ such that $\alpha(j) = \beta(i)$ and $m_i \leq n_j$.

(116) Let L be a modular upper continuous lattice with Gabriel dimension, then L has Krull dimension if and only if all $a \in L$, [a,1] has finite uniform dimension.

(117) Let $(E,\leq)$ be a nonempty ordered set. Using transfinite recursion we define the sequence of subsets $(E_\alpha)_{\alpha\geq 0}$ of E by: $E_0 = \{x \in E$, x a maximal element$\}$,

$E_\alpha = \{x \in E,\ x < y$ entails $y \in \bigcup_{\beta<\alpha} E_\beta\}$ for $\alpha > 0$. Show that $E = E_\alpha$ for some α is equivalent to $(E,\leq)$ being Noetherian.

(118) Let E be a Noetherian ordered set. Show that the set of increasing series of E is Noetherian too.

(119) Let S be the set of all semigroups with 0 which are subsemigroups of $(\mathbb{N},+)$. Prove the following statements:

a. S is a lattice having 0 and 1.
b. If $G \neq 0$ in S then there exist $m,n \in \mathbb{Z}$ suh that $G \subset \mathbb{N}m$ and $G \bigcap [n,\infty] = \mathbb{N}m \bigcap [n,\infty]$
c. S is a Noetherian lattice and Kdim S = 2.

(120) (Lemonnier [4]) Let A be a ring and let $B = A[X,\varphi,\delta]$ be the ring of skew differential polynomials in the indeterminate X, where φ is an automorphism of A and δ is a φ-derivation, i.e. $\delta(ab) = \varphi(a)\,\delta(b) + \delta(a)b$ for $a,b \in A$, and where multiplication is defined by: $Xa = \varphi(a)X + \delta(a)$, $a \in A$. Let $[A,B]$ be the set of subrings R of B such that $A \subset R \subset B$. Show that $[A,B]$ is a Noetherian latice and Kdim $[A,B] \leq 2$.

Bibliographical Comments to Chapter 3.

In the study of Noetherian commutative rings, the classical Krull dimension (the maximal length of chains of prime ideals) plays an important part but unfortunately the classical Krull dimension provides only poor information when one studies noncomutative rings.

As always, many possibilities for generalizing the notion of classical Krull dimension from the commutative to the noncommutative case present themselves, in particular when one is considering particular classes of rings. The first step to a general theory was done by Gabriel and Rentschler in [1] (1967), where they introduced the notion of finite deviation of an ordered set and start to study some applications to the study of certain classes of rings e.g. Weil algebras. Later, G. Krause extended the notion of deviation to the transfinite case (in [3], (1970)) for the lattice of submodules of a module; he also replaced the term "deviation" by "Krull dimension". In the period that followed, the theory of Krull dimension was developed further and applied to several domains of noncommutative algebra; amongst the many papers on the subject let us just mention:

T. Albu [1]; G. Krause [1],[2],[3]; B. Lemonnier [1],[2],[3],[4]; R.C. Robson [1],[2],[3]; R. Gordon [1],[2],[3],[4]; A.V. Jategaonkar [2],[3],[4]; T. Lenagan [1],[2],[3], [4],[5],[6]; T. Gulliksen [1]; C. Nastasescu [1],[2],[3],[6]; C. Nastasescu, F. Van Oystaeyen [1],[2]; N. Popescu [1],[2]; J. Fisher, C. Lanski, J. Park [1]; C. Lanski [1];

R. Gordon and J.C. Robson study in [1] and [2] the connection between the Krull dimension in the Gabriel-Rentschler-Krause sense and another notion of Krull dimension introduced by P. Gabriel in his thesis using the terminology of category theory.

In this chapter we approach Krull dimension as well as Gabriel dimension in terms of lattice theory. This allows to present a unified as well as simplified approach to a series of results otherwise requiring some theoretical machinery (and notions) which is less accessible to the reader. More precisely, in Section 3.1 we follow B. Lemonnier, cf [4], and define the Krull dimension of an ordered set. In Section 3.2. we adopt to the lattice theoretical terminology some results obtained by G. Krause, cf. [3], and R. Gordon, J. Robson [1]. In Section 3.. the results obtained by G. Krause in [1] are simplified and generalized by our use of Schreier's theorem given in Chapter 1. In Section 3.4. we follow an idea of C. Lanski [1], and define the Gabriel dimension of a modular lattice, thus avoiding the (heavy) category theoretical terminology featured in other treatments of similar material. The link between Krull dimension and Gabriel dimension presented in Section 3.5. relies on T. Lenagan's paper [1], again reinterpreted in terms of lattices.

As pointed out in the introduction, the use of lattices in this book serves several purposes, here in Chapter 3. the unifying effect of this approach is particularly evident.

Chapter 4

Krull Dimension and Gabriel Dimension of Rings and Modules.

§4.1 Definitions and Generalities.

Let R be a ring, M a left R-module. We say that M has *Krull dimension, resp. Gabriel dimension,* if the lattice L(M) has Krull dimension, resp. Gabriel dimension. The ordinal Kdim L(M), resp. Gdim L(M), will be denoted by Kdim M, resp. Gdim M, and it is called *the Krull dimension, resp. Gabriel dimension, of M.*

If L(M) is α-critical, resp. α-simple, then we say that M is *α-critical, resp. α-simple.* Note that it makes sense to talk about α-simple modules only in case α is a non-limit ordinal. The module M is said to be *critical, resp. Gabriel simple,* if it is α-critical, resp. α-simple, for some ordinal α.

For any submodule N of M we have L(M/N) $\cong$ [N,M], hence M is α-critical, resp. α-simple, if and only if Kdim M $= \alpha$ and Kdim (M/N) $< \alpha$, resp. Gdim M $=$ Gdim N $= \alpha$ and Gdim (M/N) $< \alpha$, for any nonzero submodule N of M. If ${}_RR$ has Krull, resp. Gabriel, dimension then we say that R has left Krull dimension, resp. left Gabriel dimension, and denote it by sKdim R $=$ Kdim ${}_RR$, resp. sGdim R $=$ Gdim ${}_RR$. In a similar way we define the right Krull dimension dKdim R, resp. right Gabriel dimension dGdim R. In the sequel we *shall only consider left Krull dimension*, unless otherwise stated, and we then denote it by Kdim R (similarly for Gdim R).

Now we translate some properties of Chapter 3, adding some new consequences for rings and modules.

4.1.1. (Rentschler, Gabriel [1], Gabriel [1]). If N is a submodule of M then M has Krull (resp. Gabriel) dimension if and only if N and M/N have Krull (resp. Gabriel) dimension.
Moreover: Kdim M = sup (Kdim N, Kdim M/N), Gdim M = sup (Gdim N, Gdim M/N). This property follows from Proposition 3.2.1., resp. Proposition 3.4.1., combined with L(M/N) $\cong$ [N,M].

4.1.2. If $(M_i)_{i\in I}$ is a finite (resp. arbitrary) family of R-modules then M = $\bigoplus_{i\in I} M_i$ has Krull (resp. Gabriel) dimension if and only if each M_i has Krull (resp. Gabriel) dimension. Moreover, Kdim M = $\sup_{i\in I}$\{Kdim M_i\}, (resp. Gdim M = $\sup_{i\in I}$\{Gdim M_i\}.
The property regarding Krull dimension follows from 4.1.1., for Gabriel dimension we use Proposition 3.4.2.. Note that an infinite direct sum does not have Krull dimension because it has no finite coirreducible dimension.

4.1.3. R has Krull (resp. Gabriel) dimension if and only if every finitely generated R-module (resp. any R-module) has Krull (resp. Gabriel) dimension. Moreover Kdim M $\leq$ Kdim R for each finitely generated R-module M and Gdim M $\leq$ Gdim R for any R-module M.

4.1.4. (Gordon, Robson, [1],[2]). Let R be a ring with Kdim and M an R-module with Kdim then we have Kdim M $\leq$ Kdim R even if M is not finitely generated. This follows from Corollary 3.2.10. taking into account that M = $\sum_{x\in M}$ Rx and Kdim (Rx) $\leq$ Kdim R.

4.1.5. (Michler [2], Krause [3]). If M has Kdim then Kdim M $\leq$ sup \{Kdim (M/E) + 1, E an essential R-submodule of M\}. This follows from Proposition 3.2.3.

4.1.6. If M/X has Gabriel dimension for each nonzero R-submodule X of M then M has Gabriel dimension and Gdim M $\leq 1 + \alpha$ where α = sup \{Gdim (M/X), X $\neq$ 0, X $\subset$ M\}.

4.1.7. If M has Krull (resp. Gabriel) dimension then M is an essential extension of a finite direct sum (resp. arbitrary direct sum) of critical submodules (resp. of Gabriel simple modules). This follows from Corollary 3.2.7. (resp. 3.4.2.).

4.1.8. (Gordon, Robson [2]). If M = $\sum_{i \in I} M_i$ has Kdim then Kdim M = $\sup_{i \in I}$ (Kdim M_i).

4.1.9. (Gordon, Robson [2]). If M has Krull (resp. Gabriel) dimension then Kdim M = sup {Kdim C, C a critical submodule in a factor module of M}, (resp. Gdim M = sup {Gdim S, S is a Gabriel simple submodule in a factor module of M}). Follows from Corollary 3.2.12. (resp. 3.4.4.).

4.1.10. (Gordon, Robson [2], Gabriel [1]). If M has Kdim then Kdim M $\leq$ Gdim M $\leq$ Kdim M + 1. Moreover, if M is Noetherian or if M has finite Krull dimension then Gdim M = Kdim M +1.

4.1.11. (Gordon, Robson [2]). If Kdim M = α, resp. Gdim M = α, then for any ordinal $\beta \leq \alpha$ there is a maximal submodule M_β such that Kdim $M_\beta \leq \beta$, resp. Gdim $M_\beta \leq \beta$.

If the lattice L(M) has a critical composition series (resp. simple Gabriel series of composition) then we say that M has a critical composition series (resp. a simple Gabriel series of composition).

4.1.12. Proposition. (Krause [1], Gordon [1]) Suppose that M has critical composition series (resp. simple Gabriel series of composition): $0 = M_0 \subset M_1 \subset \ldots \subset M_n = M$, $0 = N_0 \subset N_1 \subset \ldots \subset N_m = M$. Then n = m and there is a permutation Π of {1,2,...,n} such that the injective hulls $E(M_i/M_{i-1})$ and $E(N_{\Pi(i)}/N_{\Pi(i)-1})$ are isomorphic for i = 1,...,n.

Proof. We consider the case where both series are critical composition series. By Theorem 3.3.5. we have m = n. The remark following that theorem entails that there exist submodules P_i and Q_i such that $M_{i-1} \subsetneq P_i \subset M_i$, $N_{i-1} \subsetneq Q_i \subset N_i$, and

a permutation Π of $\{1,...,n\}$ such that the intervals $[M_{i-1}, P_i]$ and $[N_{\Pi(i)-1}, Q_{\Pi(i)}]$ are projective. Therefore: $P_i/M_{i-1} \cong Q_{\Pi(i)}/N_{\Pi(i)-1}$. Since M_i/M_{i-1} is an essential etension of the first and $N_{\Pi(i)}/N_{\Pi(i)-1}$ is an essential extension of the second, $E(M_i/M_{i-1}) \cong E(N_{\Pi(i)}/N_{\Pi(i)-1})$.
The other alternative in the theorem follows from Theorem 3.4.6. □

For an R-module M, the submodule $s(M) = \sum$ {c, c is a critical submodule of M of minimal Krull dimension in M} is said to be the *critical socle* of M. By transfinite induction we introduce the Loewy critical series of M, i.e. an ascending chain of submodules $s_\alpha(M)$ of M such that $s_1 = s(M)$, $s_{\alpha+1}(M)/s_\alpha(M) = s(M/s_\alpha(M))$ for a non-limit ordinal, $s_\alpha(M) = \bigcup_{\beta<\alpha} s_\beta(M)$ for a limit ordinal. If M has Krull dimension then clearly: $M = s_\gamma(M)$ for some ordinal γ. In a similar way we may define the simple Gabriel Loewy series $\{\bar{s}_\alpha(M)\}_\alpha$ which is an ascending chain of submodules of M defined as above but replacing "critical" by "Gabriel simple".

4.1.13. The R-module M has Gabriel dimension if and only if there is an ordinal γ such that $\bar{s}_\gamma(M) = M$. Furthermore, if $f \in \mathrm{Hom}_R(M,N)$, for R-modules M and N, then $f(s_\alpha(M)) \subset s_\alpha(N)$ and $f(\bar{s}_\alpha(M)) \subset \bar{s}_\alpha(N)$ for any ordinal α.

§4.2 Krull and Gabriel Dimension of Some Special Classes of Rings and Modules.

4.2.1. The Ring of Endomorphisms of a Projective Finitely Generated Module.

Let P be a finitely generated R-module and M an arbitrary R-module. Put $A = \mathrm{End}_R(P)$ and $M^* = \mathrm{Hom}_R(P,M)$; then M^* is a left A-module (see Chapter 2.10.).

4.2.1.1. Theorem. If M has Krull dimension (resp. Gabriel) then the A-module M^* has Krull dimension (resp. Gabriel). Moreover $\mathrm{Kdim}_A\, M^* \leq \mathrm{Kdim}_R\, m$ (resp. $\mathrm{Gdim}_A\, M^* \leq \mathrm{Gdim}_R\, M$).

Proof. The statements concerning the Krull dimension follow from Proposition 3.1.4. and Lemma 2.10.4.

Now suppose that M has Gabriel dimension α; we proceed by induction on α. If $\alpha = 0$ then M is semiartinian and it follows that M^* is a semiartinian A-module if we take into account Lemma 2.10.4.. Assume that the assertion holds for any R-module N with Gdim N $< \alpha$ and let M have Gdim equal to α. Using simple Gabriel Loewy series we reduce the problem to the case where M is α-simple. Since α is a non-limit ordinal, $\alpha = \beta + 1$. Consider a nonzero A-submodule Y in M^* i.e. Y $= X_N$ for some nonzero R-submodule of M (again Lemma 2.10.4.). The exactness of the sequence

$$0 \to \mathrm{Hom}_R\,(P,N) \to \mathrm{Hom}_R\,(P,M) \to \mathrm{Hom}_R\,(P,M/N) \to 0$$

together with Y $= X_N = \mathrm{Hom}_R\,(P,N)$, yields $M^*/Y \cong (M/N)^*$.

From Gdim $(M/N) < \alpha$ we derive that Gdim $(M/N)^* < \alpha$ and hence Gdim $M^*/Y \leq \beta$. Since Y is arbitrary, Gdim $M^* \leq 1 + \beta = \alpha$. □

4.2.1.2. Corollary. If P has Krull (resp. Gabriel) dimension then A has Krull (resp. Gabriel) dimension and moreover: Kdim A $\leq$ Kdim P (resp. Gdim A $\leq$ Gdim P).

4.2.1.3. Corollary. If R has Krull dimension (resp. Gabriel) and e $\in$ R is an idempotent then $M_n(R)$ and eRe have Krull (resp. Gabriel) dimension. Moreover, Kdim $M_n(R) \leq$ Kdim R and Kdim eRe $\leq$ Kdim R (resp. Gdim $M_n(R) \leq$ Gdim R, and Gdim eRe $\leq$ Gdim R).

4.2.2. Normalizing Extensions.

In this subsection we let S $= \sum_{i=0}^{n} Ra_i$ with $a_0 = 1$ be a normalizing extension of R.

4.2.2.1. Theorem. (Fisher, Lanski, Park, [1]). For an S-module M the following statements are equivalent:

a. Gdim_S M exists
b. Gdim_R M exists

In the situation considered above Gdim_S M = Gdim_R M.

Proof. Assume a., say Gdim_S M = α. If $\alpha = 0$ then the fact that Gdim_R M $\leq \alpha$ follows from Corollary 2.12.10.. Now suppose that the conclusion Gdim_R M $\leq \alpha$ is valid for all modules with Gabriel dimension strictly less than α. Using the simple Gabriel Loewy series we reduce the problem to the case where M is an α-simple S-module. Put $\alpha = \beta + 1$. Let N be an R-submodule of M maximal with the property $N^* = 0$ (see Section 2.12. also for the meaning of the notation $(-)^*$ here). If X $\supsetneq$ N is another R-submodule then $X^* \neq 0$ and therefore Gdim_S M/X^* $< \alpha$ which entails, by the induction hypothesis, that Gdim_R M/X^* $< \alpha$. Hence Gdim_R M/X $< \alpha$ and also Gdim_R M/N $\leq \alpha$. Since M embeds into
$\bigoplus_{i=0}^{n}$ M/(N:a_i), Lemma 2.12.1. entails that Gdim_R M $\leq \alpha$.

Conversely, assume Gdim_R M = γ and we will now proceed by induction to show that Gdim_S M $\leq \gamma$. If L is an R-submodule of M such that Gdim_R L $< \gamma$ then Lemma 2.12.2. yields that Gdim_R(SL) $< \gamma$ and hence Gdim_S (SL) $< \gamma$. So we may assume that M does not contain R-submodules L such that Gdim_R L $< \gamma$. Let M' be the largest S-submodule of M such that Gdim_S M' $\leq \gamma$. If M = M' then the proof is finished. Otherwise we obtain Gdim_R (M/M') $\leq \gamma$, so we put M" = M/M'. Since Gdim_R M" $\leq \gamma$ it follows that M" contains a β-simple submodule N with $\beta \leq \gamma$. Clearly it is not really restrictive to assume $\beta = \gamma$. We have SN = $\sum_{i=0}^{n} a_i$N. If a_iN $\neq 0$ then Lemma 2.12.2. together with the fact that M" does not contain R-submodules of Gabriel dimension less than γ, implies that a:N is γ-simple. Correspondingly SN is a finite sum of γ-simple modules.

Now consider a descending chain of S-submodules of SN, $P_1 \supset P_2 \supset \ldots \supset P_r \supset \ldots$. When considering the latter chain as a chain of R-modules it becomes clear that Gdim_R $(P_i/P_{i+1}) < \gamma$ for all i $\geq$ t for some t $\in \mathbb{N}$. The induction hypothesis entails that Gdim_S $(P_i/P_{i+1}) < \gamma$.

Therefore there exists an S-submodule P $\neq 0$ in SN such that Gdim_S (P/P') $< \gamma$ for all nonzero S-submodules P' of P and consequently P must be a γ-simple S-module, a contradiction. □

4.2.2.2. Corollary. S has Gabriel dimension if and only if R has, in this case: Gdim S = Gdim R.

4.2.2.3. Lemma. Let M be an S-module such that $_RM$ has Krull dimension then Kdim_S M = Kdim_R M.

Proof. By Lemma 3.1.4., Kdim_S M $\leq$ Kdim_R M. Put Kdim_S M = α. By induction on α we establish that Kdim_R M $\leq$ α. By property 4.1.7. and application of the critical Loewy series we may reduce the problem to the situation where M is an α-critical S-module. Consider N $\subset$ M, an R-submodule which is maximal with the property $N^* = 0$. If N $\subsetneq$ X $\subset$ M then $X^* \neq 0$ and hence Kdim_S $(M/X^*) < \alpha$. By the induction hypothesis Kdim_R $(M/X^*) < \alpha$ and hence Kdim_R $(M/X) < \alpha$ and Lemma 2.12.1. then yields Kdim_R M $\leq$ α . Therefore $\text{Kdim}_R(M/N) \leq \alpha$ □

4.2.2.4. Lemma. (Lemonier, [1]). If M is an S-module with Krull dimension then ${}_RM$ has Krull dimension.

Proof. If the statement is false then there exists an S-submodule N $\subset$ M such that ${}_RN$ and ${}_R(M/N)$ do not have Krull dimension. Following the lines of proof of Theorem 3.1.10. we obtain in L(${}_SM$) an ordered subset isomorphic to the set of dyadic numbers D.
Proposition 3.1.5.(5) yields the existence of an R-submodule N_0 of M such that both N_0 and M/N_0 are without Krull dimension. For any R-submodule X of M we put X^i equal to $\sum_{k=0}^{i} a_k X$, hence X^n = SX. Suppose that we have constructed for 0 $\leq$ j $\leq$ n a submodule N_j in ${}_RM$ such that both N_j^j and M/N_j^j are without Krull dimension. We claim that it is possible to construct an R-submodule N_{j+1} in M such that N_{j+1}^{j+1} and M/N_{j+1}^{j+1} are without Krull dimension.
Put P = $(N_j^j + a_{j+1}N_j)/N_j^j \cong a_{j+1}N_j/Q$ with Q = $N_j^j \bigcap a_{j+1}N_j$. We decide what N_{j+1} has to be in three different cases:

a. Kdim_R P exists. Then we put $N_{j+1} = N_j$.

b. Both P and Q are without Krull dimension. Then we put $N_{j+1} = N_j \bigcap (Q{:}a_{j+1})$. It is clear that Q = $a_{j+1}N_{j+1}$, therefore N_{j+1} does not have Krull dimension and similarly for N_{j+1}^{j+1}. Since P $\cong$ a_{j+1} N_j/a_{j+1} N_{j+1} we obtain that P is a factor module of N_j^{j+1}/N_{j+1}^{j+1}, i.e. M/N_{j+1}^{j+1} does not have Kdim.

c. Kdim_R Q exists but P does not have Krull dimension. Obviously, $a_{j+1}N_j$ cannot have Krull dimension so by Proposition 3.1.5.(5) there must exist an R-submodule Y in $a_{j+1}N_j$ such that both Y and a_{j+1} N_j/Y are without Krull dimension. Then we put $N_{j+1} = N_j \bigcap (Y{:}a_{j+1})$. Hence Y = $a_{j+1}N_{j+1}$ i.e. N_{j+1}, and also N_{j+1}^{j+1}, is without Kdim.

The exactness of the sequence:

$$0 \to Q/(N_j^j \bigcap Y) \to a_{j+1}N_j/Y \to N_j^{j+1}/N_j^j + Y \to 0$$

entails that first $N_j^{j+1}/N_j^j + Y$ cannot have Kdim and then also M/N_{j+1}^{j+1} is without Krull dimension.
Finally, putting $N = N_n^n$ constructed as above then we see that N is an S-submodule of M such that both R-modules N and M/N do not have Krull dimension. □

The foregoing lemmas may now be combined into:

4.2.2.5. Theorem. (Lemonnier [1]) For an S-module M the following statements are equivalent:

a. ${}_SM$ has Krull dimension.
b. ${}_RM$ has Krull dimension.

Moreover, under these conditions Kdim_S M = Kdim_R M.

4.2.2.6. Corollary. If M is an Artinian S-module then ${}_RM$ is an Artinian R-module.

4.2.2.7. Corollary. If S has Krull dimension then so does R and vice-versa; in this case Kdim S = Kdim R.

Let M be an R-module having Krull dimension then we say that M is K-homogeneous if Kdim M = Kdim L for all nonzero R-submodules L of M.

4.2.2.8. Corollary. If M is an α-critical S-module then ${}_RM$ is K-homogeneous and an essential extension of a direct sum of at most n α-critical R-submodules.

Proof. By Theorem 4.2.2.5. Kdim_R M $= \alpha$. Let L be a nonero R-submodule of M. By Lemma 2.12.2. we have: Kdim_R L = Kdim_R SL = Kdim_S (SL) $= \alpha$. Hence ${}_RM$ is K-homogeneous. Now let N be an α-critical R-submodule of M. Lemma 2.12.2. yields that a_iN is either zero or an α-critical R-submodule. Hence M' = SN = $\sum_{i=0}^{n} a_i$N is the sum of at most n α-critical submodules. If M' were not essential in M then there is a nonzero R-submodule L in M such that M' $\bigcap$ L = 0. But we have that Kdim_R (M' $\oplus$ L/M') = Kdim_R L = α.

On the other hand Kdim_R (M/M') =

Kdim_S (M/M') $\subset$ α and this implies that Kdim_R L $<$ α, a contradiction.

This establishes that M' is essential in ${}_RM$. Finally, the properties of the critical series, derived in Section 4.1., entail that M' is an essential extension of a direct sum of at most n α-critical modules. □

4.2.3. Rings Strongly Graded by a finite Group.

Let n be the order of G and let R = $\bigoplus_{\sigma\in G} R_\sigma$ be strongly graded by G. Consider an R-module M. The results of the forgoing section combined with those in Section 2.16. imply:

4.2.3.1. Theorem. M has Gabriel dimension if and only if ${}_{Re}M$ has Gabriel dimension and in this case Gdim_R M = Gdim_{Re} M.

4.2.3.2. Corollary. The ring R has Gabriel dimension if and only if Re has Gabriel dimension and then these dimensions are equal.

4.2.3.3. Theorem. M has Krull dimension if and only if ${}_{Re}M$ has Krull dimension and in this case Kdim_R M = Kdim_{Re} M. (the proof is as in Lemma 4.2.2.4.).

4.2.3.4. Corollary. If M is Artinian then so is ${}_{Re}M$.

4.2.3.5. Corollary. R has Krull dimension if and only if Re has Krull dimension and in this case Kdim R = Kdim Re.

4.2.3.6. Corollary. If M is α-critical then ${}_{R_e}M$ is K-homogeneous and it is an essential extension of a direct sum of at most n α-critical R_e-submodules.

4.2.4. The Ring of Invariants.

Let R be a ring and let G be a finite group of automorphisms of R with $|G| = n$. We suppose that n is invertible in R. Consider the fixed ring $R^G = \{r \in R, g(r) = r \text{ for all } g \in G\}$.

4.2.4.1. Theorem. An R-module M has Krull dimension if and only if $_{R^G}M$ has Krull dimension and in this case $\text{Kdim}_R\, M = \text{Kdim}_{R^G}\, M$.

Proof. It is obvious that $\text{Kdim}_R\, M \leq \text{Kdim}_{R^G}\, M$. By Lemma 2.17.2. we have that $\text{Kdim}_{R^G}\, M \leq \text{Kdim}_{R*G}\,(M^*)$. By Theorem 4.2.3.3. we obtain: $\text{Kdim}_{R*G}\,(M^*) = \text{Kdim}_R\, M$ and hence $\text{Kdim}_{R^G}\, M \leq \text{Kdim}_R\, M$. □

4.2.4.2. Corollary. If R has Krull dimension then R^G has Krull dimension and $\text{Kdim}\, R = \text{Kdim}\, R^G$.

Proof. By Theorem 4.2.4.1. we have $\text{Kdim}\, R = \text{Kdim}_{R^G}\, R$ and thus $\text{Kdim}\, R^G \leq \text{Kdim}\, R$. Property 4.1.4. yields: $\text{Kdim}_{R^G}\, R \leq \text{Kdim}\, R^G$, hence $\text{Kdim}\, R \leq \text{Kdim}\, R^G$. □

4.2.5. Graded Rings of Type $\mathbb{Z}$

Let $R = \bigoplus_{i\in\mathbb{Z}} R_i$ be a $\mathbb{Z}$-graded ring and $M = \bigoplus_{i\in\mathbb{Z}} M_i$ a $\mathbb{Z}$-graded R-module. If the lattice $L_g(M)$ has Krull dimension then we say that *M has graded Krull dimension* and we denote it by *gr. Kdim M* = Kdim $L_g(M)$. Whenever the lattice $L_g(M)$ is α-critical, we say that M is a *graded α-critical* module and in order to avoid confusion we will denote this by M is *gr-α-critical.* First we deal with the case of bounded gradations.

4.2.5.1. Lemma. Suppose that the graded R-module M has left (or right) bounded gradation then the following properties hold:

1. Kdim_R M = gr Kdim_R M if either one side exists.
2. If M is graded α-critical then M is α-critical.

Proof. 1. follows from Proposition 3.1.4. and Corollary 2.14.4.
2. Suppose $M_i = 0$ for all $i < n_0$ and let X be a nonzero R-submodule of M, then gr. Kdim M/X $< \alpha$. By Lemma 2.14.2. and Lemma 2.14.4. it follows that Kdim M/X $< \alpha$ and hence M is α-critical. □

4.2.5.2. Proposition. Let R be positively graded and M a graded R-module then the following properties hold:

1. Kdim_R M = gr Kdim_R M if any one side exists.
2. If M is graded α-critical then M is α-critical.

Proof. 1. For each p $\in \mathbb{Z}$, put $M_{>p} = \bigoplus_{i>p} M_i$. Each $M_{>p}$ is a graded R-submodule of M and we have that $(M/M_{>p})_i = 0$ for all $i > p$. By the Lemma Kdim M^+ = gr Kdim $M_{>p}$ and Kdim (M/$M_{>p}$) = gr Kdim (M/$M_{>p}$). By Proposition 3.2.1.: Kdim M = gr Kdim M.
2. Let X $\neq 0$ be a submodule of M. There exists a p $\in \mathbb{Z}$ such that X $\bigcap M_{>p} \neq 0$. Since M is graded α-critical, $M_{>p}$ is graded α-critical and by the lemma again, Kdim $M_{>p}$/X $\bigcap M_{>p} < \alpha$ and Kdim M/$M_{>p} < \alpha$.
The map f: L(M/X) $\rightarrow$ L($M_{>p}$/X $\bigcap M_{>p}$) $\times$ L(M/$M_{>p}$) defined by
Y/X $\mapsto$ (Y $\bigcap M_{>p}$/X $\bigcap M_{>p}$, (Y + $M_{>p}$)/$M_{>p}$) is strictly increasing (see Proposition 3.1.4.) and therefore Kdim M/X $< \alpha$ and thus M is α-critical. □

4.2.5.3. Corollary. (Gordon, Robson [1]). Let A be a ring and M an A-module. Then the A[X]-module M[X] has Krull dimension if and only if M is a Noetherian A-module. If this is the case then $\text{Kdim}_{A[X]}$ M[X] = Kdim_A M + 1. Moreover if M is α-critical then M[X] is an α+1-critical A[X]-module.

Proof. Suppose that M is not Noetherian i.e. there exists an ascending chain of A-submodules $M_0 \subset M_1 \subset ... \subset M_n \subset ...$ in M. Put U = $M_1 + M_2X + M_3X^2 + ...$, V = $M_0 + M_1X + M_2X^2 + ...$. Clearly V $\subset$ U and both are A[X]-submodules of M[X]. We have X U $\subset$ V and therefore the canonical A-module isomorphism: U/V $\cong M_1/M_0 \oplus M_2/M_1 \oplus ...$, is an isomorphism of A[X]-modules.

Consequently U/V cannot have finite coirreducible dimension and so M[X] cannot have Krull dimension as an A[X]-module, a contradiction.

In order to prove that $\mathrm{Kdim}_{A[X]}$ M[X] = 1 + Kdim_A M we may use the critical composition series of M and reduce the problem to the case where M is α-critical. Consider the strictly descending chain of A[X]-submodules of M[X]:

M[X] ⊃ XM[X] ⊃ X^2M[X] ⊃ ..., where X^iM[X]/X^{i+1}M[X] ≅ M[X]/XM[X] ≅ M, and $\mathrm{Kdim}_{A[X]}$ X^iM[X]/X^{i+1}M[X] = α. Hence $\mathrm{Kdim}_{A[X]}$ M[X] ≥ 1 + α. If mX^r, with m ≠ 0, is homogeneous in M[X] for the obvious positive gradation on M[X], then we consider the finite descending chain M[X] ⊃ XM[X] ⊃ X^2M[X] ⊃ ... ⊃ X^rM[X] ⊃ R[X] mX^r. Since X^rM[X]/R[X] mX^r ≅ M[X]/R[X]m ≅ (M/Rm)[X] we have: $\mathrm{Kdim}_{A[X]}$ X^rM[X]/R[X] mX^r ≤ α and thus: Kdim M[X]/R[X] mX^r = α. The foregoing proposition yields that M[X] is 1+α-critical. □

4.2.5.4. Corollary. (Rentschler, Gabriel [1]). If R is a left Noetherian ring then Kdim R[X] = Kdim R + 1.

4.2.5.5. Corollary. Consider T = $\begin{pmatrix} R & {}_RM_S \\ {}_SN_R & S \end{pmatrix}$ as in 2.15., where the rings R and S have Krull dimension, M is an R-module with Krull dimension and N is an S-module with Krull dimension. Then Kdim T = sup (Kdim R, Kdim S).

Proof. We see first that Kdim_R M ≤ Kdim R, Kdim_S N ≤ Kdim S. By Proposition 4.2.5.2. and Lemma 2.15.9. it follows that Kdim T ≤ sup (Kdim R, Kdim S). The converse inequality is clear. □

4.2.5.6. Lemma. Let R be a $\mathbb{Z}$-graded ring and let M be a graded R-module with graded Krull dimension, then for all i ∈ $\mathbb{Z}$: Kdim_{R_0} M_i ≤ gr Kdim_R M.

Proof. For any R_0-submodule N of M_i we have thatN = RN ⋂ M_i. Apply Proposition 3.1.4. □

If M is a graded R-module and p ∈ $\mathbb{Z}$ then we write: $M_{\geq p} = \bigoplus_{i\geq p} M_i$ and $M_{\leq p} = \bigoplus_{i\leq p} M_i$, $M_{\geq 0} = M^+$ and $M_{\leq 0} = M^-$.

4.2.5.7. Lemma. Let S be a gr-simple R-module, then:

1. $S_i = 0$ or S_i is a simple R_0-module for all $i \in \mathbb{Z}$.
2. $S_{\geq p}$ (resp. $S_{\leq p}$) with $p \geq 0$, are the only nonzero graded R^+-submodules of S^+ (resp. R^--submodules of S^-).
3. $\mathrm{Kdim}_{R^+} S^+ \leq 1$ (resp. $\mathrm{Kdim}_{R^-} S^- \leq 1$).

Proof.

1. If $S_i \neq 0$, say $x \neq 0$ is in S_i, then $Rx = S$ yields $R_0 x = S_i$ and it follows that S_i is a simple R_0-module.
2. Let M be a graded R^+-submodule of S^+ and let p be the least natural number such that $M_p \neq 0$. By 1. we have $M_p = S_p$ and on the other hand $RSp = S$ and $R^+ S_p = S_{\geq p}$ implies that $M \supset R^+ M_p = S_{\geq p}$ i.e. $M = S_{\geq p}$.
3. From 2. we may derive that S^+/M is a semisimple R_0-module of finite length for each nonzero graded R^+-submodule of S^+. Therefore S^+/M is an R^+-module of finite length and consequently gr Kdim $S^+ \leq 1$. Lemma 4.2.5.1. then allows to conclude that $\mathrm{Kdim}_{R^+} S^+ \leq 1$. □

4.2.5.8. Proposition. (cf. Nastasescu [6]) Let M be a graded Noetherian R-module with gr Kdim $M = \alpha$. Then $\mathrm{Kdim}_{R^+} M^+ \leq 1 + \alpha$, $\mathrm{Kdim}_{R^-} M^- \leq 1 + \alpha$.

Proof. By induction on α. If $\alpha = 0$ then M is gr-Artinian and hence M has a composition series where the factors are gr-simple R-modules. By Lemma 4.2.5.7. (3) the assertion follows.
If $\alpha > 0$ then M admits a critical composition series since it is Noetherian and so we reduce the problem to the case where M is a gr-critical R-module. Pick a homogeneous $x \neq 0$ in M. Then Rx is a graded submodule of M, hence gr Kdim $M/Rx < \alpha$ and the induction hypothesis entails that $\mathrm{Kdim}_{R^+} (M/Rx)^+ \leq \alpha$. It is clear that $(M/Rx)^+ = M^+/(Rx)^+$ and $R^+x \subset (Rx)^+$. Consider the exact sequence: $0 \to (Rx)^+/R^+x \to M^+/R^+x \to M^+/(Rx)^+ \to 0$. If $\deg(x) = t$ then $(Rx)^+/R^+x \cong R_{-t}x \oplus \ldots \oplus R_{-}x$ (in R_0-mod). By Lemma 4.2.5.6. it follows that $\mathrm{Kdim}_{R^+}((Rx)^+/R^+x) \leq \alpha$. Property 4.1.1. entails Kdim $M^+/R^+x \leq \alpha$ and 4.1.5. combined with Lemma 4.2.5.1. finally leads to $\mathrm{Kdim}_{R^+} M^+ \leq 1 + \alpha$. □

4.2.5.9. Theorem. If the graded Noetherian R-module M has gr $\mathrm{Kdim}_R M = \alpha$ then $\alpha \leq \mathrm{Kdim}_R M \leq 1 + \alpha$.

Proof. Direct from Proposition 4.2.5.8., Proposition 3.1.4. and Proposition 2.14.3. □

4.2.5.10. Corollary. Let R be a strongly graded ring of type $\mathbb{Z}$ such that R_0 is left Noetherian then we have:

$$\text{Kdim } R_0 \leq \text{Kdim R} \leq 1 + \text{Kdim } R_0$$
$$\text{Kdim } R_0 \leq \text{Kdim } R^+ \leq + \text{Kdim } R_0.$$

Proof. Combine the theorem with results of Section 2.16.

4.2.6. Filtered Rings and Modules.

A Ring R is *filtered* if there is an ascending chain of additive subgroups $\{F_nR, n \in \mathbb{Z}\}$ such that $1 \in F_0R$, $F_nR.F_mR \subset F_{n+m}R$ for all $n,m \in \mathbb{Z}$ and $R = \bigcup_{n\in\mathbb{Z}} F_nR$. An R-module M is *filtered* if there exists an ascending chain of additive subgroups $\{F_nM, n \in \mathbb{Z}\}$ such that $F_nR\, F_mM \subset F_{n+m} M$ for all $n,m \in \mathbb{Z}$.

We say that the filtration of M is *discrete* if $F_iM = 0$ for all $i < n_0$ for some $n_0 \in \mathbb{Z}$ The filtration of M is *separated* if $\bigcap_n F_nM = 0$; as an example: discrete filtrations are trivially separated.

The filtration of M is *exhaustive* if $M = \bigcup_n F_nM$. To a filtered ring R and a filtred R-module M we may associate: $G(R) = \bigoplus_{i\in\mathbb{Z}} F_iR/F_{i-1}R$, $G(M) = \bigoplus_{i\in\mathbb{Z}} F_iM/F_{i-1}M$, the so-called associated graded ring of R, resp. associated graded module of M. If $x \in F_pM - F_{p-1}M$ then we let x_p be the image of x in $G(M)_p$. If $a \in F_iR$, $x \in F_jM$ then we define: $a_ix_j = (ax)_{i+j}$ and this makes G(M) into a graded G(R)-module (as claimed in the terminology).

If N is a submodule of M then M/N may be filtered by putting $F(M/N) = \{N + F_iM, i \in \mathbb{Z}\}$; this filtration is the *quotient filtration.* On the other hand any R-submodule N of a filtered R-module may be filtered by putting $FN = \{N \bigcap F_iM, i \in \mathbb{Z}\}$; this filtration is called the *induced filtration.* In the latter case G(N) is a graded G(R)-submodule of G(M).

4.2.6.1. Lemma. Let M be a filtered R-module and N a submodule of M equiped with the induced filtration, then $G(M/N) \cong G(M)/G(N)$ where M/N has the quotient filtration.

Proof. From the exactness of the following sequence:

$$0 \to N \cap F_iM/N \cap F_{i-1}M \to F_iM \to \frac{N+F_iM}{N+F_{i-1}M} \to 0$$

where the maps are the canonical ones. □

4.2.6.2. Lemma. Let FM be an exhaustive filtration on the filtered R-module M. If FM is separated then M = 0 if and only if G(M) = 0.

Proof. If there exists an $x \neq 0$ in F_iM - $F_{i-1}M$ then $G(M)_i \neq 0$ hence G(M) = 0 exactly then when $F_iM = F_{i-1}M$ for all i but the latter contradicts $\bigcap_n F_n M = 0$ unless $F_i = 0$ for all i. □

4.2.6.3. Proposition. Let M be exhaustively filtered and suppose that for every submodule N the quotient filtration on M/N is separated, then:

1. If M is discrete then G(M) is a graded G(R)-module with left limited gradation.
2. The map φ: $L(M) \to L_g(G(M))$, $N \to G(N)$ is strictly increasing.

Proof. 1. Obvious.

2. Let $N \subset P$ be submodules of M. Since $G(N) \subset G(P)$ we will have G(N) = G(P) exactly then when G(P/N) = 0 (use Lemma 4.2.6.1.). The conditions of Lemma 4.2.6.2. hold for the quotient filtration on P/N so P/N = 0 and P = N follows. □

4.2.6.4. Theorem. With assumptions as in the foregoing proposition:

1. If G(M) is a Noetherian G(R)-module (rsp. Artinian) then M is a Noetherian (resp. Artinian) R-module.
2. If G(M) has $\text{Kdim}_{G(R)}$ then M has Kdim_R and moreover: Kdim M $\leq$ Kdim G(M).
3. If Kdim M = α and Kdim G(M) = α then M is α-critical if G(M) is α-critical.

Proof.

1. and 2. follow from the foregoing proposition.
2. From Lemma 4.2.6.1. and the foregoing proposition. □

Examples.

E1. Let R be a ring. A map δ: R $\to$ R is a *derivation* if $\delta(a+b) = \delta(a) + \delta(b)$ and $\delta(ab) = a\delta(b) + \delta(a)b$ for all a,b $\in$ R.
Consider the abelian group of the polynomial ring R[X] and make it into a ring by defining a multiplication as follows: for all a $\in$ R, Xa = aX + $\delta(a)$. We denote the ring just defined by R[X,δ]. On S $\in$ R[X,δ] we may consider a discrete filtration defined by F_nS = 0 if n < 0 and F_nS = {f $\in$ R[X,δ] such that $\deg_X f \leq n$}. The associated graded ring is R[X]. From the result earlier in this section we derive: if R is Noetherian (left) then R[X,δ] is left Noetherian and Kdim R $\leq$ Kdim R[X,δ] $\leq$ 1 + Kdim R.

E.2. Let R be a semi-local commutative Noetherian ring i.e. R/J(R) is semisimple. Consider an ideal I $\subset$ J(R) such that R/I is Artinian and suppose that I/I^2 is generated by t elements as an R/I-module. Put F_nR = R for n $\geq$ 0 and F_nR = I^{-n} for n < 0. It is well known that this filtration is separated (cf. N. Bourbaki, Commutative Algebra, Chap. 3). The graded ring G(R) = R/I $\oplus$ I/I^2 $\oplus$ ... $\oplus$ I^n/I^{n+1} $\oplus$ ... is a Noetherian ring and it is isomorphic to an epimorphic image of R/I $[X_1, ..., X_t]$. Theorem 4.2.6.4. then entails that Kdim R $\leq$ Kdim G(R) $\leq$ t.

E.3. Let k be a field and g a k-Lie algebra of finite length. Let U(g) be the enveloping algebra of g. Then U(g) is a left and right Noetherian ring such that Kdim U(g) $\leq$ [g:k]. Define F_n U(g) to be the k-subspace of U(g) generated by 1 and elements of the form $g_1 \cdot \ldots \cdot g_m$ for m $\leq$ n, where g_i $\in$ g. This defines a discrete filtration on U(g). By the Poincaré-Witt theorem it follows that G(U(g)) $\cong$ k$[X_1, .., X_n]$, n = [g:k]. Theorem 4.2.6.4. yields the result concerning the Krull dimension stated above.

4.2.7. Ore and Skew-Laurent Extensions. (Addendum)

In this section *R is a left Noetherian ring with finite Krull* dimension, α say. An *Ore extension* of R is a ring of the type S = R[X,δ]. If φ is an automorphism of R then R[X,φ] is the ring of *skew polynomials* over R defined by Xa = φ(a)X for all a

$\in$ R. The localization of R[X,φ] at the multiplicative set $\{X^n$, n $\geq 1\}$ is denoted by R[X,X^{-1}, φ] and it is called a *skew-Laurent extension* of R. Since R is assumed to be left Noetherian, R[X,δ], R[X,φ] and R[X,X^{-1}, φ] are left Noetherian too and by the foregoing sections the Krull dimension of these rings is either α or $1 + \alpha$. Also, in each case S is a free left (right) R-module having the powers of X as a basis. If I is a left ideal of R then S/SI $\cong$ S/S $\otimes_R$ I. Each element of S can be written in a unique way as $\sum_{i=1}^n a_i X^i$ and the image in S/SI is denoted by $(\sum^n a_i X^i)^-$.

Let us now first consider the case where S is an Ore extension of R. We say that derivations δ *and* δ*' are similar* if δ - δ' is an inner derivation of R i.e. δ - δ' = ad(a) for some a $\in$ R where ad(a)r = ar - ra for all r $\in$ R. If δ is similar to δ' then R[X,δ] $\cong$ R[X,δ'] by substituting X + a for X.

4.2.7.1. Lemma. Let R be an algebra over the rationals. If for all a $\in$ R, (δ + a)I $\not\subset$ I then SI is a maximal left ideal of S. Conversely if (δ + a)I $\subset$ I for some a $\in$ R then S/SI has infinite length as an S-module.

Proof. The first statement follows directly from K. Goodearl, Global Dimension of Differential Operator Rings II, *Trans. Amer. Math. Soc.* 209, 1975, 65-85. In order to establish the second statement we define ϕ: S/SI $\to$ S/SI, $(\sum b_i X^i)^- \mapsto (\sum (X+a) b_i X^i)^-$. If b $\in$ I then (X+a)b $\in$ SI, hence ϕ is well defined and it is clearly left S-linear. Since $\deg_X \phi(f) > \deg_X f$ we see that ϕ is injective but not surjective hence the length of S/SI has to be injective. □

If M is an S-module and M_0 an R-module such that SM_0 = M then we may filter M by putting F_nM = F_nS.M_0. In particular for M = S $\otimes_R$ N we may take M_0 to be N. One easily checks that G(S $\otimes_R$ N) $\cong$ G(S) $\otimes_R$ N. Since the filtration on S is discrete the map L(S $\otimes_R$ N) $\longrightarrow$ L_g (G(S) $\otimes_R$ N), P $\mapsto$ G(P), is injective when restricted to a chain of submodules. In the sequel we *restrict to R being an algebra over the rationals* unless otherwise stated.

From the results of the foregoing section we immediately derive that for an α-critical Noetherian R-module N we have:

Kdim (G(S) $\otimes_R$ N) = gr Kdim (G(S) $\otimes_R$ N) = Kdim N + 1,

and N $\otimes_R$ G(S) is 1 + α-critical. We have left the following proposition as an exercise (using results of Section 4.2.6.)

4.2.7.2. Proposition. Let N be a Noetherian R-module which is α-critical and let S be an Ore extension of R, then:

1. Kdim $(S \otimes_R N) \geq \alpha$
2. If $T \neq 0$ is a nonzero S-submodule of $S \otimes_R N$ then Kdim $(S/T \otimes_R N) \leq \alpha$

3. If there are no simple S-modules of finite length over R then $S \otimes_R N$ is α-critical.

Consequently, if N is simple then $N \otimes_R S$ is either simple or 1-critical.

4.2.7.3. Theorem. The assumption that all simple S-modules have finite length implies the following equivalent properties:

1. For any simple R-module N, we have that $S \otimes_R N$ is a 1-critical S-module
2. If N is a Noetherian α-critical R-module then $S \otimes_R N$ is a $\alpha+1$-critical S-module.

Each of these conditions implies that Kdim S = Kdim R + 1.

Proof. 1. Follows easily because S-submodules of $N \otimes_R S$ cannot be finitely generated over R unless they are zero so the assumption implies that $S \otimes_R N$ cannot contain simple S-modules but then we apply the final statement of the foregoing proposition.
$1° \Rightarrow 2°$. A straightforward induction on α following the lines of lemma 10 in D. Segal, [1].
$2° \Rightarrow 1°$ Obvious. The final statement is equally obvious.

4.2.7.4. Remark. If Kdim S = Kdim R + 1 then it does not necessarily follow that 1° holds. Let k be an algebraically closed field and put $R = k[Y]$ and $\delta(y) = y$, then (y) is the only maximal ideal of R stable under δ i.e. (we assume char $k = 0$) there is only one simple R-module N such that $N \otimes_R S$ is 1-critical over S but Kdim $(S/(Y)) = 1$ while Kdim S = 2. Note that a simple S-module of the form $S \otimes_R N$ may well be Artinian over R.

4.2.7.5. Theorem (McConnell, Hodges). The following statements are equivalent:

1. Simple S-modules cannot have finite R-length.
2. If N is a simple R-module then $S \otimes_R N$ is a simple S-module.
3. If N is a ρ-critical Noetherian R-module then $S \otimes_R N$ is a ρ-critical S-module.

Each of the above implies that Kdim R = Kdim S but the latter is not equivalent to the above.

Proof. 1⇒3. Follows from Proposition 4.2.7.2.

3⇒2. and 3. implies the final statement, are obvious.

2⇒1. Let P be a maximal left ideal of S with S/P being of finite length as an R-module. Then R/R $\bigcap$ P has finite length and since S/P is simple the canonical R/R $\bigcap$ S $\otimes$ P $\hookrightarrow$ S/P is surjective. By 2. the simple subquotients of R/R $\bigcap$ S $\otimes$ P are of the form S $\otimes_R$ N where N is a simple R-module. Thus S/P $\cong$ N $\otimes_R$ S for a simple R-module N. Since S $\otimes_R$ N cannot have finite R-length we reach a contradiction.
As an example consider k[X,Y] with maximal ideal M_n = (X-1,Y-n) and prime ideal P_{2n} = (X+2n), for all n $\in$ IN . Let S be the complement of the union of the M_n and P_{2n} for n $\in$ IN . If a maximal ideal M contains a P_{2n} then M $\bigcap$ S $\neq$ 0 as one easily checks. Put R = S^{-1} k[X,Y]. Then P_{2m} generates a maximal ideal of R again denoted by P_{2m}, for all m $\in$ IN . Define a derivation δ of R by δ(X) = 0, δ(Y) = 1. Then R/S $\otimes$ P_m is 1-critical for all m $\in$ IN . We have Kdim S = Kdim R = 2. This shows that the equality of Kdim S and Kdim R need not imply 1,2 or 3. □

4.2.7.6. Corollary. Let R be commutative and assume that no maximal ideals of R are stable under δ then Kdim R = Kdim S.

4.2.7.7. Corollary. If no maximal left ideal of R is stable under a derivation (similar to) δ then Kdim R = Kdim R[X,δ].

Most results mentioned above may be proved in exactly the same way for skew Laurent extensions S of R, cf. J. Connell, T. Hodges [1]; nevertheless we now proceed with skew Laurent extensions of commutative rings, following T. Hodges [1].

Let R be a commutative Noetherian ring and φ an automorphism of R then each element of S' = R[X,φ] may be written as $\sum_{i=0}^{n} a_i X^i$ in a unique way; the elements of S = R[X,X^{-1}, φ] may be written as $\sum_{i=-m}^{n} a_i X^i$ in a unique way. An ideal I of R is said to be *quasi-invariant* if there exists an n $\in$ IN such that I^{φ^n} = I and it is *invariant* if I^{φ} = I. If I is quasi-invariant then $I_0 = I^{\varphi} \bigcap ... \bigcap I^{\varphi^n}$ is invariant. A *φ-prime ideal* of R is an invariant ideal P such that I J $\subset$ P yields I $\subset$

P or J ⊂ P for invariant ideals I and J. Since the prime radical is invariant under automorphisms it follows that φ-prime ideals are semiprime. If J is a left ideal of S then the nth *leading ideal* $\gamma_n(J) = \{a \in A$, there is an element in S of the form $a_0 + a_1X + ... + a_{n-1}X^{n-1} + aX^n$, for some $a_0,...,a_{n-1} \in A\}$. The leading ideal of J is then defined to be $\gamma(J) = \bigcup_{i=0}^{\infty} \gamma_i(J)$.
Since A is Noetherian there exists an $n_0 \in \mathbb{N}$ such that $\gamma_i(J) = \gamma(J)$ for all $i \geq n_0$. Henceforth we assume that A has Krull dimension α, and we suppose that α is finite. Note that any φ-invariant multiplicatively closed set in A is an Ore set of A and φ extends in the obvious way to an automorphism of the loclization which turns out to be the skew Laurent extension of the localization of A at that set.

4.2.7.8. Proposition (Shamsuddin [1]). If A contains a quasi-invarisnt prime ideal P such that ht(P) = Kdim A then Kdim S = 1 + Kdim A.

Proof. If P is as in the statement, say $P^{\varphi^n} = P$ then P_0 is invariant. Consider the set $C(P_0)$ of elements regular modulo P_0 in A. Then $C(P_0)^{-1}$ A has only finitely many maximal ideals and all these are quasi-invariant.
Hence, Kdim $(C(P_0)^{-1} A[X,X^{-1},\varphi]) = 1 +$ Kdim $(C(P_0)^{-1}$ A) follows from the Laurent extension version of Corolary 4.2.7.6.. Since ht(P) = Kdim S $\geq$ 1 + Kdim A. Thus Kdim S = 1 + Kdim A. □

We leave the proof of the following two lemmas as an exercise (one may compare the first to Lemma 2.1 of K. Goodearl, R.B. Warfield's paper [1], the second is an easy adaption of Lemma 2.4. in R. Hart's [1]).

4.2.7.9. Lemma. If T is a left Noetherian ring and M is a compressible T-module then $M \otimes_T T[X,X^{-1},\varphi]$ is a compressible $T[X,X^{-1},\varphi]$-module.

4.2.7.10. Lemma. If P is a non quasi-invariant prime ideal of A and J is a left ideal of S such that $J \supsetneq SP$, then $J \cap A \supsetneq P$.

4.2.7.11. Lemma. Let P be a prime ideal of A and let J ⊂ I be left ideals of S such that $I/J \cong S/SP$. Then there is an $x \in \gamma(I)/\gamma(J)$ such that $\mathrm{Ann}_A\, x \subset P^{\varphi^n}$ for some

n.

Proof. Put M = (I $\bigcap$ S') + J/J and let n_0 be such that γ_{n_0}(I) = γ(I) and γ_{n_0}(J) = γ(J). Put $S'_{n_0} = \sum_{i=0}^{n_0} AX^i$ and N = (I $\bigcap$ S'_{n_0}) + J/J. Obviously N is a finitely generated A-submodule of M. Let π': S $\rightarrow$ S/SP be the canonical epimorphism and let ϕ: S/SP $\rightarrow$ I/J be an isomorphism. Then S/SP = $\bigoplus_{i\in\mathbb{Z}}$ $A\pi(X)^i$ and for large m we obtain: $\phi(1)X^m = \phi(\pi(X)^m)$ $\in$ M, and $\bigoplus_{i=m}^{\infty}$ $A\phi(\pi(X)^m)$ $\subset$ M. Since N is finitely generated we may select some k $\geq$ m large enough such that: $A\phi(\pi(X)^k)$ $\bigcap$ N = 0. Pick y $\in$ I $\bigcap$ S' such that $\tilde{y} = \phi(y) = \phi(\pi(X)^k)$; say y = $a_0 + ... + a_sX^s$, then $\tilde{y} \notin$ N implies that s > n_0. If the lemma were false then there is a $b_s \in P^{\varphi^s}$ such that $b_sa_s \in \gamma(J) = \gamma_s(J)$. Hence there exists a z = $c_0 + ... + c_{s-1}x^{s-1} + b_sa_sX^s$ $\in$ J for certain c_0 , ... , c_{s-1} $\in$ A. Consequently $y_1 = b_s^{\varphi^s}$y - z $\equiv$ $b_s^{\varphi^s}$y mod(J) and $\deg_X y_1$ < $\deg_X$y. Repeting this argument we eventually obtain y_n $\in$ I such that $\tilde{y}_n = b_{s_n}^{\varphi^{s_n}} ... b_{s_0}^{\varphi^{s_0}}\tilde{y}$ with s_0 = s, $b_{s_i} \in P^{\varphi^{k-s_i}}$ and also $\deg_X y_n$ $\leq$ n_0. But then we must have that $\tilde{y}_n$ $\in$ A$\tilde{y}$ $\bigcap$ N = 0. Thus, $b_s^{\varphi^{s_n}} ... b_{s_0}^{\varphi^{s_0}}$ $\in$ Ann_A $\tilde{y}$ = Ann_A $\phi(\pi(X)^k) = P^{\varphi^k}$. The fact that P is a prime ideal then yields that $b_{s_i}^{\varphi^{s_i}} \in P^{\varphi^k}$ for some i, contradicting the choice of b_{s_i}. □

4.2.7.12. Lemma. Let M be a finitely generated S-module. There is a chain M = $M_0 \supset .. \supset M_n = 0$ such that for each i there is a prime ideal P_i of A satisfying one of two possible conditions:

1. M_i/M_{i+1} $\cong$ S/SP_i,
2. P_i is quasi-invariant and M_i/M_{i+1} is a proper homomorphic image of S/SP_i which is torsionfree as an A/P_0-module.

Proof. Along the lines of K. Goodearl, R. Warfield [1], Lemma 2.2. □

4.2.7.13. Lemma. Let A be a φ-prime (i.e. 0 is φ-prime) with Kdim A = α. Suppose for every prime ideal P of A of height at least one that Kdim (S/SP) $\leq$ α-ht(P), then Kdim S = α.

Proof. Since A is semiprime it is sufficient to establish that Kdim (S/E) $\leq$ α - 1 for any essential left ideal E of S. Suppose E is an essential left ideal of S such that α = Kdim (S/E). There is an infinite chain of left ideals of S:

$$(*) \qquad S = J_0 \supset J_1 \supset \ldots \supset E$$

with Kdim $(J_i/J_{i+1}) = \alpha - 1$ for infinitely many i. By the foregoing lemma there is a prime ideal P_i of A satisfying either 1° or 2°. If Kdim $(J_i/J_{i+1}) = \alpha - 1$ then the hypothesis implies ht $(P_i) \leq 1$. Since A/P_i is compressible, so is S/SP_i and hence it is critical. Hence, for factors of the type mentionned in 2° of Kdim equal to α - 1 we have ht $(P_i) = 0$ and $(P_i)_0 = 0$ follows then because A is φ-prime. Consequently, in this case J_i/J_{i+1} is torsion-free as an A-module.

Let Q be the Artinian quotient ring of A i.e. the localization with respect to the set $\mathcal{C}$ of regular elements of A; we have pointed out earlier that $\mathcal{C}^{-1}S \cong Q[X,X^{-1},\varphi]$ and the latter must have Krull dimension one. Since $(\mathcal{C}^{-1}S)E$ is essential in $\mathcal{C}^{-1}S$ we obtain that if we tensor $(*)$ by $\mathcal{C}^{-1}S$ then $(J_i/J_{i+1}) \otimes_S \mathcal{C}^{-1}S$ can only be nonzero for finitely many indices i. But the latter can only be zero whenever J_i/J_{i+1} is a torsion A-module. Consequently condition 2° can only hold for finitely many indices i hence for infinitely many i we have that $J_i/J_{i+1} \simeq S/SP_i$ with $\mathrm{ht}(P_i) = 1$. The leading ideals of the J_i form an infinite chain:

$$(**) \qquad A = \gamma(J_0) \supset \gamma(J_1) \supset \ldots \supset \gamma(E)$$

By Lemma 4.2.7.11. we find $x_i \in \gamma(J_i)/\gamma(J_{i+1})$ such that Ann $x_i \subset P_i^{\varphi^{n_i}}$ for some n_i and for each i such that $J_i/J_{i+1} \simeq S/SP_i$. One easily checks that $\gamma(E)$ is essential in A. Since $\gamma(E) \subset$ Ann x_i for those P_i of height one, $P_i^{\varphi^{n_i}}$ must be minimal over $\gamma(E)$.

Let $\mathcal{D}$ be the complement of the union of the height one prime ideals minimal over $\gamma(E)$. Since all maximal ideals of $\mathcal{D}^{-1}A$ have height one we have Kdim $\mathcal{D}^{-1}A = 1$. As $\mathcal{D}^{-1}A\gamma(E)$ is essential in $\mathcal{D}^{-1}A$ we have Kdim $(\mathcal{D}^{-1}A/D\mathcal{D}^{-1}A\gamma(E)) = 0$. If we tensor $(**)$ with $\mathcal{D}^{-1}A$ then only finitely many of the B_i, where $B_i =$ $\gamma(J_i)/\gamma(J_{i+1}) \otimes_A \mathcal{D}^{-1}A$ can be nonzero. By assumption there are infinitely many indices i for which $\mathrm{ht}(P_i) = 1$ and $J_i/J_{i+1} \simeq S/SP_i$. For these i there exist $x_i \in \gamma(J_i)/\gamma(J_{i+1})$ torsion-free with respect to $\mathcal{D}$. But $B_i = 0$ if and only if $\gamma(J_i)/\gamma(J_{i+1})$ is torsion with respect to $\mathcal{D}$, a contradiction. □

4.2.7.14. Remark. In the published version of T. Hodges' paper [1] the pages 1309 and 1308 have been interchanged.

4.2.7.15. Theorem (T. Hodges [1]). With conventions as before: Kdim S = 1 + Kdim A if and only if A contains a quasi-invariant prime ideal P such that ht(P) = Kdim A.

Proof. The implication ⇐ follows from Shamsuddin's result, Proposition 4.2.7.8.

⇒. Assume that sup {ht(P), P quasi-invariant in A} < Kdim A. We then have to prove that Kdim S = Kdim A and in doing this we may restrict to the case where S is prime and A is semiprime. The proof is by induction on Kdim A = α.
If $\alpha = 0$ then A has only finitly many prime ideals so these are all quasi-invariant. The result folows then from [loc.cit.] Theorem 5.1. (the Laurent-extension version of 4.2.7.6.). Now let Kdim A = α and suppose that Kdim S > α. By the foregoing lemma there is a prime ideal P of A such that ht(P) = r $\geq$ 1 and Kdim (S/SP) > α - r. Take P such that ht(P) = r is maximal with respect to the property mentioned. We have to consider two cases.

1. P is quasi-invariant.

Since S/SP $\cong$ S/SP^{φ} we have Kdim (S/SP) = Kdim (S/SP_0). Let $\overline{\varphi}$ be the automorphism induced by φ on A/P_0, then S/SP_0 = (A/P_0) [X,X^{-1},$\overline{\varphi}$]. From ht(P) $\geq$ 1 we derive Kdim (A/P_0) < α so the induction hypothesis may be applied to conclude: Kdim (S/SP_0) = sup {Kdim (A/P_0), ht ($\overline{Q}$) + 1, where $\overline{Q}$ is a quasi-invariant prime ideal of A/P_0} $\leq$ sup {Kdim (A/P_0), ht (Q) + 1 - ht(P), where Q is a quasi-invariant prime ideal of A containing P_0} $\leq$ α - r.
The latter contradicts the choice of P.

2. P is not quasi-invariant.

Let J be a left ideal of S such that J $\underset{\neq}{\supset}$ SP. By Lemma 4.2.7.10. we have J $\bigcap$ A $\underset{\neq}{\supset}$ P, so we obtain a chain of ideals of A: A = N_0 $\supset$ N_1 $\supset$ ... $\supset$ N_t = J $\bigcap$ A, where N_i/N_{i+1} $\cong$ A/Q_i for some prime ideals Q_i of A properly containing P. Extending this chain to S yields: S = SN_0 $\supset$ SN_1 $\supset$ ... $\supset$ SN_t = S (J $\bigcap$ A), where SN_i/SN_{i+1} $\cong$ S/SQ_i. Since ht (Q) > ht (P) = r, the maximality assumption on r entails that we have the inequalities: Kdim (S/SQ_i) $\leq$ α - ht (Q_i) < α - r. Thus Kdim (S/J) $\leq$ Kdim (S/S(J $\bigcap$ A)) < α - r.
Therefore all proper factors of S/SP have Krull dimension less that α - r whereas we must have Kdim (S/SP) $\leq$ α - r, a contradiction. □

4.2.8. Affine P.I. Algebras (Addendum).

Even though affine P.I. algebras need not have Krull dimension a result of Gordon, Small cf [1] states that they have Gabriel dimension, consequently all modules over an affine P.I. algebra have Gabriel dimension. Note that the classical Krull dimension of an affine P.I. ring defined in terms of chains of prime ideals does exist. For the general theory of P.I. rings we refer to C. Procesi [1], actually the chapter on P.I. rings in P.M. Cohn's book [1] or L. Rowen [1] will suffice for our needs in this section.

4.2.8.1. Theorem (Gordon, Small [1]). **If R is a P.I. ring with the ascending chain condition on semiprime ideals and such that the prime radical of every factor ring is nilpotent then Gdim R exists and it satisfies Gdim R = cl Kdim R + 1.**

Proof. Every prime factor ring of R is a prime P.I. ring hence it is a Goldie ring an also bounded (cf. Section 6.3 and 5.5.). Put cl Kdim R = α. By the ascending chain condition semiprime ideals are finite intersections of prime ideals and as the prime radical of R is nilpotent we see that $P_1 \cdot \ldots \cdot P_n = 0$ for some prime ideals of R. In $P \supset P_1 \supset P_1P_2 \supset \ldots \supset 0$ the ith-factor module is an R/P_i-module and if P_i is maximal the assumptions imply that R/P_i is simple Artinian. The theorem is thus valid for $\alpha = 0$.

More generally R will have Gdim if each R/P_i has Gdim. From $P_1 \cdot \ldots \cdot P_n = 0$ it is clear that there are finitely many minimal prime ideals in R, thus cl Kdim R = cl Kdim (R/P) for some prime ideal P of R. In case R has Gdim then

Gdim R = Gdim (R/P_i) for some i. Consequently we may assume R to be a prime ring such that the theorem holds for every factor ring of cl Kdim < α.

Now let J be a left ideal of R such that R/J is a torsion R-module, then J is essential and as such it contains a nonzero ideal I. By the argument above I contains a product of prime ideals each of which contains I, hence cl Kdim (R/I) < α and therefore Gdim R/I = cl Kdim (R/I) + 1 or Gdim (R/J) < $1 + \alpha$. Each proper factor module of a cyclic uniform left ideal of R is a cyclic torsion module, hence it has Gabriel dimension less than $\alpha + 1$. By Goldie's theorem R embeds in a direct sum of cyclic uniform left ideals. Thus, Gdim R exists and it does not exceed $1 + \alpha$. On the other hand cl Kdim(R/P) + 1 = Gdim (R/P) for any prime ideal P $\neq$ 0 of R.

Since Gdim R is a nonlimit ordinal and Gdim (R/P) < Gdim R, it follows that $1 + \alpha \leq$ Gdim R by the properties of Gabriel dimension. □

4.2.8.2. Remark. The assumptions of the theorem are equivalent to:

1. P satisfies the ascending chain condition on prime ideals
2. For any proper ideal J of R there is afinite set of prime ideals containing J such that their product is contained in J.

4.2.8.3. Corollary. Affine P.I. rings have finite Gabriel dimension.

Proof. Envoking some results of C. Procesi, [1] Corollary 2.2 p.106 and Corollary

4.5. p.116, it becomes clear that affine P.I. rings have finite classical Krull dimension and that they satisfy the ascending chain condition on semiprime ideals so that the theorem applies. □

For the definition of an affine P.I. ring it is common to use P.I. algebras over a field i.e. P.I. rings generated as a k-algebra by a finite number of elements but all the result proved and used above remain valid if k is only assumed to be a commutative Noetherian ring.

4.2.8.4. Proposition. With assuptions as in Theorem 4.2.8.1. we have: if M is a finitely generated faithful R-module then Gdim M = Gdim R holds in the following cases:

1. R is prime.
2. M is Noetherian.

Proof. cf. Gordon, Small [1] Proposition 2.2., p. 1296. □

After some considerations concerning the Gelfand-Kirillov dimension one may proceed to obtain first a corollary stating that cl Kdim B $\leq$ cl Kdim A whenever B is a subalgebra of an affine P.I. algebra A, as well as:

4.2.8.5. Theorem. Let S and R be affine P.I. k-algebras over a field k and let M be an S-R-bimodule such that ${}_SM$ is faithful and M_R is finitely generated. Then Gdim S $\leq$ Gdim R.

4.2.8.6. Corollary. Let R and S be as in the theorem and M an S-R-bimodule finitely generated and faithful over R and over S, then Gdim R = Gdim S.

§4.3 Exercises.

(121) Let R be a prime ring. Prove that the following statements are equivalent:

1. R is a domain with Kdim R = α,
2. R is a (left) Ore domain and Kdim R = α,
3. R is α-critical as a left R-module.

(122) Show that R is left Artinian if and only if R has Krull dimension and it is a semi-primary ring.

(123) (Gordon, Robson [1]) An R-module M is Noetherian if and only if each of its factor modules has a critical composition series. Hint: if M is α-critical then M[X] is 1+α-critical and apply Corollary 4.2.5.3.

(124) (Gordon, Robson [1]) Any ordinal number α may be written in a unique way as $\alpha = \omega^{\beta_1} n_1 + ... + \omega^{\beta_r} n_r$ where $n_j \in \mathbb{N}$ and the β_j are ordinals suh that $\beta_1 > \beta_2 > ... > \beta_r \geq 0$. Let M be an R-module whose lattice L(M)$^{\circ}$ is well-ordered with the associated ordinal equal to α. Show (by transfinite recursion on α) that Kdim M = β_1.

(125) If k R = k[X_1, ... ,X_N, ...] is a polynomial ring in infinitly many indeterminates over the field k then R does not have Kdim.

(126) Let R be graded of type $\mathbb{Z}$, M a gr-simple R-module. Then either M is simple or M is 1-critical (cf. Nâstâsescu, [6]).

(127) Let R be strongly graded of type G, M a graded R-module.
Prove that gr Kdim_R M = Kdim_{R_e} M_e.

(128) Let R be strongly graded by a polycyclic-by-finite group G with normal series $\{e\} = G_0 \subset G_1 \subset ... \subset G_s$ = G. If R_e is left Noetherian and Kdim$R_e = \alpha$ then $\alpha \leq$ Kdim R $\leq \alpha$ + h(G) (C. Nâstâsescu, F. Van Oystaeyen [2]) where h(G) is Hirch number of group G.

(129) Let A be a left Noetherian ring with Kdim A = α. If G is as in exercise 128. then $\alpha \leq$ Kdim A[G] $\leq \alpha$ + h(G) where h(G) is Hirch number of group G.

(130) Let A be a ring, M a left A-module. Let A[[X]] be the ring of formal series with coefficients in A and M[[X]] the left A[[X]]-module M[[X]] = $\{\sum_{i=0}^{\infty} M_i X^i,\ M_i \in$ M$\}$. Establish the following claims:

1. M[[X]] has Kdim if and only if M is a Noetherian A-module.

2. If M is an α-critical Noetherian A-module then M[[X]] is an $\alpha+1$-critical Noetherian A[[X]]-module.
3. If M is a Noetherian A-module then M[[X]] is a Noetherian A[[X]]-module and $\text{Kdim}_{A[[X]]}$ M[[X]] = 1 + Kdim_A M.
4. If A is left Noetherian then Kdim A[[X]] = 1 + Kdim A.

(131) (Gordon, Robson [2]) Let the R-module M have Gdim M = α + n where n $\in$ $\mathbb{N}$ and $\alpha = 0$ or a limit ordinal. Then for the R[X]-module M[X] we obtain: α + n + 1 $\leq$ $\text{Gdim}_{R[X]}$ M[X] $\leq$ α + 2n if n > 0 and $\text{Gdim}_{R[X]}$ M[X] = α if n = 0.

(132) Let k be a field, M the faithful k[X]-module k[X,X^{-1}]/k[X] then $\text{Gdim}_{k[X]}$ M = 1 < Gdim k[X] = 2.

(133) Let M be a finitely generated faithful module over a ring R with Gdim, then M has a cyclic submodule C such that: Gdim R = Gdim (R/Ann C) (Gordon, Small [1] Lemma 2.1. p. 1295).

(134) Let k and M be as in exercise 133. and consider the ring R = $\begin{pmatrix} k & M \\ 0 & k[X] \end{pmatrix}$. Then R is a P.I. ring and N = $\begin{pmatrix} k & M \\ 0 & 0 \end{pmatrix}$ is a faithful cyclic R-module. Prove that R has Kdim and show that Gdim N < Gdim R (note: M is Artinian over k[X]!).

(135) (J. Okninski [1]) Let A be a commutative ring, S a commutative semigroup. We denote by rk(S) the supremum of the torsion free ranks of all groups <T> which are fraction groups of 0-cancellative homomorphic images T of S. Prove that the following statements hold:

a. rk(S) = supremum of rk(T) where T is a free subsemigroup of S.
b. cl dim A[S] = sup {cl Kdim A[$< S_p >$], P a prime ideal of A[S]}
(here $< S_p >$ is the group of fractions of the semigroup S_p obtained from S by taking the congruence induced in S by P).
c. cl Kdim A[S] < ∞ if and only if cl dim A < ∞ and rk(S) = ∞.
d. If A is Noetherian ring then cl Kdim A[S] = cl Kdim A + rk(S).
e. If Kdim A[S] exists then:

1) Kdim A exists;
2) rk(S) < ∞;
3) if rk(S) > 0 then A is Noetherian;
4) if rk(S) = 0 then S is finite

f. If the semigroup S satisfying the ascending chain condition on principal ideals and if Kdim A[S] exists then S is finitely generated and
Kdim A[S] = Kdim A + rk(S)

Bibliographical Comments to Chapter 4.

The first paragraph of this chapter provides a module theoretic translation for the results of Chapter 3. Some of the main contributions to the calculus of Krull and Gabriel dimension we have drawn from the following papers:

J. Bit David [1]; J. Fisher, C. Lanski, J. Park [1]; R. Gordon, J.C. Robson [1],[2]; C. Lanski [1]; B. Lemonnier [3],[4]; M. Lorenz, D. Passman [1]; C. Năstăsescu [1],[6]; R. Rentschler, P. Gabriel [1], C. Năstăsescu, F. Van Oystaeyen [2],[3]; T. Hodges [1]; T. Hodges, J. McConnell [1]; K. Goodearl, R. Warfield [1]; R. Hart [1]; A. Shamsuddin [1]; K. Goodearl [1]; D. Segal [1]; R. Gordon, L. Small [1].

A sporadic reference to bounded prime rings does appear but facts concerning these rings will be treated in more detail in Chapter 5 (Section 5.5.) and also in Chapter 6, 6.3., 6.4..

For the general theory of P.I. algebras we referred to C. Procesi [1]. In P.I. theory many particular techniques for studying these rings are available and the use of Gabriel dimension is not very frequent as far as we know. However, the final section of the foregoing chapter is only meant to give an example of a very nice class of rings having Gabriel dimension, e.g. the affine P.I. algebras, whereas the Krull dimension need not be defined. As Gordon and Small pointed out in [1] it is indeed somewhat paradoxical that the Gabriel dimension is then so closely related to the classical Krull dimension.

Chapter 5

Rings with Krull Dimension.

§5.1 Nil Ideals.

The aim of this section is to establish that a nil ideal of a ring with Krull dimension is nilpotent. We start with some lemmas:

5.1.1. Lemma. (T. Lenagan [4]). Consider an R-module M and an ascending chain of R-submodules $0 = A_0 \subset A_1 \subset A_2 \subset \ldots$, such that $M = \bigcup_{i=0}^{\infty} A_i$. If there exists a descending chain of submodules: $M = M_0 \supset M_1 \supset M_2 \supset \ldots$, such that $M_i \not\subset M_{i+1} + A_j$ for every i,j, then M cannot have Krull dimension.

Proof. We reduce the proof of the lemma to that of Lemma 3.2.8. in the following way. Put $B = \sum_{i=0}^{\infty} (A_i \bigcap M_i)$. In M/B we consider the chains:

$$0 = (B + A_0)/B \subset (B + A_1)/B \subset (B + A_2)/B \subset \ldots,$$
$$M/B = (M_0 + B)/B \supset (M_1 + B)/B \supset (M_2 + B)/B \supset \ldots,$$

putting $M_i' = (M_i + B)/B$ and $A_i' = (B + A_i)/B$.
Suppose that $M_i + B \subset M_{i+1} + A_j + B$. If $j \leq i$ then $B \subset A_i + M_{i+1}$ and $A_j \subset A_i$ entail $M_i \subset M_{i+1} + A_i$, a contradiction. If $i < j$ then $B \subset A_j + M_{j+1} \subset A_j + M_{i+1}$ and $A_i \subset A_j$ yield $M_i \subset M_{i+1} + A_j$, again a contradiction.

Consequently, $M'_i \not\subset M'_{i+1} + A'_j$ for every i,j. Now we may apply Lemma 3.2.8.. □

5.1.2. Lemma. Let R be a ring with Krull dimension and let I be a left ideal contained in the Jacobson radical J(R). If L^2 = L then L = 0.

Proof. Consider an arbitrary left R-module M with Krull dimension, α = Kdim M say. By transfinite recursion we will show that LM = 0 and hence L = 0 will follow. If $\alpha = 0$ then M is Artinian. Using the Loewy series of M combined with the fact that LS = 0 for every simple R-module S (and also using L^2 = L) one arrives at LM = 0. Assume now that the claim has been established for any R-module N such that Kdim N < α. If Kdim M = α then we reduce the problem to the case where M is an α-critical R-module by passing to terms in the critical Loewy series. For a nonzero R-submodule M' of M we then have Kdim (M/M') < α and thus LM ⊂ M'. But since M' is arbitrary it follows that LM is a simple R-module.
However L ⊂ J(R) then yields L (LM) = 0, i.e. LM = 0 as claimed. □

5.1.3. Lemma. In a ring with Krull dimension any left critical nil ideal L satisfies $L^2 = 0$.

Proof. For a nonzero a ∈ L we consider φ_a: L → L, x ↦ xa. Clearly,
Ker $\varphi_a = l_R(a) \bigcap L$. Since a is nilpotent, $l_R(a) \bigcap L \neq 0$ and hence Ker $\varphi_a \neq 0$. Consequently Kdim (Im φ_a) < α.
Since L is critical, we must have Im $\varphi_a = 0$, or La = 0 and $L^2 = 0$ follows. □

5.1.4. Lemma. Let N be a non-nilpotent ideal of a ring R which has Krull dimension. In N one can construct an ascending chain of ideals $0 = A_0 \subset A_1 \subset A_2 \subset$ ..., such that each A_i is nilpotent but $\bigcup_{i=0}^{\infty} A_i$ is not nilpotent.

Proof. Since N has Krull dimension as a left R-module, it follows from Lemma 5.1.3. that N contains a left ideal I such that $L^2 = 0$. From (LR)(LR) = L^2R = 0 it follows that N contains an ideal J such that $J^2 = 0$. Applying Zorn's lemma, we obtain an ideal I_1 in N which is maximal with respect to the property: $I_1^2 = 0$.
Repeating this argument for the ring R/I_1 we arrive at the ascending chain 0 =

$I_0 \subset I_1 \subset I_2 \subset ... \subset I_n \subset ...$ such that each I_{i+1}/I_i is an ideal in R/I_i maximal with the property $(A_{i+1}/A_i)^2 = 0$. Clearly, every ideal I_i is nilpotent. If I = $\bigcup_{i=0}^{\infty} I_i$ is nilpotent then I = I_n for some n.
If I $\neq$ N then we obtain a nil ideal N/I in R/I and by Lemma 5.1.3. this means that N/I contains an ideal A/I of R/I such that $(A/I)^2 = 0$. But then $I_{n+1} \neq I_n$, a contradiction, hence I is not nilpotent. □

5.1.5. Remark. The foregoing lemmas also hold for rings having Gabriel dimension.

5.1.5. Theorem. (Gordon, Robson [1], Lenagan [3]). Let R be a ring having Krull dimension. If N is a nil ideal of R then N is nilpotent.

Proof. Suppose that N is not nilpotent. By Lemma 5.1.4., we may construct an ideal A = $\bigcup_{i=0}^{\infty} A_i$ contained in N and such that each A_i is a nilpotent ideal but A is not nilpotent. Since N has Krull dimension we may apply Lemma 5.1.1. and so we obtain numbers n and j such that $N^n \subset N^{n+1} + A_j$. In the ring R/$A_j$ we obtain: $(N/A_j)^n = (N/A_j)^{n+1}$. Because N/A $\subset$ J(R/A_j) we may apply Lemma 5.1.2. to conclude that $(N/A_j)^n = 0$, hence $N^n \subset A_j$. That N is nilpotent follows because A_j is. □

5.1.6. Corollary. The prime radical, rad R, of a ring R which has Krull dimension is a nilpotent ideal.

§5.2 Semiprime Rings with Krull Dimension.

Certain dimension properties of semiprime rings lead to structural results e.g. A. Goldie and L. Small's result stating that a semiprime ring that has Krull dimension is necessarily a left Goldie ring (see Corollary 5.2.4.). We start here in a somewhat more general situation which will allow us to consider Gabriel dimension instead of Krull dimension.

A nonzero R-module M is said to be *monomorphic* if for any submodule N of M and any nonzero f $\in$ Hom_R(N,M) it follows that f is a monomorphism. If an R-module M has the property that each nonzero submodule of M contains a (nonzero) monomorphic submodule then we say that the R-module *M has enough monomorphic submodules.*

5.2.1. Lemma. An α-critical or α-simple R-module M is monomorphic.

Proof. Suppose that M is α-critical; let $f \neq 0$ be an R-linear map $N \to M$, where N is a nonzero R-submodule of M.
Then $0 \neq \mathrm{Im}\, f \cong N/\mathrm{Ker}\, f$. If $\mathrm{Ker}\, f \neq 0$ then $\mathrm{Kdim}\, \mathrm{Im}\, f < \alpha$ because N is α-critical, but as Im f is α-critical too this leads to a contradiction.
In conclusion, $\mathrm{Ker}\, f = 0$ and M is monomorphic. In a similar way one treats the case where M is an α-simple R-module. □

5.2.2. Lemma. Let L be a nonzero left ideal of the semiprime ring R. If L is monomorphic as a left module then there is an $a \in L$ such that $l_R(a) \bigcap L = 0$.

Proof. Since $L^2 \neq 0$ there is an $a \in L$ such that $La \neq 0$. Define φ_a: $L \to L$ by $x \mapsto xa$. Clearly $\varphi_a \neq 0$ but since $\mathrm{Ker}\, \varphi_a = 0$ it follows that $0 = l_R(a) \bigcap L$. □

5.2.3. Theorem. Let R be a semiprime ring such that ${}_RR$ has finite coirreducible dimension and it has enough monomorphic submodules, then R is a left Goldie ring.

Proof. Suppose first that the singular radical $Z(R) \neq 0$; then Z(R) contains a monomorphic left ideal L. If $a \in L$ is nonzero then $l_R(a)$ is an essential left ideal of R but by the foregoing lemma, $l_R(a) \bigcap L = 0$. Hence $L = 0$, contradiction i.e. Z(R) = 0.
Let E be any essential left ideal of R and let $L_1 \subset E$ be a monomorphic left ideal. By the foregoing lemma there is an $a_1 \in L_1$ such that $l_R(a_1) \bigcap L_1 = 0$. Since $l_R(a_1) \bigcap E \neq 0$ there is a monomorphic left ideal $L_2 \subset l_R(a_1) \bigcap E$ and by the foregoing lemma there is an $a_2 \in L_2$ such that $l_R(a_2) \bigcap L_2 = 0$. Clearly: $L_1 \bigcap L_2 = 0$, $l_R(a_1 + a_2) \bigcap (L_1 \bigoplus L_2) = 0$. Since $l_R(a_1 + a_2) \bigcap E \neq 0$, we find a monomorphic left ideal $L_3 \subset l_R(a_1 + a_2) \bigcap E$. Pick $a_3 \in L_3$ such that $l_R(a_3) \bigcap L_3 = 0$. We now obtain:

$L_3 \bigcap (L_1 \bigoplus L_2) = 0$ and $l_R(a_1 + a_2 + a_3) \bigcap (L_1 \bigoplus L_2 \bigoplus L_3) = 0$. Continuing this argumentation leads to left ideals $L_1, ..., L_m$ and elements $a_i \in L_i$ such that $l_R(a_1 + ... + a_m) \bigcap (L_1 \bigoplus ... \bigoplus L_m) = 0$. If the coirreducible dimension of ${}_R R$ equals n then there exists an $r \leq n$ such that $l_R(a_1 + ... + a_n) = 0$. Put $a = a_1 + ... + a_r$. Since $l_R(a) = 0$ and $Z(R) = 0$ we see that a is a regular element.
So we have constructed, in a given essential left ideal E a regular element a of R i.e. R is a left Goldie ring. □

5.2.4. Corollary. (A. Goldie, L. Small, [1]; R. Gordon, J.C. Robson [1]). A semiprime ring with Krull dimension is a left Goldie ring

5.2.5. Corollary. A semiprime ring with Gabriel dimension and finite coirreducible dimension is a left Goldie ring.

5.2.6. Corollary. If R has Krull dimension then we have: Kdim R = Kdim (R/rad R).

5.2.7. Corollary. If R is a semiprime ring with Krull dimension then Kdim R = α = sup {1 + Kdim (R/L), L an essential left ideal of R}.

Proof. If L is an essential left ideal then L contains a regular element, a say. From $Ra^n/Ra^{1+n} \cong R/Ra$ it follows that Kdim (Ra^n/Ra^{1+n}) = Kdim (R/Ra) $\geq$ Kdim (R/L).
Because the chain $Ra \supset Ra^2 \supset Ra^3 \supset ...$ is strictly increasing, Kdim R > Kdim (R/L) and hence Kdim R $\geq \alpha$. By the property 4.1.5., the equality follows. □

5.2.8. Corollary. Let R be a semiprime ring with Krull dimension and let $\{L_\alpha\}_{\alpha \in \bigwedge}$ be the set of critical left ideal of R; then Kdim R = sup {Kdim L_α, $\alpha \in \bigwedge$}.

Proof. There are critical left ideals $L_1.,,,.L_n$ such that $L = L_1 \bigoplus ... \bigoplus L_n$ is essential in R. Corollary 5.2.6. entails that Kdim R > Kdim (R/L). On the other hand, from Kdim R = sup {Kdim (R/L), Kdim L} it follows that Kdim R = Kdim

L = sup {Kdim L_j, j = 1,...,n} = sup {Kdim L_α, $\alpha \in \Lambda$}. □

5.2.9. Corollary. If L is a critical left ideal of a prime ring R with Krull dimension, then Kdim R = Kdim L.

Proof. Since R is a left Goldie ring it has a simple Artinian total ring of fractions $Q_{cl}(R)$, i.e. ${}_RR$ is an essential extension of a finite direct sum, $\bigoplus_{i=1}^{n}$ U_i, where each U_i is a coirreducible R-module and all are isomorphic. Pick some U_{i_0} and let it contain a critical left ideal J. Then ${}_RR$ is an essential extension of a direct sum $\bigoplus_{i=1}^{n}$ J_i where each $J_i \cong$ J, i = 1,...,n. By Corollary 5.2.7. we conclude that Kdim R = Kdim J.
Since L $\bigcap$ ($\bigoplus_{i=1}^{n}$ J_i) $\neq 0$, there exist nonzero left ideals L' $\subset$ L, J' $\subset$ J such that L' $\cong$ J'.
Finally, the equality: Kdim J = Kdim J' = Kdim L' = Kdim L, yields Kdim R = Kdim L. □

5.2.10. Corollary. Let R be a prime ring with Krull dimension. If M is a finitely generated R-module, the following statements are equivalent:

1. Kdim M < Kdim R.
2. M is singular, i.e. M = Z(M).

Proof. Assume that Kdim M < Kdim R and Z(M) $\neq$ M. So there is an x $\in$ M such that l_R(x) is not an essential left ideal and hence there exists a critical left ideal I such that I $\bigcap$ l_R(x) = 0. Therefore, M contains a submodule isomorphic to I. From Corollary 5.2.9. we retain: Kdim M $\geq$ Kdim I = Kdim R, a contradiction.
Conversely, assume that M is singular. Since M is finitely generated by assumption, there exists a cyclic submodule N of M such that Kdim M = Kdim N. Put N = Rx. Since l_R(x) is an essential left ideal there is a c $\in$ R which is regular and such that cx = 0. Define f: R $\rightarrow$ N, $\lambda \mapsto \lambda$x. Then Ker f = l_R(x) and N $\cong$ R/L_R(x). Thus, N is a factor module of the R-module R/Rc.
Applying Corollary 5.2.7., we obtain: Kdim (R/Rc) < Kdim R and consequently: Kdim M < Kdim R. □

5.2.11. Corollary. Let R be a semiprime ring with Kdim R = α.If M is an α-critical

R-module, then M contains a submodule isomorphic to an α-critical ideal of R.

Proof. We use the notations of the proof of Corollary 5.2.9.. If LM = 0 then there exists a nonzero morphism f: $R/L \to M$. Hence α = Kdim M = Kdim Im f $\leq$ Kdim (R/L) < Kdim R = α, contradiction. So, LM $\neq 0$. There must then exist a j, $1 \leq j \leq n$, such that $L_j M \neq 0$,. For some x $\in$ M, I_j x $\neq 0$.
Let g: $L_j \to M$, $\lambda \mapsto \lambda x$, then Ker g = 0 (because otherwise: α = Kdim M = Kdim Im g = Kdim L/Ker g < α, a contradicion) i.e. g is monomorphic. □

5.2.12. Corollary. Let R be a ring with Krull dimension and M a critical R-module. Then Kdim R = Kdim M if and only if M contains a nonzero submodule isomorphic to a left ideal of R.

Proof. An easy consequence of Corollary 5.2.9. and 5.2.11. □

§5.3 Classical Krull Dimension of a Ring.

For a ring we let Spec R be the set of prime ideals of R. By transfinite recursion we define on Spec R the following filtration: $(\text{Spec } R)_0$ is the set of maximal ideals of R. If $(\text{Spec } R)_\beta$ has been defined for any $\beta < \alpha$ then $(\text{Spec } R)_\alpha$ is defined by: $\{P \in \text{Spec } R$, for all $Q \in \text{Spec } R$ such that $P \subsetneq Q$ we have $Q \in \bigcup_{\beta<\alpha} (\text{Spec } R)_\beta\}$.
So we obtain an ascending chain:

$$(\text{Spec } R)_o \subset (\text{Spec } R)_1 \subset \ldots \subset (\text{Spec } R)_\alpha \subset \ldots \qquad (*)$$

5.3.1. Proposition. The following statements are equivalent:

1. $(\text{Spec } R)_0 \neq \phi$.
2. The set $\bigcup_{\alpha \geq 0} (\text{Spec } R)_\alpha$ satisfies the ascending chain condition.
3. There exists an ordinal α such that Spec R = $(\text{Spec } R)_\alpha$ if and only if R satisfies the ascending chain condition for prime ideals.

Proof. Easy. □

A ring is said to have *classical Krull dimension* if there exists an ordinal α such that Spec R = $(\text{Spec R})_\alpha$. The least ordinal with that property is said to be *the classical Krull dimension of* R and it is denoted by clKdim R.

5.3.2. Proposition. 1. Let φ: R $\to$ S be a surjective ring morphism and φ_*: Spec S $\to$ Spec R the induced map, $\varphi_*(Q) = \varphi^{-1}(Q)$ for Q $\in$ Spec S. Then, for all α, $\varphi_*\ ((\text{Spec S})_\alpha) \subset (\text{Spec R})_\alpha$.

2. P $\in$ $(\text{Spec R})_\alpha$ if and only if R/P has classical Krull dimension and clKdim R/P $\leq \alpha$.

Proof. 1. By transfinite recursion on α.
2. Follows from 1. in a straightforward way. □

5.3.3. Proposition. If the ring R has Krull dimension α then R has classical Krull dimension and cl.Kdim R $\leq \alpha$.

Proof. If $\alpha = 0$ then R is a left Artinian ring. Since a prime ideal in an Artinian ring is a maximal ideal it follows that cl.Kdim R = 0.
Now suppose that we have established the assertion for all rings with Krull dimension less than α and proceed to prove the claim for R, a ring with KdimR = α. If P $\in$ Spec R then Kdim R/P $\leq \alpha$. If P $\subsetneq$ Q then Q/P is a nonzero ideal in R/P, hence it is an essential left ideal in R/P. By Corollary 5.2.7. it follows that Kdim (R/Q) < Kdim (R/P), or Kdim (R/Q) < α. By the assumption cl.Kdim R/Q < α. Now Proposition 5.3.2. entails that Q $\in$ $(\text{Spec R})_\beta$ with $\beta < \alpha$.
Consequently, P $\in$ $(\text{Spec R})_\alpha$ and hence $(\text{Spec R})_\alpha$ = Spec R, so cl.Kdim R $\leq \alpha$. □

§5.4 Associated Prime Ideals.

Consider an R-module M. We say that a prime ideal P $\in$ Spec R is *associated to M* if there is a nonzero submodule N of M such that P = Ann N = Ann N' for any nonzero submodule N' of N. By Ass(M) we denote the set of all prime ideals of R associated to M. In case R is a commutative ring it is well known that P $\in$ Ass(M) if and only if there exists an x $\in$ M, x $\neq$ 0, such that P = l_R(x).

We now aboard the study of associated prime ideals for modules over rings with Krull dimension.

5.4.1. Proposition. We have the following general properties:

1. If we consider an exact sequence of R-modules,

$$0 \longrightarrow M' \xrightarrow{f} M \xrightarrow{g} M" \longrightarrow 0$$

   then: Ass(M') $\subset$ Ass(M) $\subset$ Ass(M') $\bigcup$ Ass(M").
2. For a directed family $(M_i)_{i\in I}$ of R-submodules of M such that M = $\bigcup_{i\in I} M_i$, then Ass(M) = $\bigcup_i$ Ass(M_i).
3. If M = $\bigoplus_{i\in I} M_i$ then Ass(M) = $\bigcup_i$ Ass(M_i).
4. If M is an essential extension of the R-submodule M' then Ass(M') = Ass(M).

Proof. 1. The inclusion Ass(M') $\subset$ Ass(M) is obvious. Let P be in Ass(M) then there exists a submodule N of M such that P = Ann N = Ann N' for any nonzero submodule N' of N. If N $\bigcap$ M' $\neq$ 0 then P = Ann(N $\bigcap$ M') = Ann (X) for any R-submodule X in N $\bigcap$ M', hence P $\in$ Ass(M'). If N $\bigcap$ M' = 0 then there exists a monomorphism 0 $\to$N $\to$ M" and this establishes that P $\in$ Ass(M") and the claim follows.

2. From 1. it is clear that $\bigcup_{i\in I}$ Ass(M_i) $\subset$ Ass(M). If P is in Ass(M) then there exists a nonzero submodule N $\subset$ M such that P = Ann N = Ann N' for any submodule N' in N. But there exists an i $\in$ I such that N $\bigcap$ M_i $\neq$ 0, consequently:
P $\in$ Ass(N $\bigcap$ M_i) $\subset$ Ass(M_i), and the inclusion Ass(M) $\subset$ $\bigcup_{i\in I}$ Ass(M_i) is then obvious.

3. From 1. and 2..

4. Obvious, by definition. □

5.4.2. Lemma. Let J be an ideal of a ring R with Krull dimension, then there exist prime ideals $P_1, ..., P_n$ in Spec R such that J $\supset$ $P_1 .... P_n$, J $\subset$ P_i for all i = 1,...,n.

Proof. If we write rad(I) for the intersection of prime ideals contining I then rad(I)/I is the prime radical of the ring R/I. Applying Corollary 5.1.6. to R/J yields that $\mathrm{rad}(J)^k \subset J$. On the other hand, since R/rad(J) is a semiprime left Goldie ring it follows that there is a finite number of prime ideals $P_1, ..., P_s$ such that rad(J) = $P_1 \bigcap ... \bigcap P_s$. From rad(J) $\supset P_1 P_2 ... P_s$ it then follows that $(\mathrm{rad}(J))^k$ contains a finite product of prime ideals and therefore J does too. □

5.4.3. Proposition. Let R be a ring with Krull dimension. If M is a nonzero uniform R-module then Ass(M) = {P} for some prime ideal P of R.

Proof. From Lemma 5.4.2. it follows that there is a nonzero submodule N of M and a prime ideal Q of R such that QN = 0. By the ascending chain condition on prime ideals we know that there is a prime ideal P maximal with respect to the property PN_0 for some nonzero submodule N_0 of M. If N_0' is a nonzero submodule of N_0 then P $\subset$ Ann N_0' and Lemma 5.4.2. entails the existence of a prime ideal Q $\supset$ Ann N_o' annihilating a nonzero submodule of N_0'. By the maximality assumption: P = Q and P = Ann N_0' or P $\in$ Ass(M) follows.
The fact that M is uniform allows to apply Proposition 5.4.1. and to deduce that Ass(M) = {P}. □

5.4.4. Corollary. Let M be a nonzero module over a ring R with Krull dimension, then Ass(M) $\neq \phi$.

Proof. Pick x $\neq$ 0 in M. Since Rx has Krull dimension it has finite coirreducible dimension and therefore it must contain a nonzero coirreducible submodule. The claim now folows the proposition. □

For an injective indecomposable R-module Q we let [Q] be the isomorphism class of Q and refer to this as the *type* of the injective module Q. If x $\neq$ 0 in Q then $l_R(x)$ is an irreducible left ideal and hence Q $\cong$ E(R/ l_R(x)). The class of all types of indecomposable injectives is a set which we denote by $\mathcal{E}$(R).

5.4.5. Corollary. Let M be a nonzero module over a ring R with Krull dimension.

Then Ass(M) = {P} if and only if $r_M(P)$ is an essential submodule of M such that P contains each ideal that annihilates a nonzero submodule of M.

Proof. If $0 \neq N \subset M$ then Ass(N) $\neq \phi$ i.e. Ass(N) = {P} and there exists a nonzero submodule N' in N such that P = Ann N'. From PN' = 0, i.e. $N' \subset r_M(P)$, it follows that $r_M(P) \bigcap N \neq 0$, showing that $r_M(P)$ is essential. If J is an ideal annihilating the nonzero submodule K of M then Ass(K) = {P} entails that P = Ann(K') where $0 \neq K' \subset K$. Obviously JK' = 0 yields $J \subset P$. Conversely, look at $Q \in$ Ass(M). By the hypothesis, $Q \subset P$. On the other hand there exists a nonzero submodule $N \subset M$ such that Q = Ann N = Ann N' for any submodule $N' \neq 0$ of N. From $N \bigcap r_P(M) \neq 0$ it follows that $Q = \text{Ann}\,(N \bigcap r_P(M))$ and hence $P \subset Q$ i.e. P = Q. □

5.4.6. Lemma. Let P be a prime ideal of a ring with Krull dimension. There exists an indecomposable injective Q_P such that $E(R/P) \cong Q_P^n$ where n is the coirreducible dimension of the ring R/P.

Proof. Since R/P has a simple Artinian classical ring of fractions, it is an essential extension of n coirreducible submodules which are all isomorphic. Let X be one of the latter submodules. If $Q_P = E(X)$ then $E(R/P) \cong Q_P^n$. □

By Lemma 5.4.6. it is possible to define a map:

$$\varphi\colon \text{Spec } R \to \mathcal{E}(R),\ P \mapsto Q_P.$$

By Proposition 5.4.3. it is possible to define a map:

$$\Psi\colon \mathcal{E}(R) \to \text{Spec } R,\ Q \mapsto P_Q$$

where P_Q is the unique prime ideal of R for which Ass(Q) = $\{P_Q\}$. It is now a direct consequence of Lemma 5.4.6. that:

5.4.7. Proposition. If R has Krull dimension then $\Psi \circ \varphi$ is the identity of Spec R In particular φ is injective.

A ring R with Krull dimension such that φ is a bijection is said to be *a ring with enough prime ideals.*

§5.5 Fully Left Bounded Rings with Krull Dimension.

Let R be ring with Krull dimension. We say that R is *left bounded* if essential left ideals of R contain nonzero ideals. If for every prime ideal P of R, R/P is left bounded, then R is said to be *fuly left bounded.* Clearly: a commutative ring is fully left bounded. Note also that the property of being fully left bounded is inherited by epimorphic images.

5.5.1. Lemma. In a ring R with Krull dimension there exists a prime ideal P such that Kdim R = Kdim R/P.

Proof. Let $P_1, ..., P_n$, be the prime ideals of R such that $P_n.....P_1 = 0$, the existence of which is garanteed by Lemma 5.4.2.. Considering the chain: R $\supset P_1 \supset P_2P_1 \supset ... \supset P_nP_{n-1} ... P_1 = 0$, we see that: Kdim R = sup (Kdim R/P_1, Kdim P_1/P_2P_1, ...). Since the module $P_{i-1}.....P_1/P_iP_{i-1}.....P_1$ is an R/P_i-module we have: Kdim $(P_{i-1}.....P_1)/(P_iP_{i-1}.....P_1) \leq$ Kdim R/P_i. Therefore: Kdim R $\leq$ sup {Kdim R/P_i, i = 1,...,n}, and this leads to Kdim R = sup {Kdim R/P_i, i = 1,...,n}. The latter implies the existence of an i such that Kdim R = Kdim R/P_i. □

5.5.2. Theorem. **(G. Krause [2]).** For a fully left bounded ring R with Krull dimension we have that Kdim R = cl.Kdim R.

Proof. The lemma and Proposition 5.3.3. allow to assume that R is a prime ring. Put Kdim R = α and let $\beta < \alpha$. By Corollary 5.2.7. there exists an essential left ideal L such that $\beta \leq$ Kdim R/L. Since R contains a nonzero ideal J, KdimR/L $\leq$ Kdim R/J < Kdim R. Lemma 5.5.1. implies that there exists a prime ideal P containing J such that Kdim R/J = Kdim R/P. Hence $\beta \leq$ Kdim R/P < α.
Now we may use transfinite recursion on α. If $\alpha = 0$ then R is simple
Artinian and hence cl.Kdim R = 0. Let R be a prime ring such that Kdim R = α. If cl.Kdim R $\neq \alpha$ then cl.Kdim R < α (cf. Proposition 5.3.3.). So there exists a prime ideal P of R such that cl.Kdim R $\leq$ Kdim R/P < α. The induction hypothesis yields: cl.Kdim R/P = Kdim R/P. The fact that R is a prime ring entails cl.Kdim R > cl.Kdim R/P what consitutes a contradiction. □

5.5.3. Corollary. If R is a fully left bounded ring with Krull dimension, α say, then there is a prime ideal P of R, such that cl.Kdim R/P = β, for each $\beta \leq \alpha$.

Proof. For $\beta = \alpha$ the statement is a consequence of Lemma 5.5.1., and Theorem 5.5.2.. If $\beta < \alpha$ then there exists a prime ideal P suh that $\beta \leq$ cl.Kdim R/P $< \alpha$. Proposition 5.3.1. enables us to choose P such that it is maximal with respect to the property $\beta \leq$ cl.Kdim R/P $< \alpha$. If $\beta <$ cl.Kdim R/P then a similar argument allows us to select a prime ideal Q $\supsetneq$ P such that $\beta \leq$ cl.Kdim R/Q < cl.Kdim R/P. The maximality property of P then forces P = Q, a contradiction. □

5.5.4. Corollary. For a commutative ring with Krull dimension cl.Kdim R = Kdim R.

5.5.5. Corollary. If F is a field then cl.Kdim F$[X_1, ..., X_n]$ = n.

5.5.6. Theorem. A fully left bounded ring R with Krull dimension has enough prime ideals.

Proof. Let Q be an indecomposable injective R-module and assume that Ass(Q) = {P}, E(R/P) $\cong Q_P^n$. If we establish that Q $\cong Q_P$ i.e. Q contains a coirreducible submodule isomorphic to an irreducible submodule of R/P, then it will follow that Ψ: Spec R $\rightarrow \mathcal{E}$(R) is surjective, as desired. Clearly we may reduce to the case where P = 0, i.e. R is a prime ring such that Ass(Q) = {(0)}. Obviously Q then contains a critical submodule M such that Ann M = Ann M' = 0 for any nonzero submodule M' of M. We may suppose that M $\cong$ R/L or some left ideal L of R. If L is essential then it contains a nonzero ideal J but then JM = 0 leads to the contradiction
J $\subset$ Ann M.
Consequently L is not essential, so we may find a nonzero left ideal L_1 such that L $\bigcap L_1$ = 0, i.e. there is a monomorphism 0 $\rightarrow$ L$_1$ $\rightarrow$ M. The latter implies that M contains a coirreducible submodule isomorphic to a nonzero left ideal. □

An R-module M $\neq$ 0 is said to be *compressible* if for each nonzero submodule N of M there exists a monomorphism from M to N.

5.5.7. Proposition. Let M be a compressible module over a ring R with Krull dimension, then:

1. M is critical.
2. Ann M is a prime ideal.
3. $\mathrm{End}_R(M)$ is a left Ore domain.

Proof. 1. M contains a critical module C, so there exists a monomorphism $0 \to M \to C$ and therefore M is critical.

2. If $N \neq 0$ is a submodule of M then Ann M = Ann N = P. If I and J are ideals such that $IJ \subset P$ then $IJM = 0$. If $JM \neq 0$ then $I \subset \mathrm{Ann}\, JM = \mathrm{Ann}\, M$, hence $I \subset P$; consequently P is a prime ideal.

3. Every nonzero $f \in \mathrm{End}_R(M)$ is a monomorphism because M is critical, hence $\mathrm{End}_R(M)$ is a domain. On the other hand, if f and y are nonzero R-endomorphisms of M, then $N = \mathrm{Im}\, f \bigcap \mathrm{Im}\, g \neq 0$ yields that there exists a monomorphism $0 \longrightarrow M \xrightarrow{h} N$. Note that $f(f^{-1}(N)) = N = g(g^{-1}(N))$ and consider the diagram below obtained by restricting f and g to $f^{-1}(N)$ and $g^{-1}(N)$ respectively:

$$\begin{array}{ccc} M & \xrightarrow{\beta} & g^{-1}(N) \\ \alpha \downarrow & \searrow h & \downarrow g' \\ f'(N) & \xrightarrow[f']{} & N \end{array}$$

here f' and g' are isomorphisms and $\alpha = (f')^{-1} \circ h$, $\beta = (g')^{-1} \circ h$. Since $f \circ \alpha = f' \circ \alpha = h = g \circ \beta = g' \circ \beta$, $\alpha f = \beta g$ follows and this establishes that $\mathrm{End}_R(M)$ is a left Ore domain. □

5.5.8. Corollary. Every nonzero module over a fully left bounded ring with Krull dimension contains a compressible submodule.

Proof. M contains a critical submodule C; let $\mathrm{Ass}(C) = \{P\}$. There exists a submodule D of C such that P = Ann D = Ann D' for all nonzero submodules D' of D. Since D is an R/P-module it is not restrictive to assume that P = 0, or that R is a prime ring, while Ann D = Ann D' for any nonzero submodule D' of D.
The assumption on R entails that D contains a submodule D_0 isomorphic to a nonzero left ideal L of R. Consider a left ideal $J \neq 0$ in L. Then $LJ \neq 0$. Pick $a \in J$ such that

La $\neq 0$.
Since L is critical, say α-critical, we obtain Kdim J = Kdim L = α. The nonzero homomorphism φ_a: L $\rightarrow$ J, $\lambda \mapsto \lambda$a, is a monomorphism. Hence L is compressible and D_0 is a compressible R-module. □

§5.6 Examples of Noetherian Rings of Arbitrary Krull Dimension.

The idea of this section is to use certain families of a subset of indeterminates in order to contruct commutative Noetherian domains having for their dimensions any given ordinals.

Let T be an arbitrary set. A family $\mathcal{G}$ of subsets of T is said to be a *gang* if: T $= \bigcup_{g \in \mathcal{G}}$ G, G $\in \mathcal{G}$ and G' $\subset$ G yields G' $\in \mathcal{G}$, ascending chains in $\mathcal{G}$ are stationary. These conditions imply that each G $\in \mathcal{G}$ is a finite set. The third condition allows to define: $\mathcal{G}_0 = \{$G $\in \mathcal{G}$, G is a maximal element of $\mathcal{G}\}$, supposing that $\mathcal{G}_\beta$ has been defined for every $\beta < \alpha$ then: $\mathcal{G}_\alpha = \{$G $\in \mathcal{G}$, all G' $\in \mathcal{G}$ containing G are in $\bigcup_{\beta < \alpha} \mathcal{G}_\beta\}$. This gives rise to the ascending chain $\mathcal{G}_0 \subset \mathcal{G}_1 \subset \ldots \subset \mathcal{G}_\alpha \subset \mathcal{G}_{\alpha+1} \subset \ldots$. There exists an ordinal α such that $\mathcal{G} = \mathcal{G}_\alpha$ and the least such ordinal is said to be the Krull dimension of $\mathcal{G}$ (denoted by Kdim $\mathcal{G} = \alpha$). Clearly Kdim $\mathcal{G} = 0$ if and only if T = ϕ.

5.6.1. Proposition. (R. Gordon, J.C. Robson). For any ordinal α there is a set T and a gang $\mathcal{G}$ on T such that Kdim $\mathcal{G} = \alpha$

Proof. If $\alpha = 0$ then T = ϕ, $\mathcal{G} = \phi$. If $\alpha > 0$ is a nonlimit ordinal then $\alpha = \beta + 1$ for some β. Let U be a set, $\mathcal{H}$ a gang on U with Kdim $\mathcal{H} = \beta$. Let x $\notin$ U and put: T = U $\cup$ {x}, $\mathcal{G} = \{$G $\subset$ T such that G = H or G = H $\cup$ {x} for some H $\in \mathcal{H}\}$. Obviously $\mathcal{G}$ is a gang on T such that Kdim $\mathcal{G} = \alpha$. If α is a limit ordinal then there is a family of ordinals $(\beta_i)_{i \in I}$ with $\beta_i < \alpha$ having α as the sup of this family. Let T_i be any set with a gang $\mathcal{G}_i$ such that Kdim $\mathcal{G}_i = \beta_i$. Put T = $\amalg_{i \in I} T_i$, $\mathcal{G} = \amalg_{i \in I} \mathcal{G}_i$. It is easy enough to check that $\mathcal{G}$ is a gang on T and that Kdim $\mathcal{G} = \alpha$.

For a prime ideal P of a commutative ring R, The *height* of P is the least upper-bound of $\{n \in \mathbb{N}$, there is a strictly increasing sequence of prime ideals of R: $P_0 \subset P_1 \subset \ldots \subset P_n = P\}$ and it is denoted by: ht(P). If ht(P) is finite then P is said to have *finite height* and we write ht(P) $< \infty$. Otherwise P is of infinite height and we then write ht(P) $= \infty$. Obviously, in a commutative Noetherian ring, if P has finite height then ht(P) $\leq$ cl.Kdim R. The following is equally obvious:

5.6.2. Lemma. Let P be a prime ideal of a commutative ring R. Then PR $[X_\alpha, \alpha \in I]$ is a prime ideal in the polynomial ring R $[X_\alpha, \alpha \in I]$.

5.6.3. Proposition. Let R = F $[X_\alpha, \alpha \in I]$ be the ring of polynomials over the field F. For a prime ideal P of R, ht(P) $< \infty$ if and only if P is finitely generated.

Proof. For a finite subset S of I we let $R_{(S)}$ be the ring F $[X_\alpha, \alpha \in S]$. Clearly R $= \bigcup_{S \subset I} R_{(S)}$. If P is finitely generated then there is a finite subset S of I such that $P \subset R_{(S)}$ and then it is clear that ht(P) $\leq$ Card(S).
Conversely, suppose that ht P $< \infty$. Since $(P \bigcap R_{(S)})R$ is a prime ideal contained in P for all infinite subsets S of I it follows that there is a finite subset S_0 in I such that for every finite set $I \supset S \supset S_0$: $(P \bigcap R_{(S_0)})R = (P \bigcap R_{(S)})R$. Clearly P = $(P \bigcap R_{(S_0)})R$ follows and since $P \bigcap R_{(S_0)}$ is finitely generated in $R_{(S_0)}$, P is finitely generated in R. □

5.6.4. Lemma. If $P_1, \ldots, P_n$ are prime ideals of R, a commutative ring, and I is an ideal of R contained in $P_1 \bigcup \ldots \bigcup P_n$ then $I \subset P_j$ for some $j \in \{1,\ldots,n\}$.

Proof. We may suppose that $P_i \subsetneq P_j$ for all $i,j \in \{1,\ldots,n\}$. If $I \not\subset P_i$ for all i, then $I \sqcap_{i \neq j} P_j \not\subset P_i$ and so we may select $a_i \in I \sqcap_{i \neq j} P_j$ such that $a_i \notin P_i$. Now a = $a_1 + \ldots + a_n$ is in I but evidently $a \notin \bigcup_{i=1}^n P_i$, contradiction. □

5.6.5. Lemma. (Cohen). A commutative ring R such that all of its prime ideals are finitely generated is Noetherian.

Proof. If R were not Noetherian then there exists an ideal I maximal in the set of ideals which cannot be finitely generated (using Zorn's lemma). If I is not prime then we may find ideals A,B $\supsetneq$ I such that AB $\subset$ I. the choice of I entails that R/A and R/B are Noetherian rings. Since B is also finitely generated, B/AB is a finitely generated R/A-module hence it is Noetherian. From I/AB $\subset$ B/AB it then follows that I/AB is finitely generated as an R-module. Now the fact that AB is finitely generated as well forces I to be finitely generated, contradiction. □

Let R = F[T] be the ring of polynomils with indeterminates in the set T, where F is a field. If $\mathcal{G}$ is a gang on T we may look at the multiplicatvely closed set in R: $S(\mathcal{G}) = F[T] - \bigcup_{G\in\mathcal{G}} GF[T]$, where GF[T] is the prime ideal generated in F[T] by the elements of G. Let F{$\mathcal{G}$} be the ring of fractions of F[T] with respect to S($\mathcal{G}$). With these notations we arrive at:

5.6.6. Theorem. (R. Gordon, J.C. Robson [1]). The ring F{$\mathcal{G}$} is a Noetherian domain and clKdim F{$\mathcal{G}$} $\geq$ Kdim $\mathcal{G}$.

Proof. That F{$\mathcal{G}$} is a domain is evident. If I is an ideal of F[T] such that I $\bigcap$ S($\mathcal{G}$) = ϕ then we claim that I $\subset$ GF[T] for some G $\in$ $\mathcal{G}$. Pick $a_1 \in$ I. There is a finite T_1 in T such that $a_1 \in F[T_1]$. Put $S_1 = F[T_1] - HF[T_1]$ where
H = $\bigcup$ {G $\in$ $\mathcal{G}$, G $\subset$ T_1}. It is clear that S_1 = S $\bigcap$ F[T_1]. Since S_1^{-1} F[T_1] is semilocal we apply Lemma 5.6.4. to derive the existence of a G $\in$ $\mathcal{G}$, G $\subset$ T_1, such that I $\bigcap$ F[T_1] $\subset$ GF[T_1]. Let $G_{11},\ldots,G_{1n_1}$ be the elements of $\mathcal{G}$ contained in T_1 and minimal with respect to the above property. If I $\subsetneq \bigcup_{i=1}^{n_1} G_{1i}$ F[T], let $a_2 \in$ I be outside of the latter union. there exists a finite $T_2 \subset$ T, $T_1 \subset T_2$, such that $a_2 \in F[T_2]$. As before, choose $G_{21},\ldots,G_{2n_2}$ from $\mathcal{G}$ and contained in T_2 such that they are minimal with respect to I $\bigcap$ F[T_2] $\subset$ GF[T_2]. It is obvious that for every i there exists a j such that $G_{2i} \supset G_{1j}$. In case I $\subsetneq \bigcup_{i=1}^{n_2} G_{2i}$ F[T] we reproduce the foregoing argument. The strictly increasing sequence
$\bigcup_{i=1}^{n_1} G_{1i}$ F[T] $\subsetneq \bigcup_{i=1}^{n_2} G_{2i}$ F[T] $\subsetneq \ldots$, induces a strictly increasing sequence of elements of $\mathcal{G}$ ($G_{1j} \subset G_{2i} \subset \ldots$) and hence the process terminates. If k $\geq$ 1 is such that I $\subset \bigcup_{i=1}^{n_k} G_{ki}$ F[T] then by Lemma 5.6.4. it follows that: I $\subset G_{ki}$ F[T] for a certain i $\in$ {1,...,n_k}. Now if P is a prime ideal of F[T] such that P $\bigcap$ S($\mathcal{G}$) = ϕ then there is a G $\in$ $\mathcal{G}$ such that P $\in$ GF[T]. Hence ht(P) $< \infty$ and then P is finitely generated in view of Propsition 5.6.3. For a prime ideal Q of F{$\mathcal{G}$} there exists a prime ideal P of F[T] such that P $\bigcap$ S($\mathcal{G}$) = ϕ and Q = PF[T]. Consequently, each prime ideal of F{$\mathcal{G}$} is finitely generated. because of Lemma 5.6.5., we may conclude that F{$\mathcal{G}$} is Noetherian.

Obviously, the map $G \to GF\{\mathcal{G}\}$ defines an injection from $\mathcal{G}$ to Spec $F\{\mathcal{G}\}$, therefore cl.Kdim $F\{\mathcal{G}\} \geq$ Kdim $\mathcal{G}$. □

5.6.7. Corollary. (Gulliksen, Gordon, Robson). For any ordinal α there exists a commutative Noetherian domain R such that cl.Kdim R $= \alpha$.

Proof. We have noted in the beginning of this section that there is a set T and a gang $\mathcal{G}$ on T such that Kdim $\mathcal{G} = \alpha$. From the theorem it follows that $F\{\mathcal{G}\}$ is Noetherian and cl.Kdim $F\{\mathcal{G}\} \geq \alpha$.
Equality will follow from Corollary 5.5.3. □

§5.7 Exercises.

(136) Let R be a ring with Gdim R $= \alpha$. Prove that for all $\beta \leq \alpha$, β a non-limit ordinal, there exists a β-simple module.

(137) Prove that a ring R with Gabriel dimension is a domain if and only if ${}_RR$ is Gabriel simple.

(138) Let $p \subsetneq q$ be prime ideals of a commutative ring with Gabriel dimension. Show that Gdim $R/q <$ Gdim R/p.

(139) Let M be an R-module for a commutative ring R. Show that the following statements are equivalent:

1. Gdim M $\leq \alpha$.
2. For any submodule $M' \subsetneq M$ we have that: Ass$(M/M') \neq \phi$ is in $(\text{Spec } R)_{\alpha-1}$ if α is a non-limit ordinal and in $(\text{Spec } R)_{\alpha}$ if α is a limit ordinal.

(140) If R is a semiprime ring with Gabriel dimension and finite coirreducible dimension then Gdim R is a non-limit ordinal.

(141) If R is a fully left bounded ring with left Krull dimension then Gdim R = Kdim R + 1.

(142) Let ${}_RM_S$ be an R-S-bimodule and let T = $\begin{pmatrix} R & M \\ O & S \end{pmatrix}$. If the rings R and S are both fully left bounded with Krull dimension and if ${}_RM$ has Krull dimension, then T is a fully left bounded ring with left Krull dimension.

(143) Let R be a left Noetherian rring and let $a_1, a_2, ..., a_n$ be central elements of R contained in J(R), then: Kdim R $\leq$ Kdim (R/Ra_1 + ... + Ra_n) + n.

(144) Let R be a commutative ring with cl.Kdim R = α + n, where n $\in$ IN and α is either 0 or a limit ordinal. Prove that R[X] has classcal Krull dimension and that α + n + 1 $\leq$ cl.Kdim R[X] $\leq$ α + 2n + 1.

(145) If M is agraded module over a $\mathbb{Z}$ -graded ring R, then every P $\in$ Ass(M) is a graded prime ideal.

(146) Let M be an A-module. Prove: $\mathrm{Ass}_{A[X]}$ (M[X]) = {P[X], P $\in$ Ass(M)}.

(147) A commutative ring R such that its lattice of ideals is totally ordered has Krull dimension if and only if it has Gabriel dimension.

(148) Let R be a valuation ring (cf. Bourbaki, Commutative Algebra, chap. 6) with valuation group $\mathbb{Z}^n$, n $\geq$ 1, ordered lexicographically. Show that Kdim R = n.

Bibliographical Comments to Chapter 5.

In the first section we have established that a nilideal in a ring with Krull dimension is nilpotent. For the proof of this we follow T. Lenagan, cf [4]. The result may be generalized (cf. R. Gordon, J.C. Robson [1],[3]) as follows: a nilsubring of a ring with Krull dimension is nilpotent.

For some important contributions to the study of rings with Krull dimension we referred to:

A. Goldie, L. Small [1]; R. Gordon, J.C. Robson [1],[2]; R. Gordon [1]; A.V. Jategaonkar [2]; T. Lenagan [1],[3],[4],[5].

Chapter 6

Krull Dimension of Noetherian Rings. The Principal Ideal Theorem.

§6.1 Fully Left Bounded Left Noetherian Rings.

In this and consequent sections we study some properties of fully left bounded and fully (left and right) bounded rings.

6.1.1. Theorem. (G. Krause [2]). For a left Noetherian ring R the following statements are equivalent:

1. R is fully left bounded.
2. R has enough prime ideals.

Proof. 1 ⇒ 2. cf. Theorem 5.5.6.
2 ⇒ 1. First we establish that property 2. is inherited by epimorphic images. If I is an ideal of R, let $\overline{Q}$ be an indecomposable injective R/I-module. Since $\overline{Q}$ is a uniform R-module, the injective hull E($\overline{Q}$) is again indecomposable.
If P ∈ Spec R is such that $\mathrm{Ass}_R(\mathrm{E}(\overline{Q})) = \{\mathrm{P}\}$ then $\mathrm{Ass}_{R/I}(\overline{Q}) = \{\mathrm{P}\}$. By the hypothesis: $E_R(\mathrm{R/P}) \cong Q_P^n$, where $Q_P \cong E_R(\overline{Q})$. Now $r_{E_R(R/P)}(\mathrm{I}) = \{ x \in E_R(\mathrm{R/P}), \mathrm{I}x = 0\}$ is the injective hull of R/P in R/I-mod. On the other hand: $r_{E_R(R/P)}(\mathrm{I}) \cong r_{Q_P^n}(\mathrm{I}) \simeq Q^n$. Thus $E_{R/I}(\mathrm{R/P}) \cong \overline{Q}^n$ and therefore R/I has enough prime ideals.

By these remarks we may reduce the problem to the case where R is a prime ring. Suppose that I is an essential left ideal, I $\neq$ R. Then R/I contains a critical submodule $M_1 = I_1/I$ such that Ann $M_1 = P_1$ and Ass $(M_1) = \{P_1\}$. If $I_1 \neq$ R then we proceed in the same way till we have obtained a chain I $\subset I_1 \subset I_2 \subset ... \subset I_n \subset ...$, such that I_k/I_{k-1} is a critical module, for very k ≥ 1, with Ann $M_k = P_k$, Ass $(M_k) = \{P_k\}$. Since R is left Noetherian, R $= I_n$ for some n. However: Kdim $M_k \leq$ Kdim $(R/I_{k-1}) <$ Kdim R. The hypotheses imply: $E(R/P_k) \cong Q^t_{pk}$, some t, where $Q_{P_k} \cong E(M_k)$. From this we derive that Kdim M_k = Kdim R/P_k and hence Kdim $R/P_k <$ Kdim R. Since R is assumed to be a prime ring, we must have $P_k \neq 0$ for all k ≥ 1. Hence: $P_n P_{n-1} \cdots P_1 \neq 0$ but $P_n P_{n-1} \cdots P_2 \subset$ I i.e. R is left bounded. □

Any commutative Noetherian ring is fully (left and right) bounded. More generally, any left Noetherian subring of a ring of matrices over a commutative ring can be shown to be fully left bounded. In particular any ring R which is finitely generated as a module over its Noetherian centre Z(R) is also a fully bounded ring. Any simple Noetherian ring which is not Artinian will definitely be not fully (left) bounded because it contains essential left ideals which cannot contain a nonzero ideal. Let us construct a rather natural example of such a ring.

Let k be a field of characteristic zero (!) and let $k[\underline{Y}]$ be the polynomial ring in the indeterminates $\underline{Y} = \{Y_1, ..., Y_n\}$. Let $\underline{\delta}$ be the set of derivations $\frac{\partial}{\partial Y_1}, ..., \frac{\partial}{\partial Y_n}$ and form the ring of differential polynomials: $k[\underline{Y}]\ [\underline{X},\underline{\delta}]$ where $X_iY_i - Y_iX_i = 1$, $X_jY_i - Y_iX_j = 0$ for i $\neq$ j, $X_iX_j - X_jX_i = 0$ for all i,j. It is not hard to verify the following relations:

$$[X_i, X_i^\alpha Y_i^\beta] = \beta\, X_i^\alpha Y_i^{\beta-1},\ [Y_i,\ X_i^\alpha Y_i^\beta] = -\alpha X_i^{\alpha-1} Y_i^\beta,$$

where [-,-] denotes the usual Lie-brackets. The ring $A_n(k)$ constructed above is called the nth-Weyl algebra; it is a filtered ring (by total degree in the X_i and the Y_i) with associated graded ring $k[X_1, ..., X_n, Y_1, ..., Y_n]$. Hence $A_n(k)$ is a Noetherian domain. Suppose I $\neq 0$ were an ideal of $A_1(k)$; say u $\neq 0 \in$ I, u $= \sum_{\alpha,\beta} U_{\alpha\beta} X_1^\alpha Y_1^\beta$ with $u_{\alpha\beta} \in$ k.
From $[X_1,u] \in$ I and $[Y_1,u] \in$ I it follows that I $\bigcap$ k $\neq 0$ (one uses char(k) = 0 here) and hence I = $A_1(k)$ or $A_1(k)$ is a simple ring. In a similar (inductive) way it can be shown that $A_n(k)$ is a simple Noetherian Ore domain, for each n $\in \mathbb{N}$.
Thus cl.Kdim $A_n(k) = 0$. Since $A_1(k)$ is obviously not Artinian (use the associated graded ring or direct computation to verify this) we also have Kdim $A_1(k) \neq 0$ (in fact a calculation of the global dimension (see later) will show that Kdim $A_n(k)$ = n). Clearly $A_n(k)$ is neither left nor right bounded.

§6.2 The Reduced Rank of a Module.

Unless otherwise stated *R will be a semiprime left Noetherian ring throughout this section.* Let S = S_{reg} be the set of regular elements of R and let Q be its classical ring of fractions, $S^{-1}R = Q$. By Goldie's theorems, Q is a semisimple Artinian ring. For a finitely generated left R-module M, $Q(M) = S^{-1}M$ is a finitely generated Q-module and therefore Q(M) has finite length as a Q-module. We define $\rho_R(M)$ to be the length $l_Q(Q(M))$ and refer to this number as the *reduced rank of M.*

6.2.1. Proposition. Let M be a finitely generated left R-module, then:

1. If N is a submodule of M then we have additivity for ρ, $\rho(M) = \rho(N) + \rho(M/N)$.
2. The R-module M is torsion if and only if $\rho(M) = 0$.

Proof. 1. By localisation we obtain the exact sequence of Q-modules:

$$0 \to Q(N) \to Q(M) \to Q(M/N) \to 0.$$

Applying Corollaries 2.4.4. yields the result.
2. Obvious, cf. Section 2.8. □

For any left Noetherian ring R the prime radical is nilpotent and R/rad R is semiprime left Noetherian. So if M is a finitely generated R-module, repeated application of the proposition yields:
$\rho_R(M) = \sum_{i=0}^{k-1} \rho_{R/rad\, R}\, [(\text{rad } R)^i\, M/(\text{rad } R)^{i+1}\, M]$, a number that may be regarded as the *reduced rank of* M over the left Noetherian (not necessarily semiprime) ring R, cf. A. Goldie [4].
If J is an ideal of R we put $\mathcal{C}(J) = \{\ a \in R,\ \overline{a}\ \in R/J$ is a regular element $\}$; in particular $\mathcal{C}$ is the set of regular elements of R.

6.2.2. Theorem. (Chatters a.o. [1]). Let M be a finitely generated R-module over a left Noetherian ring and let $N \subset M$ be a submodule, then:

1. $\rho_R(M) = \rho_R(N) + \rho_R(M/N)$.
2. $\rho_R(M) = 0$ if and only if for all $x \in M$ there exists an $a \in \mathcal{C}(\text{rad } R)$ such that $ax = 0$.

Proof. 1. Put rad R = I and let k be the smallest number such that $I^kM = 0$. The proof is by induction on k. For k = 1, the statement results from Proposition 6.2.1.. Let k > 1. Since $I^{k-1}(IM) = 0$ and $I^{k-1}(IM \bigcap N) = 0$, we apply the induction hypothesis: $\rho_R(IM) = \rho_R(IM \bigcap N) + \rho_R(IM/(N \bigcap IM))$, and
$\rho_R(N \bigcap IM) = \rho_R(N \bigcap IM/IN) + \rho_R(IN)$.
Therefore: $\rho_R(IM) = \rho_R(IM/N \bigcap IM) + \rho_R(N \bigcap IM/IN) + \rho_R(IN)$.
From Proposition 6.2.1. we then obtain: $\rho_R(M/IM) = \rho_R(M/IM + N) + \rho_R(IM + N/IM) = \rho_R(M/IM + N) + \rho_R(N/N \bigcap IM)$.
Envoking $\rho_R(M) = \rho_R(M/IM) + \rho_R(IM)$ we obtain:

$$\rho_R(M) = [\rho_R(M/IM + N) + \rho_R(IM + N/IM)] + [\rho_R(IM/N \bigcap IM) + \rho_R(N \bigcap IM/IN) + \rho_R(IN)] \qquad (*)$$

The definition of $\rho_R(M/N)$ implies: $\rho_R(M/N) = \rho_R(M/IM + N) + \rho_R(IM + N/N)$. Again using Proposition 6.2.1. and the fact that $I(N/IN) = 0$ we get: $\rho_R(N/IN) = \rho_R(N \bigcap IM/IN) + \rho_R(N/N \bigcap IM)$. In combination with (*) this leads to the relation: $\rho_R(M) = \rho_R(M/N) + \rho_R(N/IN) + \rho_R(IN)$. The definition of $\rho_R(N)$ then leads to $\rho_R(N) = \rho_R(N/IN) + \rho_R(IN)$ and finally to: $\rho_R(M) = \rho_R(M/N) + \rho_R(N)$.

2. If ∈ M and $\rho_R(M) = 0$ then $\rho_{R/I}(I^iM/I^{i+1}M) = 0$ for all $i \geq 0$. By the foregoing proposition there exist $a_1.a_2, ..., a_k \in \mathcal{C}(I)$ such that $a_i.a_{i-1}.....a_1x = 0$ in $I^{i-1}M/I^iM$, i.e. $a_ka_{h-1}...a_1x = 0$. Putting $a = a_ka_{k-1}...a_1$, then: $a \in \mathcal{C}(I)$, $ax = 0$. For the converse implication it suffices to apply Proposition 6.2.1. to obtain: $\rho_{R/I}(I^iM/I^{i+1}M) = 0$ for all $i \geq 0$, or $\rho_R(M) = 0$. □

The foregoing theorem yields an easy proof of a result of L. Small:

6.2.3. Theorem. Let R be a left Noetherian ring. If $\mathcal{C}(0)$ equals $\mathcal{C}(\text{rad } R)$ then R is a left order in a left Artinian ring.

Proof. Pick $a \in R$, $s \in \mathcal{C}(0)$. Since $R \cong Rs$ we see that $\rho_R(Rs) = \rho_R(R)$ and thus by Theorem 6.2.2. we obtain $\rho_R(R/Rs) = 0$. By the same theorem, there exists a $t \in \mathcal{C}(\text{rad } R) = \mathcal{C}(0)$ such that $ta \in Rs$. Hence $\mathcal{C}(0)$ satisfies the left Ore conditions and therefore we may consider the left ring of fractions $Q = \mathcal{C}(0)^{-1}R$. If $J \subsetneq I$ are left ideals of Q then $J \bigcap R \subsetneq I \bigcap R$ and these are $\mathcal{C}(0)$-saturated (i.e. if for $r \in R$, $sr \in J \bigcap R$ with $s \in \mathcal{C}(0)$, then $r \in J \bigcap R$).
Therefore $\rho_R(I \bigcap R/J \bigcap R) \neq 0$ or $\rho_R(I \bigcap R) > \rho_R(J \bigcap R)$. It is now evident that this leads to the conclusion that Q is a left Artinian ring. □

6.2.4. Remark. The converse of Theorem 6.2.3. is valid too: if R is a left order in a left Artinian ring then $\mathcal{C}(0) = \mathcal{C}(\text{rad R})$.

§6.3 Noetherian Rings Satisfying Condition H.

A left Noetherian ring R is said to be an *H-ring*, or to *satisfy condition H*, if for every finitely generated R-module M there exist $x_1, ..., x_n \in$ M such that Ann(M) = Ann $\{x_1, ..., x_n\}$.

Clearly, a commutative Noetherian ring R is an H-ring. More generally every R-algebra S which is finitely generated as an R-module will be a Noetherian H-ring. Indeed, let M be a finitely generated S-module and take an x ∈ M. Put $M_x = \{y \in M, l_S(x).y = 0\}$. Obviously, M_x contains x and it is an R-submodule of M; moreover it is also clear that $l_S(x) = l_S(M_x)$. If x,y ∈ M then $l_S(M_x + M_y) = l_S(M_x) \bigcap l_S(M_y) = l_S(x) \bigcap l_S(y)$. Since M is a Noetherian R-module there exist finitely many $x_1, ..., x_n \in$ M such that: $M = M_{x_1} + ... + M_{x_n}$. Then: $Ann_S(M) = \bigcap_{i=1}^n l_S(M_{x_i}) = \bigcap_{i=1}^n l_S(x_i)$ and consequently, S is an H-ring.

Perhaps the choice of the simple examples given bove suggests a relation between the class of H-rings and the class of fully left bounded rings. This relation does exist and it is the topic of the following results.

6.3.1. Proposition. An H-ring is fully left bounded.

Proof. Since it is clear that epimorphic images of H-rings are H-rings we may assume that R is a prime ring. Let L be an essential left ideal of R and put M = R/L. By the hypothesis: Ann(M) = $\bigcap_{i=1}^n l_R(x_i)$ for some finite set $\{x_1, ..., x_n\} \subset$ M. If a_i represents x_i in R, i = 1,...,n, then $l_R(x_i) = (L:a_i)$. The latter is an essential left ideal since L is such and therefore Ann(M) is also essential hence a nonzero ideal. Because Ann(M) ⊂ L it follows that R is bounded. □

Next we provide some results that will eventually lead (in Theorem 6.3.6.) to the complete equivalence of the notions of H-rings and left Noetherian fully bounded rings.

6.3.2. Lemma. Let R be a prime left Noetherian fully bounded ring. If M is a nonzero finitely generated R-module then $\rho_R(M) = 0$ implies that $0 \notin \text{Ass}(M)$.

Proof. For $P \in \text{Ass}(M)$ there exists a nonzero submodule $N \subset M$ such that $P = \text{Ann}(N) = \text{Ann}(N')$ for all nonzero submodules N' of N. Pick $x \neq 0$ in N. Since $\rho_R(M) = 0$, $\rho_R(N) = 0$ and hence there exists a regular element $c \in R$ such that $cx = 0$ (cf. Proposition 6.2.1.). Since Rc is essential, and using the assumption that R is fully left bounded, it follows that $Ix = 0$ for some nonzero ideal I of R contained in Rc. Hence $I(Rx) = 0$ or $I \subset P$ and therefore $P \neq 0$. □

For any R-module M, let Z(M) be the singular radical of M. We say that M is *singular* if $Z(M) = M$ and we say that M is non-singular when $Z(M) = 0$.

6.3.3. Proposition. Let R be a semiprime left Noetherian ring and let M be an R-module, then:

1. If M is uniform then it is either singular or non-singular.
2. If M is finitely generated and singular while R is a fully left bounded prime ring then M cannot be faithful.

Proof. 1. Suppose that $Z(M) \neq 0$, then $M/Z(M)$ is singular. If $x \in M$ then $(Z(M) : x)$ is an essential left ideal hence there exists a regular element $c \in R$ such that $c \in (Z(M) : x)$, i.e. $cx \in Z(M)$. Since $l_R(cx)$ is essential, there is a regular element d in R such that $d(cx) = 0$. From $dc \in l_R(x)$ we may conclude $x \in Z(M)$.
2. Let $x_1, ..., x_n$ be generators of M. Since R is left bounded and each $l_R(x_k)$ being essential we obtain that $l_R(x_k)$ contains a nonzero ideal I_k, $k = 1,...,n$. Now $I_k x_k = 0$ entails $I_k(Rx_k) = 0$. Put $I = \bigcap_{k=1}^n I_k$; then $I \neq 0$ since R is prime. On the other hand $I(Rx_k) = 0$ for all $k = 1,...,n$, yields $I \subset \text{Ann } M$. □

6.3.4. Corollary. If R is a left Noetherian fully left bounded prime ring and M is a faithful finitely generated uniform R-module, then $\text{Ass}(M) = \{(0)\}$.

Proof. By the foregoing proposition it follows that M is nonsingular. Pick $x \neq 0$ in M. Then $l_R(x)$ is not an essential left ideal and there exists a nonzero left ideal K such that K $\bigcap$ $l_R(x) = 0$. So we arrive at a monomorphism

$$0 \longrightarrow K \xrightarrow{\alpha} M.$$

Since M is uniform we obtain: Ass(K) = Ass(M) and thus Ass(K) = {(0)}. □

6.3.5. Corollary. f M is a faithful finitely generated uniform R-module and R is a left Noetherian fully left bounded ring then $r_R(P)$ is an essential left ideal, where {P} = Ass(M).

Proof. If I is a nonzero left ideal and Q ∈ Ass(I) then there exists a nonzero left ideal J ⊂ I such that Q = Ann(J). From JM ≠ 0, Ass(JM) = {P} follows and Q ⊂ Ann(JM). If λ ∈ Ann(JM) then λJM = 0 entails λJ = 0, or λ ∈ Q. This establishes Q = Ann (JM) and this in turn entails that JM is a faithful uniform R/Q-module. In view of Corollary 5.4.5. we may conclude that Ass(JM) = {Q}. It follows that Q = P, PJ = 0 and J ⊂ $r_R(P)$. Therefore $r_R(P) \bigcap I \neq 0$ and we have proved that $r_R(P)$ is an essential left ideal. □

We are now ready to give the proof of:

6.3.6. Theorem (G. Cauchon [1]). For a left Noetherian ring the following assertions are equivalent:

1. R is fully left bounded.
2. R has enough prime ideals.
3. R is an H-ring.

Proof. 1.⇔2. cf. Theorem 6.1.1.
3.⇒1. cf. Proposition 6.3.1.
1.⇒3. Let M be a finitely generated R-module. Since M is (left) Noetherian it has finite Goldie (uniform) dimension and consequently there exist irreducible submodules $N_1, ..., N_r$ of M such that $0 = N_1 \bigcap ... \bigcap N_r$ and also a canonical monomorphism $M \rightarrow \bigoplus_{i=1}^r M/N_i$. Put $M_i = M/N_i$; then Ann M = $\bigcap_{i=1}^r$ Ann M_i and the M_i are uniform. Suppose, for each i, that there exist $y_{i_1}, ..., y_{i_m} \in M_i$ such that Ann $(M_i) = \bigcap_{t=1}^m l_R(y_{i_t})$ (we may assume that m is the same for all i). Let $x_{i_t} \in M$ represent y_{i_t}, then Ann M = $\bigcap_{i=1}^r \bigcap_{t=1}^m l_R(x_{i_t})$.

This enables us to reduce the problem to the situation where M is uniform itself, and up to replacing R by R/Ann M we may also ssume that M is a faithful R-module. Let Ass (M) = {P}. By the foregoing corollary $r_R(P)$ is an essential left ideal. Pick $x_1, ..., x_n \in M$ such that the number s, s = $\rho_{R/P}$ $(r_R(P) \bigcap l_R(x_1, ..., x_n))$, is minimal. This means that s = $\rho_{R/P}$ $(r_R(P)) \bigcap l_R(x_1, ..., x_n, x)$ holds for every x $\in$ M.

From Theorem 6.2.2 we retain:

$\rho_{R/P}$ $(r_R(P) \bigcap l_R(x_1, ..., x_n) \,/\, r_R(P) \bigcap l_R(x_1, x_n, x)) = 0$. Now the module $(r_R(P) \bigcap l_R(x_1, ..., x_n))x = N$ is isomorphic to a quotient module of $r_R(P) \bigcap l_R(x_1, ..., x_n) \,/\, r_R(P) \bigcap l_R(x_1, ..., x_n, x)$ and hence $\rho_{R/P}(N) = 0$.

If N $\neq 0$ then application of Lemma 6.3.2. learns that P $\notin$ Ass (N). However, from N $\subset$ M it follows that Ass (N) = Ass (M) = {P}, contradiction.

Hence N = 0 and therefore: $r_R(P) \bigcap l_R(x_1, ..., x_n) \subset \text{Ann}(M) = 0$. Finally, $l_R(x_1, ..., x_n) = 0$ may be derived from the fact that $r_R(P)$ is an essential left ideal.

□

6.3.7. Corollary. For a left Noetherian fully bounded ring and a faithful finitely generated R-module M we have Kdim M = Kdim R.

Proof. The theorem establishes that Ann (M) = $l_R(x_1) \bigcap ... \bigcap l_R(x_n)$, for certain $x_1, ..., x_n \in M$, hence there exists a monomorphism:

$$0 \to R \to R/l_R(x_1) \oplus ... \oplus R/l_R(x_n) \cong Rx_1 \oplus ... \oplus Rx_n.$$

Therefore Kdim R $\leq$ Kdim M. The converse inequality holds, so the claim follows.

□

6.3.8. Lemma. If R is left Noetherian then Z(R) is nilpotent.

Proof. Put I = Z(R). There is an n $\in \mathbb{N}$ such that $l(A^n) = l(A^{n+1}) = ...$. If $A_{n+1} \neq 0$, say $aA^n \neq 0$ for some a$\in A$ and we may assume that a is such that $l_R(a)$ is maximal. For any b$\in A$, $l_R(b)$ is essential hence there is a c $\in$ R such that ca $\neq$ 0 and cab = 0. Since $l_R(a) \subset l_R(ab)$ and c $\notin l_R(a)$ it follows that $l_R(a) \subsetneq l_R(ab)$ and thus $abA^n = 0$. In the latter equality b is arbitrary, hence $aA^{n+1} = 0$ or a $\in$ $l(A^n) = l(A^{n+1})$. Obviously this leads to a contradiction:$aA^n = 0$. □

6.3.9. Theorem. Let R be a Noetherian (left and right) ring which is fully left bounded, then: $\bigcap_{n=1}^{\infty} J(R)^n = 0$.

Proof. Consider a $\in$ R and let I_a be the left ideal maximal with respect to a $\notin I_a$. Obviously $\bigcap_{a\in R} I_a = 0$. Put $I_a^* = \text{Ann}_R\ (R/I_a)$. Then I_a^* is an ideal which is maximal with respect to the property of being contained in I_a. By definition, I_a is irreducible i.e. R/I_a is uniform. Moreover $Ra + I_a$ is the intersection of all left ideals of R strictly containing I_a, thus $Ra + I_a/I_a$ is the socle of R/I_a. Since R/I_a is uniform, the socle is a simple R-module and thus its annihilator P is a prime ideal. Using the properties of an H-ring (and evoking Theorem 6.3.6.) we may conclude that R/P is simple Artinian and that there exist $b_1, ..., b_n \in R$ such that $I_a^* = (I_a : b_1) \bigcap ... \bigcap (I_a : b_n)$. The monomorphisms $R/(I_a : b_i) \to R/I_a$ may be combined into a monomorphism $R/I_a^* \to R/I_a$ and therefore $\text{Ass}\ (R/I_a^*) = \{P\}$. The latter entails that the ideal $K = \{t \in R/I_a^*,\ Pt = 0\} = r_{R/I_a^*}(P)$ is an essential left ideal of R/I_a^* (Corollary 5.4.5.). Because R/P is simple Artinian, K is the left socle of T and because K is a left R/I_a^*-module of finite length we must have that $K\ J(R/I_a^*)^n = K\ J(R/I_a^*)^{n+1} = ...$, for some $n \in \mathbb{N}$. Applying Nakayama's lemma yields $K\ J(R/I_a^*)^n = 0$ i.e. $J(R/I_a^*)^n$ is in the singular radical $Z(R/I_a^*)$. By the lemma it follows that $J(R/I_a^*)$ is nilpotent i.e. $J(R)^m \subset I_a^* \subset I_a$ for some m. We have established that for all a $\in$ R, $\bigcap_{n\geq 1} J(R)^n \subset I_a$, hence $\bigcap_{n=1}^{\infty} J(R)^n = 0$.
□

The two-sided Noetherian condition in the foregoing theorem is essential; this may be verified by looking at the following example. Let $\mathbb{Z}_{(p)}$ be the local ring at the prime number $p \in \mathbb{Z}$. The ring $R = \begin{pmatrix} \mathbb{Q} & \mathbb{Q} \\ 0 & \mathbb{Z}_{(p)} \end{pmatrix}$ is a left Noetherian fully left bounded ring. It is not hard to see that $J(R) = \begin{pmatrix} 0 & \mathbb{Q} \\ 0 & p\mathbb{Z}_{(p)} \end{pmatrix}$ and $\begin{pmatrix} 0 & \mathbb{Q} \\ 0 & 0 \end{pmatrix}$ are in $\bigcap_{n=1}^{\infty} J(R)^n \neq 0$.

§6.4 Fully Bounded Noetherian Rings.

By a fully bounded Noetherian ring we mean a ring satisfying both the left and right conditions. For a fully bounded ring the left Krull dimension equals the right Krull dimension and it also equals the classical Krull dimension. The content of this section is based on Jategaonkar's results in [4].

6.4.1. Lemma. Consider rings R and S together with an R-S-bimodule ${}_RM_S$. Suppose that R is left Noetherian fully left bounded and that ${}_RM$ is finitely generated. Then, Kdim_R M equals the Krull dimension of the lattice L(${}_RM_S$) of all subbimodules of ${}_RM_S$.

Proof. Put μ(M) = Kdim L(${}_RM_S$). Proposition 3.1.4. yields that μ(M) $\leq$ kdim_R (M). We establish the equality by induction on μ(M). If μ(M) = -1 then M = 0 and Kdim_R M =-1.
Assume that the lemma holds for all R-S-bimodules ${}_RL_S$ such that ${}_RL$ is finitely generated and μ(L) < μ(M) = α. Suppose μ(M) < Kdim_R M i.e. there exists an R-submodule N of M such that Kdim_R (M/N) $\not<$ α. Since ${}_RM$ is Noetherian we may select an R-submodule N_1 of M maximal with respect to the property Kdim (M/N_1) $\not<$ α. If Kdim (M/N_1) > α then there exists a strictly descending chain, $(P_i/N_1)_{i\geq 1}$, of submodules of M/N_1 such that Kdim_R (P_i/P_{i+1}) $\geq$ α for all i $\geq$ 1 (cf. Proposition 3.1.5.). Then Kdim_R (M/P_i) $\geq$ α for some $P_i \supsetneq N_1$ would contradict the choice of N_1, i.e. Kdim (M/N_1) = α. Put I = Ann_R (M/N) and M_1 = IM. Clearly: I is an ideal and M_1 $\in$ L(${}_RM_S$). It is equally obvious that both M/N_1 and M/M_1 are faithful R/I-modules. Corollary 6.3.7. entails:
Kdim_R (M/M_1) = Kdim (R/I) = Kdim (M/N_1) = α. By the induction hypothesis we arrive at μ(M/M_1) = α.
On the other hand, Kdim_R (M/M_1) = α and Kdim_R (M) > α imply that Kdim_R (M_1) > α. Also, $\mu(M_1)$ $\leq$ μ(M) = α.
The assumption $\mu(M_1)$ < α contradicts the induction hypothesis, therefore $\mu(M_1)$ = α and we have found a submodule M_1 of M such that $\mu(M_1)$ = μ(M/M_1) = α < Kdim_R M. Starting the same procedure at M_1 instead of M we arrive at a strictly descending chain: M = M_0 $\supset$ M_1 $\supset \dots \supset$ M_n $\supset \dots$, of R-S-subbimodules of M such that $\mu(M_i/M_{i-1}) = \alpha$ for all i. This entails that μ(M) > α, a contradiction.
□

6.4.2. Theorem. Let R and S be Noetherian fully bounded rings. If there eists an R-S-bimodule ${}_RM_S$ such that ${}_RM$ is faithful and finitely generated while M_S is also faithful and finitely generated then Kdim R = Kdim S.

Proof. Apply the lemma and Corollary 6.3.7. □

6.4.3. Lemma. Let R be a Noetherian fully bounded ring. Consider prime ideals p and q of R and a finitely generated uniform R-module M and a proper submodule N such that $p = l_R(N)$, $N = r_M(p)$, $q\,M \subset N$ and such that M/N is a nonsingular R/q-module. Then Kdim R/p = Kdim R/q.

Proof. Since N is a faithful uniform R/p-module it is nonsingular (cf. Proposition 6.3.3.), hence each nonzero submodule of N is a faithful R/p-module. Let $M' \subset M$ be such that $pM' \neq 0$ i.e. $M' \not\subset N$. Obviously, $r_{M'}(p) = N \bigcap M' \subsetneq M'$.

We obtain: $p = \text{Ann}\,(N \bigcap M')$ and $qM' \subset M' \bigcap N$.

From $M'/N \bigcap M' \cong M' + N/N \hookrightarrow M/N$ it follows that $M'/N \bigcap M'$ is a nonsingular R/q-module. Clearly: $\text{Ann}_R(M') \supset \text{Ann}_R(M)$.

Consider a submodule M' of M such that $pM' = 0$ and $\text{Ann}_R(M')$ maximal with respect to this property. Up to replacing M by M' we may suppose from the start that $\text{Ann}_R(M) = \text{Ann}_R(M')$ for any submodule M' of M with the property $pM' \neq 0$. Then put $I = \text{Ann}_R(M)$. Obviously $I \subset p$. If we had $I \not\subset q$ then $I(M/N) = 0$ entails that M/N cannot be a singular R/q-module. Therefore we must have $I \subset q$, or $I \subset p \bigcap q$. Suppose that $I = p \bigcap q$. Then, from $pM \neq 0$, we obtain that

$pM \bigcap N \neq 0$. But $q(pM \bigcap N) = 0$ because $qp \subset I$. Since $pM \bigcap N$ is necessarily a faithful R/p-module, thus $q \subset p$ and $I = q$ follows. Consequently M is a uniform faithful R/q-module and as such it is nonsingular. So any nonzero submodule of M must be a faithful R/q-module and in particular $q = \text{Ann}\,(N) = p$ or $I = p$. The latter leads to $pM = 0$, a contradiction.

Hence we have to assume that $I \subsetneq p \cap q$. Then put $X = (p \bigcap q)/I$ which is clearly an R-bimodule. Let $\underline{a}$ be the annihilator of X considered as a left R-module, $\underline{b}$ the annihilator of X considered as a right R-module. From $pq \subset I$ we obtain that $p \subset \underline{a}$, $q \subset \underline{b}$. Now $\underline{a}(p \cap q) \subset I$ yields $\underline{a}(p \cap q)\,M = 0$ while $(p \cap q)\,M \neq 0$ also yields

$(p \cap q)\,M \bigcap N \neq 0$. However the latter submodule is a faithful R/p-module, i.e.

$\underline{a} \subset p$ and thus $\underline{a} = p$. Similarly, we prove

$(p \cap q)\underline{b}\,M = 0$ and if $\underline{b}M \not\subset N$ then $I = \text{Ann}(\underline{b}M)$ leads to the contradiction

$p \cap q \subset I$. So we must have $\underline{b}M \subset N$, i.e. $\underline{b}(M/N) = 0$. The latter yields $\underline{b} \subset q$ nd $\underline{b} = q$.

It follows that X is a faithful R/p-R/q-bimodule so by Theorem 6.4.2. the result now follows. □

6.4.4. Corollary. Let M be a finitely generated left module over a fully bounded Noetherian ring R. Then M is a critical module if and only if $\text{Ann}_R\,M$ is a prime ideal and M is a uniform nonsingular $R/\text{Ann}_R M$-module.

Proof. First assume the latter condition, then we have: Kdim_R M = Kdim R/Ann_R M = α (cf. Corollary 6.3.7.). For a nonzero M' ⊂ M we see that M/M' is singular as M' is esential in M. From Corollary 5.2.0. we retain that Kdim M/M' < α and therefore M is α-critical.

Conversely, suppose that M is α-critical. Put Ass (M) = {p} and N = r_M(p). Clearly, l_R(N) = p and thus N is a uniform faithful R/p-module, i.e. N is a nonsingular R/p-module and thus Kdim_R N= Kdim R/p. If M/N ≠ 0 we let

g∈Ass(M/N), i.e. there is a submodule L/N ⊂ M/N such that q = Ann_R (L/N) and L/N is a critical module. Consequently L/N is a nonsingular R/q-module in view of Proposition 6.3.3.. On the other hand: Kdim R/q = Kdim L/N < Kdim M ≤ Kdim N = Kdim R/p. Lemma 6.4.3. leads to the contradiction Kdim R/p = Kdim R/q. So we have M/N = 0 and thus Ann M = p. □

6.4.5. Corollary. Let M be a finitely generated left module over a fully bounded Noetherian ring. Let $0 = M_0 \subset M_1 \subset \ldots \subset M_n$ = M be a critical composition series for M, α_i = Kdim_R (M_i/M_{i+1}). Then M contains an α_i-critical submodule for all i = 1,...,n.

Conversely, if M contains an α-critical submodule then there is an i, 1 ≤ i ≤ n, such that $\alpha = \alpha_i$.

Proof. We have $\alpha_1 \leq \alpha_2 \leq \ldots \leq \alpha_n$. The assertion that M contains an α_i-critical submodule for all i = 1,...,n holds for n = 1. Suppose that n > 1 and proceed by induction. By the induction hypothesis, M/M_1 contains an α_i-critical submodule for all i = 1,...,n. Let $L_i/M_1 \subset M/M_1$ be an α_i-critical submodule, where i ≥ 2. Clearly we may suppose that $\alpha_i > \alpha_1$. By Corolary 6.4.4.:

p = $\mathrm{Ann}_R(M_1)$ and q = Ann_R (L_i/M_1) are prime ideals.

Thus Kdim_R M = Kdim R/p and Kdim_R (L_i/M_1) = Kdim R/q and this implies Kdim R/p < Kdim R/q or $r_{L_i}(\mathrm{p}) = M_1$. Lemma 6.4.2. yields that L_i cannot be uniform, so M_1 cannot be an essential submodule in L_i and there must exist a nonzero submodule $N_1 \subset L_i$ such that $M_1 \bigcap N_1 = 0$. Hence there exists a monomorphism $0 \rightarrow N_1 \rightarrow L_i/M_1$ and thus N_1 is α_i-critical. Consequently, M contains an α_i-critical submodule.

For the second statement we start from the assumption that M contains an α-critical submodule of M, N say. Let k be the largest integer, 0 ≤ k ≤ n such that N $\bigcap M_k$ = 0. Thus N $\bigcap M_{k+1} \neq 0$, i.e. N contains a nonzero N' which is isomorphic to a submodule of M_{k+}/M_k. Thus α = Kdim N = Kdim N' = Kdim $M_{k+1}/M_k = \alpha_{k+1}$, what proves the claim. □

6.4.6. Corollary. Let M be a finitely generated left module over a Noetherian fully bounded ring. Let $0 = M_0 \subset M_1 \subset ... \subset M_n$ = M be a critical composition series such that Kdim $M_i/M_{i-1} = \alpha$, for all i = 1,...,n. If E is a finitely generated R-module such that it is an essential extension of M then every critical composition series of E has the property that its factors are α-critical.

Proof. Direct from Corollary 6.4.5. □

6.4.7. Corollary. Let M be a finitely generated left module over a fully bounded ring R. If s(M) is an essential submodule then M has finite length.

6.4.8. Lemma. Let R be a semiprime Noetherian ring and let M be a uniform nonsingular finitely generated left R-module then M is isomorphic to a left ideal of R.

Proof. Let Q be the classical left (and right) ring of fractions of R, i.e. Q = S^{-1}R, S the set of (left) regular elements of R. Let φ_M: M $\rightarrow$ Q(M) = S^{-1}M be the canonical morphism. Since M is nonsingular we have Ker $\varphi_M = 0$. Since M is uniform, Q(M) is a uniform Q-module and thus Q(M) is a simple Q-module (because Q is a semisimple Artinian ring) and also it is isomorphic to a minimal left ideal of Q. This allows to assume from the start that M is a finitely generated left R-submodule of Q. Since Q is the right ring of fractions too, there exists an s $\in$ S such that Ms $\subset$ R. The map φ: M $\rightarrow$ ${}_RR$, $\mapsto$ xs, is clearly a monomorphism.

□

6.4.9. Theorem. Let R be a Noetherian fully bounded ring. A finitely left R-module M is critical if and only if M is a compressible module.

Proof. If M is compressible then it is critical because of Proposition 6.5.7.. Conversely, suppose that M is critrical. Corollary 6.4.4. and Lemma 6.4.8. entail that M is isomorphic to a critical left ideal of R/Ann (M). However, a critical left ideal is necessarily compressible. □

Recall that a left R-module M is said to be *cotertiary* if Ass(M) consists of one element only and a submodule L of M is *tertiary* if M/L is cotertiary. By the results of section 5.4. every uniform module is cotertiary and irreducible submodules are tertiary.
If M is finitely generated then we say that a submodule N of M has a *tertiary decomposition* if $N = M_1 \cap ... \cap M_n$, where each M_i is tertiary, such that Ass $(M/M_i) \neq$ Ass (M/M_j) for $i \neq j$ and such that the intersection is irredundant.

6.4.10. Proposition. (Lesieur, Croisot [1]). Every submodule N of a finitely generated R-module M has a tertiary decomposition. If $N = M_1 \cap ... \cap M_n = N_1 \cap ... \cap N_n$ are tertiary decompositions of N then m = n and the sets of associated prime ideals {Ass (M/M_i) i = 1,...,m} and {Ass (M/N_j), j = 1,...,n} coincide.

Proof. It is clear that N can be written as an irredundant intersection $N = M_1 \cap ... \cap M_n$ of irreducible submodules an thus we may take M_P to be the intersection of the M_i for which Ass $(M/M_i) = \{P\}$. Then $N = \bigcap M_p$ is the desired decomposition. Consider the monomorphism $M/N \to M/M_1 \oplus ... \oplus M/M_m$. By the irredundancy hypothesis, M/N intersects each M/M_i non-trivially, in L_i say. From the existence of the monomorphisms:
$L_1 \oplus ... \oplus L_m \to M/N \to M/M_1 \oplus ... \oplus M/M_m$ we obtain
$\bigcup_i$ Ass $(L_i) \subset$ Ass $(M/N) \subset \bigcup_i$ Ass (M/M_i). Since Ass $(L_i) \neq \phi$ it follows that Ass $(L_i) =$ Ass (M/M_i) and then also: Ass $(M/N) = \bigcup_i$ Ass (M/M_i). □

Recall that a submodule N of an R-module M is said to be *primary* if it is tertiary and $P^nM \subset N$, where $\{P\} =$ Ass (M/N), for some $n \in \mathbb{N}$. Primary decompositions are defined just like the tertiary decompositions. Obviously, a left ideal I will be primary if and only if it is tertiary and it contains a power of the associated prime ideal i.e. if and only if $aRb \subset I$ with $b \notin I$ yields $a \in$ rad (I) (the intersection of the prime ideals containing I). It is equally evident that Ass (R/I) equals rad(I) whenever I is primary.

Let R be any ring; a finitely generated R-module M is an *Artin-Rees module* if for every ideal J of R and every submodule N of M the following property holds: for every $n \in \mathbb{N}$ there exists an $m \in \mathbb{Z}$ such that $J^mM \cap N \subset J^nN$. An ideal I of R is said to satisfy the *left Artin-Rees property* if for any finitely generated left R-module M and any submodule N of M, and any $n \in \mathbb{N}$, there exists an $m \in \mathbb{N}$ such that $I^mM \cap N \subset I^nN$.

6.4.11. Lemma. Let M be a finitely generated left R-module with submodules L $\subset$ N $\subset$ M. There exists a submodule M' of M such that N $\bigcap$ M' = L and Ass(N/L) = Ass(M/M').

Proof. If L = N then we may take M' = M, so let us assume that N $\neq$ L. Decompose L tertiary in M as L = $M_1 \bigcap ... \bigcap M_n$. Whenever $M_i \bigcap$ N $\neq$ N, $M_i \bigcap$ N is tertiary in N and we obtain a tertiary decomposition in N : L = $(M_1 \bigcap N) \bigcap ... \bigcap (M_r \bigcap N)$, up to reordering. Put M' = $M_1 \bigcap ... \bigcap M_r$. Then L = M' $\bigcap$ N and Ass(M/M') = $\bigcup_{i=1}^r$ Ass(M/M_i) = $\bigcap_{i=1}^r$ Ass(N/$M_i \bigcap$ N) = Ass(N/L). □

6.4.12. Proposition. A finitely generated left R-module M is an Artin-Rees module if and only if the tertiary submodules are primary.

Proof. First assume that M satisfies the Artin-Rees condition and let N be a tertiary submodule with P = Ass(M/N). Putting M' = $\{m \in M, Pm \subset N\}$ we obtain that P = Ann(M'/N). Since Ass(L/N) = {P} for every nonzero L/N in M/N, the maximality of M' with respect to the property P = Ann(M'/N) entails that M'/N is essential in M/N. The Artin-Rees property yields: $P^nM \bigcap M' \subset PM' \subset N$ for some n $\in \mathbb{N}$, but then $P^nM \subset N$ by essentiality of M'/N in M/N. In conclusion N is primary.
Conversely, suppose that submodules of M which are tertiary are necessarily primary and let I be any ideal of R, N a submodule of M, and n $\in \mathbb{N}$. Put L = I^nN. Since I^n annihilates N/L we have I $\subset$ P for every P $\in$ Ass(N/L). By the lemma we may find a submodule M' of M such that L = N $\bigcap$ M' and Ass(N/L) = Ass(M/M'). Put M' = $M_1 \bigcap ... \bigcap M_r$, a tertiary decomposition of M' in M and let Ass(M/M_i) = $\{P_i\}$. Then $P_i \in$ Ass(M/M') = Ass(N/L) and I $\subset P_i$ by the foregoing remark. Since M_i is primary in M by hypothesis we arrive at $P_i^tM \subset M_i$ for some t. Hence there is an integer t such that $I^tM \subset \bigcap_{i=1}^r P_i^tM \subset \bigcap_{i=1}^r M_i$ = M' and so consequently: $I^tM \bigcap N \subset M' \bigcap N = L = I^nN$ establishing the Artin-Rees property. □

The following theorem is an easy combination of the foregoing results:

6.4.13. Theorem. Let R be a left Noetherian ring. The following statements are equivalent:

1. Every irreducible left ideal is primary.
2. Every tertiary submodule of a finitely generated left R-module is primary in M.
3. R is an Artin-Rees module as a left R-module.
4. All finitely generated left R-modules are Artin-Rees modules.
5. All ideals of R satisfy the left Artin-Rees property.
6. If E is an indecomposable injective left R-module with associated prime ideal P then for every $x \in E$, $P^n x = 0$ for some number n.

A ring satisfying one of the conditions in the theorem is called a *left classical ring*. Commutative Noetherian rings as well as Azumaya algebas over these are (left) classical rings.

6.4.14. Theorem. (Krull Intersection Theorem). Let R be a left classical ring, I an ideal of R and M a finitely generated left R-module, then $\bigcap_n I^n M =$ $\{m \in M, (1\text{-}a)m = 0 \text{ for some } a \in I\}$.

Proof. If $m = ma$ for some $a \in I$ then $= ma^t$ for all $t \in \mathbb{N}$ and thus $m \in \bigcap_n I^n M$. On the other hand, for an $m \in \bigcap_n I^n M$ we obtain $Rm \subset I^n M$ for all $n \in \mathbb{N}$. By the Artin-Rees property it follows that the I-adic topology of Rm is trivial i.e. $Im = Rm$ and consequently there is an $a \in I$ such that $m = am$. □

6.4.15. Theorem. For a semilocal fully bounded Noetherian ring R, J(R) has the (left) Artin-Rees property.

Proof. Let N be a submodule of a finitely generated left R-module. Consider $M' \subset M$ which is maximal with respect to the property $M' \bigcap N = J(R)^n N$ for some given $n \in \mathbb{N}$. Then M/M' is an essential extension of $N/J(R)^n$ with $J(R)^n (N/J(R)^n N) = 0$. Now R is semilocal, hence $R/J(R)$ is a semisimple Artinian ring, and therefore $N/J(R)^n N$ is a left R-module of finite length. By Corollary 6.4.7., M/M' has finite length and so there is a $t \in \mathbb{N}$ such that $J(R)^t (M/M') = 0$. This leads to: $J(R)^t M \bigcap N \subset J(R)^n N$, i.e. the Artin-Rees property holds for J(R). □

§6.5 Krull Dimension and Invertible Ideals in a Noetherian Ring.

Let R be a ring and I an ideal of R. Consider an overring T of R such that there is an R-submodule J of T such that IJ = JI = R; then I is said to be invertible (in T) and we write $J = I^{-1}$. It is not hard to see that for a prime left Noetherian ring R every ideal I is such that I^{-1} may be viewed as a subbimodule of Q, the classical ring of fractions of R.

For example, if $a \neq 0$ is normalizing in R, i.e. aR = Ra, then I = aR contains a regular element s because a prime left Noetherian ring is a prime Goldie ring; in particular a is regular and I is invertible with $I^{-1} = Ra^{-1} \subset Q$.

Throughout this section I will be an invertible ideal of R and we may consider $\bigcup_{n=0}^{\infty} I^{-n}$ as the overring T.

6.5.1. Lemma. If I is an invertible ideal of a left Noetherian ring R then $T = \bigcup_{n=0}^{\infty} I^{-n}$ is left Noetherian and Kdim T $\leq$ Kdim R.

Proof. The ring $R(I) = \bigoplus_{n\in\mathbb{Z}} I^n$ is strongly $\mathbb{Z}$ -graded and T is an epimorphic image of R(I). Since R is left Noetherian, R(I) is left Noetherian too (2.6), and so is T. Let L be a left ideal of T, then $L = T(L \bigcap R)$. Indeed, $T(L \bigcap R) \subset L$ is obvious and for the converse, if $a \in L$, then $a \in I^{-n}$ for some $n \geq 0$ and thus $I^n a \subset L \bigcap R$, yielding $a \in I^{-n}(I^n a) \subset T(L \bigcap R)$.
The mapping $L \to L \bigcap R$ is strictly increasing $L_s(T) \to L_s(R)$ an therefore Kdim T $\leq$ Kdim R. □

6.5.2. Lemma. If L is a left ideal in the left Noetherian ring R then $I^{2n} \bigcap L \subset I^n L$ for some $n > 0$ (i.e. I satisfies the Artin-Rees property) and in particular $I^m \bigcap L \subset IL$ for some $m \in \mathbb{Z}$.

Proof. The ascending chain $L \subset I^{-1}(I \bigcap L) \subset I^{-2}(I^2 \bigcap L) \subset \ldots$ terminates, hence $I^{-n}(I^n \bigcap L) = I^{-n-1}(I^{n+1} \bigcap L) = \ldots$, hence $I^{-n}(I^n \bigcap L) = I^{-2n}(I^{2n} \bigcap L)$ and thus: $I^{2n} \bigcap L \subset I^n(I^n \bigcap L) \subset I^n L$. □

6.5.3. Lemma. If I is an invertible ideal in a ring with Krull dimension R, then Kdim R/I < Kdim R.

Proof. Consider left ideals of R such that $I \subset L \subsetneq L' \subset R$. Since I is invertible: $I^{n+1} \subset I^n L \subsetneq I^n L' \subset I^n$. The map φ: $L_s(R/I) \to L(I^n/I^{n+1})$, $A/I \mapsto I^n A/I^{n+1}$, is strictly increasing. Therefore $\text{Kdim}_R\,(I^n/I^{n+1}) \geq$ Kdim R. Since there exists an infinite strictly descending chain $R \supset I \supset I^2 \supset \ldots$, it follows that Kdim R > Kdim R/I. □

6.5.4. Lemma. Suppose that R is a left Noetherian ring and that $I \subset J(R)$ is an invertible ideal then for every left ideal L of R we have: $L = \bigcap_{n=1}^{\infty} (L + I^n)$.

Proof. If the assertion in false we may choose A to be maximal among left ideals where the assertion fails. Put $A' = \bigcap_{n=1}^{\infty} (A + I^n) \supsetneq A$. If B is a left ideal properly contining A the B' = B and thus $A' \subset B' = B$. From Lemma 6.5.2. it follows that there is an integer m such that $A' \bigcap I^m \subset IA' \subset A$.
Therefore: $A = A + (A' \bigcap I^m) = A' \bigcap (A + I^m) = A'$, contradiction. □

For any invertible ideal I of R we may construct the abelian group $G_I^*(R) = \bigoplus_{n\in\mathbb{Z}} I^n/I^{n+1}$. For $x \in I^n - I^{n+1}$ we let x_n be the "image" of x in the part I^n/I^{n+1}. Defining $a_n.b_m = (ab)_{m+n}$ defines the structure of a $\mathbb{Z}$-graded ring on $G_I^*(R)$ (in fact $G_I^*(R)$ is the associated graded ring of the I-adic $\mathbb{Z}$-filtration on R).Clearly $G_I^*(R)$ is strongly graded and its positive part is $G_I(R) = \bigoplus_{n\geq 0} I^n/I^{n+1}$ (the Rees ring of R).

6.5.5 Lemma. If R/I is left Noetherian then $G_I^*(R)$ and $G_I(R)$ are left Noetherian. Moreover, Kdim R $\leq$ Kdim $G_I^*(R) \leq$ Kdim (R/I) + 1, Kdim (R/I) $\leq$ Kdim $G_I(R)$ $\leq$ Kdim (R/I) + 1.

Proof. Apply Proposition 4.2.5.8. and Corollary 4.2.5.10. □

6.5.6. Theorem (cf. T. Lenagan [2]). If $I \subset J(R)$, in a left Noetherian ring R, is an invertible ideal then Kdim R = Kdim (R/I) + 1.

Proof. From Lemma 6.5.3. we retain: Kdim (R/I) + 1 $\leq$ Kdim R. Lemma 6.5.5. yields that Kdim G_I(R) $\leq$ Kdim (R/I) + 1. Now apply Theorem 4.2.6.4. □

We say that R is an *N-ring* if for each P $\in$ Spec R, any nonzero ideal of R/P contains a nonzero normal element. For example, if R is a left Noetherian P.I. ring then it is an N-ring (cf. E. Formanek or L. Rowen [1]).

6.5.7. Theorem (T. Lenagan [2]). Let R be a left Noetherian N-ring. If Kdim (R/J(R)) = α then Kdim R = α + n for some positive integer n.

Proof. Let I be an ideal of R, maximal such that R/I does not satisfy the assertion (assuming it fails for I = 0) and replace R by R/I. Let Kdim R = β. There is a P $\in$ Spec R such that Kdim R = Kdim (R/P); if P $\neq$ 0 then R/P satifies the assertion of the theorem. Put J(R/P) = K/P, then we have: β = Kdim [(R/P)/J(R/P)] + n = Kdim (R/K) + n, for some n $\in \mathbb{N}$. From J(R) $\subset$ K, Kdim (R/K) $\leq$ Kdim (R/J(R)) = α and thus $\alpha \leq \beta \leq \alpha$ + n or $\beta = \alpha$ + m for some m $\in \mathbb{N}$ follows. If P = 0 then R is prime; if J(R) = 0 the proof is finished.
If P = 0 and J(R) $\neq$ 0 then there exists an a $\in$ J(R), a $\neq$ 0, which is a normal element. Applying Theorem 6.5.6. to I = aR = Ra we obtain
Kdim R = Kdim (R/Ra) + 1. But J(R)/Ra is the Jacobson radical of R/Ra and therefore the hypothesis yields: Kdim (R/Ra) = α + n for some n $\in \mathbb{N}$.
Consequently Kdim R = α + n + 1 leads to a contradiction. □

We say that R is *semi-local* whenever R/J(R) is semisimple.

6.5.8. Corollary. Let R be a semi-local left Noetherian ring. If R is an N-ring then Kdim R is a finite number.

6.5.9. Example. Let R be a left Noetherian semi-local PI ring, then Kdim R is finite (in particular: if R is a commutative Noetherian ring).

§6.6 The Principal Ideal Theorem.

In this section R is left Noetherian, I is an invertible ideal of R and T = $\bigcup_{n=0}^{\infty} I^{-n}$. If A,B are left R-submodules of T then I(A $\cap$ B) = IA $\cap$ IB and I^{-1}(A $\cap$ B) = I^{-1}A $\cap$ I^{-1}B. Indeed, A $\cap$ B $\subset$ A, A $\cap$ B $\subset$ B yields I(A $\cap$ B) $\subset$ IA $\cap$ IB and similarly I^{-1} (IA $\cap$ IB) $\subset$ I^{-1}IA $\cap$ I^{-1}IB = A $\cap$ B; thus IA $\cap$ IB $\subset$ I(A $\cap$ B) and this proves the first equality (the second follows in the similar way).

6.6.1. Lemma. If P $\in$ Spec R does not contain I then:

1. IP = I $\cap$ P = PI and I^{-1}P = PI^{-1}.
2. I + P/P is invertible in R/P.

Proof. 1. Obviously I^{-1}(I $\cap$ P) is an ideal of R and then from I(I^{-1}(I $\cap$ P)) $\subset$ P and I $\not\subset$ P we may conclude that I^{-1}(I $\cap$ P) $\subset$ P, i.e. I $\cap$ P = IP. The equality PI = I $\cap$ P is established similarly. From IP = PI we derive that P = I^{-1}PI or PI^{-1} = I^{-1}P.
2. The ideals I^n, n $\geq$ 1, are invertible and contained in P. From 1. we retain that I^nP = I^n $\cap$ P and I^{-n}P = PI^{-n}. On the other hand: P = I^{-n}(I^nP) = I^{-n} (P $\cap$ I^n) = I^{-n}P $\cap$ R, hence TP $\cap$ R = ($\bigcup_{n=1}^{\infty}$ I^{-n})P $\cap$ R = $\bigcup_{n=1}^{\infty}$ (I^{-n}P $\cap$ R) = P. Similarly, TP = PT follows. Hence PT is an ideal of T. From TP $\cap$ R = P it follows that φ: R/P $\to$ T/TP is an injective ring homomorphism. Now we calculate: (I + TP) (I^{-1} + TP) = II^{-1} + ITP + TPI^{-1} + $(TP)^2$ $\subset$ R + TP + TI^{-1}P = R + TP + TP = R + TP. The latter establishes that (I + P)/P is invertible and its inverse is I^{-1} + PT/PT. □

6.6.2. Lemma (A. Chatters, A. Goldie a.o. [1]). If B $\subset$ A are finitely generated R-submodules of T such that IA $\subset$ B, then $\rho_{R/I}$ (A/B) = $\rho_{R/I}$ (IA/IB).

Proof. Put N/I = rad (R/I). Since N/I is nilpotent, N^n $\subset$ I for some n $\in$ $\mathbb{N}$ From $(I^{-1}NI)^n$ $\subset$ I, I^{-1}NI $\subset$ N and NI $\subset$ IN follow. Similarly IN $\subset$ NI, and thus IN = NI. By definition of $\rho_{R/I}$ we may reduce the problem to the case where N(A/B) = 0 i.e. where NA $\subset$ B. Because of IN = NI we then obtain N(IA) $\subset$ IB. The map φ: L(A/B) $\to$ L(IA/IB), C/B $\mapsto$ IC/IB is a lattice isomorphism. The equality $\rho_{R/I}$

$(A/B) = \rho_{R/I}$ (IA/IB) will result if we establish that φ respects torsion with respect to the multiplicative system of regular elements of R/N (i.e. L((A/B)/t(A/B)) $\cong$ L((IA/IB)/t(IA/IB)).)
We go on to show that if A/B is a R/N torsion module then JA/JB is too, where J = I or J = I^{-1}. Since IN = NI, I^nN = NI^n and JN = NJ. So if t $\in$ T then tY $\subset$ N if and only if Yt $\subset$ N. Let a $\in$ JA and denote: K = {r $\in$ R, ra $\in$ JB} $\supset$ N. To prove that K $\bigcap$ $\mathcal{C}$(N) $\neq$ ϕ, where $\mathcal{C}$(N) is the set of r $\in$ R hich are regular modulo N, is equivalent to establish that K/N is an essential left ideal of R/N.
Suppose K $\bigcap$ K' = N for some left ideal K' containing N. Pick s $\in$ J^{-1}K'. From J^{-1}K' $\subset$ J^{-1} we derive that s $\in$ J^{-1}. But sa $\in$ J^{-1}JA = A. Since we have assumed that A/B is torsion there is a c $\in$ $\mathcal{C}$(N) such that csa $\in$ B, i.e. Jcsa $\subset$ JB. Now Jcs $\subset$ JJ^{-1}K' = K' entails: Jcs $\subset$ K' $\bigcap$ K = N and thus: csJ $\subset$ N. But then sJ $\subset$ J^{-1}K'J $\subset$ R and c $\in$ $\mathcal{C}$(N) yields sJ $\subset$ N, hence Js $\subset$ N. Since s $\in$ J^{-1}K' is arbitrary we arrive at J(J^{-1}K') $\subset$ N or K' $\subset$ N, proving that R/N is an essential extension of K/N as desired. □

Recall that *P has height n*, ht(P) = n, if there exists a chain of prime ideals $P_0 \subsetneq P_1 \subsetneq P_2 \subsetneq \dots \subsetneq P_n$ = P which cannot be refined properly.

6.6.5. Theorem (cf. Chatters, Goldie, Hajarnavis, Lenagan [1]). Let I be a proper invertible ideal of the left Noetherian ring R and let P $\in$ Spec R be minimal such that I $\subset$ P then ht(P) $\leq$ 1.

Proof. Suppose ht(P) $\geq$ 2, i.e. let there be a chain of prime ideals $P_0 \subsetneq P_1 \subsetneq P_2$ = P. By Lemma 6.6.3., I + P_0/P_0 is an invertible ideal of R/P_0. Clearly P/P_0 is minimal over I + P_0/P_0 and ht(P/P_0) $\geq$ 2. Replacing R by R/P_0 if necessary we may assume that P_0 = 0 and R is prime.
Let a $\in$ P_1 be a regular element. By Lemma 6.5.1. there is an n such that
Ra $\bigcap$ I^{2n} $\subset$ I^na. Since P is minimal over I^n we may replace I by I^n, i.e. we may assume that Ra $\bigcap$ I^2 $\subset$ Ia. Put N/I = rad(R/I) and we will write ρ(-) for any finitely generated R/I-module meaning the reduced rank.
Then: $\rho(I^2$ + Ra/Ia + $I^2)$ = $\rho(I^2$ + IA + Ra/Ia + $I^2)$ = ρ(Ra/Ra $\bigcap$ (Ia + I^2)) = ρ(Ra/Ia + (Ra $\bigcap$ I^2)) = ρ(Ra/Ia) = ρ(R/I) (because Ra/Ia $\cong$ R/I).
By Lemma 6.6.2. we also have: ρ(I + Ra/I^2 + Ia) = ρ(I + Ra/I) + ρ(I/I^2 + Ra) = ρ(I + Ra/I) + ρ(R/I + Ra) = ρ(R/I).
Combining all this yields: ρ(I + Ra/I^2 + Ra) = 0. Hence if x $\in$ I then there is a c $\in$ $\mathcal{C}$(N) such that cx $\in$ I^2 + Ra $\subset$ I^2 + P_1. Therefore cx $\in$ I^2 + P_1 $\bigcap$ I. Lemma

6.6.2. and $P_1 \bigcap I = P_1 I = IP_1$ yields $cx \in I^2 + P_1 I$ and thus: $cxI^{-1} \subset I + P_1$. Now $xI^{-1} \subset R$ and $c \in \mathcal{C}(N) \subset \mathcal{C}(P)$ (note: P/N is a minimal prime ideal of R/N), hence $xI^{-1} \subset P$.
Finally, since $x \in I$ was arbitrary, $II^{-1} \subset P$ and P = R results but the latter is a contradiction. □

6.6.6. Theorem (Principal Ideal Theorem; Jategaonkar [2]). Let a be a normal element in the left Noetherian ring R and assume taht a is not a unit of R. If $P \in \text{Spec } R$ is minimal such that $a \in P$ then $ht(P) \leq 1$.

Proof. Let $P_0 \in \text{Spec } R$ be minimal in R and contained in P. If $P = P_0$ then ht(P) = 0. If $P \neq P_0$ then $a \notin P_0$ and we may replace R by R/P_0 i.e. we may assume that R is prime. However, in this case I = Ra = aR is invertible and we may apply Theorem 6.6.5. □

6.6.7. Corollary. Let P be a prime ideal of a left Noetherian ring R. If, for every prime ideal Q of R such that $Q \subsetneq P$, P/Q contains a normal element of R/Q, then ht(P) is finite.

Proof. Suppose ht(P) is infinite. By the Noetherian property we may assume that $ht(P/I) < \infty$ in R/I for every nonzero ideal $I \subset P$. In this case R is a prime ring. Since $P \neq 0$ there is an $a \in P$ which is normal. Consider the chain

$$P = P_0 \supset P_1 \supset \ldots \supset P_n \tag{*}$$

and a chain of prime ideals

$$P = Q_0 \supset Q_1 \supset \ldots \supset Q_n \tag{**}$$

such that $a \in Q_{n-1}$. To construct this we take a chain of type (**) where the number of terms containing a is maximal. If j is maximal such that $a \in Q_j$ and j < n-1 let $\overline{a}$ be the image of a in $\overline{R} = R/Q_{j+2}$.
Obviously, $\overline{a}$ is normal in $\overline{R}$ and therefore $\overline{a}$ is a regular element of $\overline{R}$. Put $\overline{Q}_j = Q_j/Q_{j+z}$. Then $ht(\overline{Q}_j) \geq z$. Theorem 6.6.6. yields the existence of an ideal $\tilde{Q}$ of $\overline{R}$ such that $\overline{a} \in \tilde{Q} \subsetneq \overline{Q}_j$. Replacing Q_{j+1} by the preimage of $\tilde{Q}$ (under the canonical $R \to \overline{R}$) we do obtain a chain of type (**) where the number of terms containing a is strictly larger than j, contradicting the choice of j! Hence if ht(P) were infinite

then ht(P/Ra) is infinite in R/Ra, contradicting our assumptions. □

6.6.8. Example. A left Noetherian P.I. ring satisfies the hypothesis of the above Corollary hence for all $P \in \text{Spec } R$, $\text{ht}(P) < \infty$.

§6.7 Exercises.

(149) Let R be a fully bounded Noetherian (left and right) ring. If I is an ideal of R then Kdim $({}_R I)$ = Kdim (I_R).

(150) (G. Cauchon [1]). Let R be a left Noetherian H-ring and let M be a left R-module. If Ass(M) consists of minimal prime ideals of R then the set of annihilators of subsets of M satisfies the ascending chain condition.

(151) (T. Lenagan [7]). Let I be an ideal in a left and right Noetherian ring R. Then ${}_R I$ is Artinian if and only if I_R is Artinian.

(152) For a semi-local fully bounded Noetherian ring, Kdim R $< \infty$.

(153) Let R be a left and right Noetherian ring such that Kdim R ≤ 1 then $\bigcap_{n=1}^{\infty} J(R)^n = 0$.

Bibliographical Comments to Chapter 6.

Condition (H) was introduced by P. Gabriel in [1], 1962. He established that a left Noetherian ring R satisfying condition (H) has enough prime ideals. Theorem 6.3.6. gives new necessary and sufficient conditions for a ring to have enough prime ideals i.e. a left Noetherian ring has enough prime ideals exactly when it satisfies condition (H). This result is due to G. Cauchon [1] but we follow the proof given by A.W. Chatters, A. Goldie, Hajarnavis, T. Lenagan in [1], using the notion of reduced rank (cf. A. Goldie [1]).

Using the properties of the reduced rank we gave a less complicated proof of a theorem of L. Small [1]. As a consequence of 6.3.6. the Jacobson conjecture is valid for fully left bounded left and right Noetherian rings. This result of G. Cauchon [1] extends an earlier result of Jategaonkar [4].

Our proof of Theorem 6.3.8. follows the one given by A. Goldie in [4]. Example 6.3.9. was given by I. Herstein in [1], 1965, it shows how the Jacobson conjecture fails if one only allows one-sided conditions (left Noetherian fully left bounded). The Jacobson conjecture for left and right Noetherian rings in general remains unsettled at this moment.

Section 6.4. contains several of Jategaonkar's results stemming from [4]. Section 6.5. contains part of Lenagan [2] but we have been able to simplify T. Lenagan's proofs by using strongly graded rings (in fact generalized Rees rings). The final section deals with Jategaonkar's principal ideal theorem for noncommutative rings (proved in a different way in [2], 1974). The proof we included is similar to the one given by A. Chatters, A. Goldie, Hajernavis, T. Lenagan in [1].

Chapter 7

Relative Krull and Gabriel Dimensions.

§7.1 Additive Topologies and Torsion Theories.

Throughout R is a ring with identity and R-mod denotes the class of all left R-modules.

7.1.1. Definition. A *(left) linear topology* on R is a nonempty set $\mathcal{J}$ of left ideals of R satisfying the following conditions:

a. If $I \in \mathcal{J}$ and J is a left ideal of R containing I then $J \in \mathcal{J}$.
b. If $I, J \in \mathcal{J}$ then $I \cap J \in \mathcal{J}$.
c. If $I \in \mathcal{J}$ then $(I : a) = \{ x \in R, xa \in I\} \in \mathcal{J}$ for all $a \in R$.

We say that $\mathcal{J}$ is an *additive topology* on R if it is a linear topology satisfying the extra condition:

d. If $J \in \mathcal{J}$ and I is a left ideal of R such that $(I : a) \in \mathcal{J}$ for all $a \in J$ then $I \in \mathcal{J}$.

7.1.2. Definition. A nonempty subclass $\mathcal{A}$ of R-mod is said to be *closed* if each direct sum of objects of $\mathcal{A}$ is in $\mathcal{A}$ and for each exact sequence of type:

$$0 \longrightarrow M' \xrightarrow[f]{} M \xrightarrow[g]{} M'' \longrightarrow 0 \qquad (*)$$

with M $\in \mathcal{A}$ we have that M',M" $\in \mathcal{A}$. We say that $\mathcal{A}$ is a *hereditary torsion class* if it is closed and for an exact sequence of type (*) with M',M" $\in \mathcal{A}$ we have that M $\in \mathcal{A}$.

To a linear topology $\mathcal{J}$ on R we may associate the closed subclass $\mathcal{A}$ in R-mod, $\mathcal{A}_F = \{M \in \text{R-mod}, l_R(x) \in \mathcal{J} \text{ for all } x \in M\}$. It is clear that $\mathcal{A}_F$ is a hereditary torsion class when $\mathcal{F}$ is an additive topology. On the other hand, suppose we are given a nonempty subclass $\mathcal{A}$ of R-mod and define $\mathcal{F}_\mathcal{A} = \{L \text{ left ideal of R}, R/L \in \mathcal{A}\}$.

7.1.3. Proposition. 1. If $\mathcal{A}$ is closed then $\mathcal{F}_\mathcal{A}$ is a linear topology.
2. If $\mathcal{A}$ is a hereditary torsion class then $\mathcal{F}_\mathcal{A}$ is an additive topology.

Proof. 1. The conditions a. and b. are easily verified. Consider I $\in \mathcal{F}_A$ and x $\in$ R together with the R-linear φ_x: $R \to R/I$, $a \mapsto \overline{ax}$ the kernel of which is Ker $\varphi_x = (I : x)$. So φ_x induces an injective morphism $R/(I : x) \hookrightarrow R/I$, hence $(I : x) \in \mathcal{F}_\mathcal{A}$. Thus $\mathcal{F}_\mathcal{A}$ satisfies condition c of Definition 7.1.1. as well.
2. Consider J $\in \mathcal{F}_\mathcal{A}$ and a left ideal I of R such that $(I : x) \in \mathcal{F}_\mathcal{A}$ for all x $\in$ J. Clearly $R/I + J \in \mathcal{A}$.We claim: $(I + J)/I \cong J/I \bigcap J \in \mathcal{A}$. Let $\overline{x} \in J/I \bigcap J$, ($\overline{x} = x + I \bigcap J$, x $\in$ J). Then $l_R(\overline{x}) = ((I \bigcap J) : x) = (I : x)$. But $(I : x) \in \mathcal{F}_\mathcal{A}$, hence $R\overline{x} \cong R/l_R(\overline{x}) \in \mathcal{A}$. Since $\mathcal{A}$ is closed under direct sums it follows that $I/I \bigcap J \in \mathcal{A}$ and thus $(I + J)/I \in \mathcal{A}$. From the exact sequence:
$0 \to (I + J)/I \to R/I \to R/(I + J) \to$, 0 it follows that $R/I \in \mathcal{A}$ or $I \in \mathcal{F}_\mathcal{A}$. □

7.1.4. Theorem. The correspondences $\mathcal{F} \mapsto \mathcal{A}_\mathcal{F}$, $\mathcal{A} \mapsto \mathcal{F}_\mathcal{A}$, are inverse to each other and they induce a bijective correspondence between the set of additive topologies on R and set of hereditary torsion classes in R-mod.

Proof. One easily verifies that $\mathcal{F}_{\mathcal{A}_\mathcal{F}} = \mathcal{F}$ and $\mathcal{A} = \mathcal{A}_{\mathcal{F}_\mathcal{A}}$ for a given linear topology $\mathcal{F}$, resp. a closed subclass $\mathcal{A}$ of R-mod. □

Let $\mathcal{F}$ be an additive topology on R; a left R-module in $\mathcal{A}_\mathcal{F}$ is said to be $\mathcal{F}$-torsion. For every left R-module X there exists a unique largest submodule $t_\mathcal{F}$ (X) of X which is $\mathcal{F}$-torsion; this submodule is called the *$\mathcal{F}$-torsion submodule* *of X*. The functor $t_{\mathcal{F}}$ in R-mod is called the *torsion radical* associated to $\mathcal{F}$. A left R-module M such that $t_{\mathcal{F}}$ (M) = 0 is said to be *$\mathcal{F}$-torsion free*. Since $\mathcal{A}_{\mathcal{F}}$ is closed under extensions it follows that, for each left R-module X the left R-module $X/t_{\mathcal{F}}$ (X) is $\mathcal{F}$-torsion free. A left R-module which is both $\mathcal{F}$-torsion and $\mathcal{F}$-torsion free is necessarily the zero-module.

Some Examples of Additive Topologies.

E.1. Let S be a multiplicatively closed subset of R. Put $\mathcal{F}_S = \{$L left ideal of R, for all $x \in R$ there is an $s \in S$, $s \in (L : x)\}$. The hereditary torsion class associated to this additive topology is $\mathcal{A}_S = \{$left R-modules M, for all $x \in M$ there exists an $s \in S$ with $sx = 0\}$.
Moreover, if S satisfies the left Ore conditions then we obtain: $\mathcal{F}_S = \{$L left ideal of R, $L \cap S \neq \phi\}$.

E.2. By the results of Chapter 1 we may conclude that the class of all semiartinian left R-modules is a hereditary torsion class, called the Dickson torsion class. The corresponding additive topology is $\mathcal{F}_D = \{$L left ideal of R, R/L is a semiartinian R-module$\}$.

E.3. Let Q be an injective left R-module and put, $\mathcal{A}_Q = \{$M a left R-module, Hom_R (M,Q) = 0$\}$.
It is easily seen that $\mathcal{A}_Q$ is a hereditary torsion class and the associated topology is $\mathcal{F}_Q = \{$L left ideal of R, Hom_R (R/I, Q) = 0$\}$. When Q is the injective envelope of some M, the $\mathcal{F}_Q$ may be characterized as follows:

7.1.5. Proposition. $\mathcal{F}_Q = \{$L left ideal of R, for all $a \in R$ and $x \neq 0$ in M, $(I : a) \not\subset l_R(x)\}$, where $Q \cong E_R$ (M).

Proof. Consider $L \in \mathcal{F}_Q$, $x \neq 0$ in M and $a \in R$ such that $(L : a) \subset l_R(x)$. From $(L : a) \in \mathcal{F}_Q$ it follows that $l_R(x) \in \mathcal{F}_Q$ and hence there exists a nonzero $f \in \text{Hom}_R$ $(R/l_R(x), Q)$ what contradicts $R/l_R(x) \in \mathcal{A}_Q$.
Conversely, if Hom_R (R/L, E(M)) $\neq 0$ for some left ideal L such that for all $x \neq 0$ in M and $a \in R$ we have $(L : a) \not\subset l_R(x)$, then there is a morphism
f: $R/L \to E(M)$ which is nonzero. Since M is essential in E(M) there is an $\overline{a} \in R/L$, with $a \in R$ representing $\overline{a}$, such that $x = f(\overline{a}) \neq 0$. Now $(L : a) \subset l_R(x)$ yields a contradiction.

E.4. A left ideal L of R is said to be *dense* if for each a $\in$ R, r_L (L : a) = 0 i.e. (L : a) $\lambda = 0$ entails $\lambda = 0$. Taking M = $_R R$ in E.3. we obtain the additive topology consisting of the dense left ideals of R which is associated to Q = E($_R R$). This topology will be called Lambek's topology.

E.5. Let I be an ideal of R and put $\mathcal{F}_I = \mathcal{F}_{E(R/I)}$. When R/I is a semiprime left Goldie ring we put $\mathcal{C}$(I) equal to the set of c $\in$ R such that $\bar{c} \in$ R/I is a regular element of R/I. From Proposition 2.8.7. we retain: $\mathcal{C}$(I) = {c $\in$ R, xc $\in$ I implies x $\in$ I}.

7.1.6. Proposition. Let I be an ideal of R such that R/I is a semiprime left Goldie ring, then $\mathcal{F}_I$ = {L left ideal of R, (L : x) $\bigcap$ $\mathcal{C}$(I) $\neq \phi$ for all x $\in$ R}.

Proof. In view of Proposition 7.1.5. we have: $\mathcal{F}_I$ = {L left ideal of R, (L : x) $\not\subset l_R(\bar{y})$ for all x $\in$ R, y $\in$ R with y $\notin$ I}. Clearly if (L : x) $\bigcap$ $\mathcal{C}$(I) for all x $\in$ R, then L $\in \mathcal{F}_I$. Conversely, consider L $\in \mathcal{F}_I$, x $\in$ R. Let π: R $\to$ R/I be the canonical epimorphism and put J = π (L : x). Since (L : y) $\bar{y} \neq 0$ for all y $\notin$ I it follows that there is a $\lambda \in$ (L : y) such that $\lambda\bar{y} \neq 0$. But from λy $\in$ L we derive that $\overline{\lambda y} \neq 0$ and $\overline{\lambda y} \in$ J, hence J is an essential left ideal of R/I. Since the latter is a semiprime left Goldie ring there is a regular element $\bar{c}$ in R/I with $\bar{c} \in$ J and then it follows also that (L : a) $\bigcap$ $\mathcal{C}$(I) $\neq \phi$. □

Next we assume that R is $\mathbb{Z}$ -graded; R = $\bigoplus_n R_n$, h(R) = $\bigcup_n R_n$.

7.1.7. Definition. A nonempty set $\mathcal{H}$ of graded left ideals of R (the set of all graded left ideals of R will be denoted by L_g(R)) is a *graded filter* (or topology) on R if it satisfies:

1. If I $\in \mathcal{H}$ and I' $\in L_g$(R) such that I $\subset$ I' then I' $\in \mathcal{H}$.
2. If $I_1, I_2 \in \mathcal{H}$ then $I_1 \bigcap I_2 \in \mathcal{H}$.
3. If I $\in \mathcal{H}$ then (I : x) $\in \mathcal{H}$ for all x $\in$ h(R).
4. If $I_1 \in \mathcal{H}$ and (I : x) $\in \mathcal{H}$ for all x $\in I_1 \bigcap$ h(R) = h(I_1), then I $\in \mathcal{H}$.

If M = $\bigoplus_{n\in\mathbb{Z}} M_n$ is a graded left R-module then we define M(m) = $\bigoplus_{n\in\mathbb{Z}} M_{n+m}$ to be the *m-suspension* or *m-shifted module* of M.

It is clear how the definition of a localizing class or hereditary torsion class may now be rephrased in terms of graded left R-modules i.e. in R-gr; we leave the exact wording of these modified definitions to the reader. A localizing subclass of R-gr is said to be *rigid* if contains all shifted modules of any module it contains. To a graded filter $\mathcal{H}$ there corrseponds a rigid localizing subclass of R-gr, say $T_{\mathcal{H}}$ (easy to check!). Elements of $T_{\mathcal{H}}$ are said to be gr-$\mathcal{H}$-torsion (or just $\mathcal{H}$-torsion if no confusion is possible). For an $M \in$ R-gr we let $t_{\mathcal{H}}(M)$ be the largest graded R-submodule of M which is still in $T_{\mathcal{H}}$. If $t_{\mathcal{H}}(M) = 0$ then M is said to be gr-$\mathcal{H}$-torsion free (or just $\mathcal{H}$-torsion free).

For a more elaborated treatment of rigid localizing classes and localization of graded rings we refer to c. Năstăsescu, F. Van Oystaeyen [3]; here we restrict to the bare necessities.

§7.2 The Lattices $C_{\mathcal{F}}(M)$ and $C^g_{\mathcal{H}}(M)$.

Fix notations as follows: $\mathcal{F}$ is an additive topology on R with associated localizing class $T_{\mathcal{F}}$ and torsion radical $t_{\mathcal{F}}$.

The *$\mathcal{F}$-closure* or *$\mathcal{F}$-saturation* of a submodule N of the left R-module M is defined to be the submodule of M given by: $N^{\sim} = \{x \in M, (N : x) \in \mathcal{F}\}$. It is very easy to prove:

7.2.1. Proposition. For submodules N and N' of M we have:

1. $N \subset N^{\sim}$.
2. $N^{\sim}/N = t_{\mathcal{F}}(M/N)$ and $M/N^{\sim}$ is $\mathcal{F}$-torsion free.
3. If $N \subset N'$ then $N^{\sim} \subset (N')^{\sim}$.
4. $(N \bigcap N')^{\sim} = N^{\sim} \bigcap (N')^{\sim}$.
5. $(N^{\sim})^{\sim} = N^{\sim}$
6. $(N + N')^{\sim} = (N^{\sim} + (N')^{\sim})^{\sim}$.

7.2.2. Definition. A submodule N of M is said to be *$\mathcal{F}$-closed* or *$\mathcal{F}$-saturated* if $N = N^{\sim}$. If $K \subset N$ are submodules of M such that $K = \{x \in N, (K : x) \in \mathcal{F}\}$ then we say that *K is $\mathcal{F}$-closed in N*. Write $C_{\mathcal{F}}(M) = \{N \subset M, N = N^{\sim}\}$.

The set $C_{\mathcal{F}}(M)$, ordered by inclusion, is a lattice. Indeed, if N, N' $\in C_{\mathcal{F}}(M)$ then $N \cap N'$ and $(N + N')^{\sim} \in C_{\mathcal{F}}(M)$, thus we may define the lattice-operations in $C_{\mathcal{F}}$ as follows: $N \wedge N' = N \cap N'$, $N \vee N' = (N + N')^{\sim}$.

7.2.3. Proposition. The lattice $C_{\mathcal{F}}(M)$ is a complete upper-continuous modular lattice; $t_{\mathcal{F}}(M)$ and M may be taken for 0,1 respectively.

Proof. For a family $\{N_i, i \in \mathcal{F}\}$ in $C_{\mathcal{F}}(M)$ it is easily seen that $\bigcap \{N_i, i \in \mathcal{F}\} \in C_{\mathcal{F}}(M)$ and this is the infemum of that family, whilst $(\sum_{i \in J} N_i)^{\sim}$ is its supremum. Let H,K,L $\in C_{\mathcal{F}}(M)$ with $H \subset K$. Then $K \wedge (H \vee L) =$ $K^{\sim} \cap (H + L)^{\sim} = (K \cap (H + L))^{\sim} = (H + (K \cap L))^{\sim} = H \vee (K \wedge L)$ (we did use modularity of the lattice of all submodules of M). That $C_{\mathcal{F}}(M)$ is upper-continuous is easily verified. □

7.2.4. Proposition. For a submodule N of M we denote $a = N^{\sim}$ in $C_{\mathcal{F}}(M)$. The lattice $C_{\mathcal{F}}(N)$ is isomorphic to $[0,a]$ and the lattice $C_{\mathcal{F}}(M/N)$ is isomorphic to $[a,1]$.

Proof. Define φ: $C_{\mathcal{F}}(N) \to [0,a]$, $X \mapsto x = X^{\sim}$. If X,Y $\in C_{\mathcal{F}}(N)$ satisfy $\varphi(X) = \varphi(Y)$ then $X^{\sim} = Y^{\sim}$. Clearly the $\mathcal{F}$-closure in N is obtained by intersecting N with the $\mathcal{F}$-closure in M, hence it follows from $X^{\sim} = Y^{\sim}$ in M that X = Y because the latter are $\mathcal{F}$-closed. Consequently φ is injective. Conversely, let $b \in [0,a]$. Then $0 \leq b \leq a$ and b = B for some B $\in C_{\mathcal{F}}(M)$. It is clear that $B \cap N \in C_{\mathcal{F}}(N)$ and since $N^{\sim}/N$ is $\mathcal{F}$-torsion also $B/B \cap N$ is $\mathcal{F}$-torsion, hence $(B \cap N)^{\sim} = B$ in M. Consequently φ is surjective and thus bijective.
We now calculate: $\varphi(X \wedge Y) = \varphi(X \cap Y) = (X \cap Y)^{\sim} = X^{\sim} \cap Y^{\sim} =$ $\varphi(X) \wedge \varphi(Y)$, $\varphi(X \vee Y) = (X \vee Y)^{\sim} = (X + Y)^{\sim} = (X^{\sim} + Y^{\sim})^{\sim} =$ $(\varphi(X) + \varphi(Y))^{\sim} = \varphi(X) \vee \varphi(Y)$.
Define Ψ: $C_{\mathcal{F}}(M/N) \to [a,1]$, $X/N \mapsto X$. It is easily verified that Ψ also defines an isomorphism of lattices. □

7.2.5. Corollary.

1. An N $\in C_{\mathcal{F}}(M)$ is irreducible if and only if it is an irreducible submodule of M.
2. If M is $\mathcal{F}$-torsion free and N $\in C_{\mathcal{F}}(M)$ then N is uniform in $C_{\mathcal{F}}(M)$ if and only if N is a uniform submodule of M.

Proof. 1. Put N = P $\bigcap$ Q for submodules P and Q of M. Then N = $N^\sim$ = (P $\bigcap$ Q)$^\sim$ = $P^\sim \bigcap Q^\sim = P^\sim \bigwedge Q^\sim$ yields N = $P^\sim$ or N = $Q^\sim$ because N is irreducible in $C_{\mathcal{F}}$(M). Therefore N = P or N = Q and N is irreducile as a submodule. The converse implication is clear.

2. Consider nonzero submodules P and Q of N. We have: $P^\sim \subset N^\sim$ = N and similarly $Q^\sim \subset$ N. Since M has no $\mathcal{F}$-torsion it follows that $P^\sim \neq 0$ and $Q^\sim \neq 0$. Therefore, $P^\sim \bigwedge Q^\sim \neq 0$, hence (P $\bigcap$ Q)$^\sim \neq 0$ and also P $\bigcap$ Q $\neq 0$. This proves that N is a uniform submodule of M if it is a uniform element of $C_{\mathcal{F}}$(M). The converse is obvious. □

We now include some examples, related to the examples of additive topologies given in Section 7.1.

E.1. Let S be a multiplicatively closed set such that R satisfies the left Ore conditions with respect to S. Let $\mathcal{F}_S$ = {L left ideal of R, L $\bigcap$ S $\neq \phi$} (see E.1. in foregoing section). The associated hereditary torsion class is $\mathcal{T}_S$ = {M ∈ R-mod, for every x in M there is an 0 in S such that sx = 0}. Let t_S be the torsion radical corresponding to $\mathcal{F}_S$, i.e. for M ∈ R-mod, t_S(M) = {m ∈ M, sm = 0 for some s ∈ S}. If N is a submodule of M then $N^\sim$ = {m ∈ M, sm ∈ N for some s ∈ S}.

E.2. Let Q be an injective left R-module and let $\mathcal{F}_Q$ be as in E.3. in Section 7.1. Write C_Q(R) = $C_{\mathcal{F}_Q}$(R). The elements of C_Q(R) are just the left annihilators of nonempty subsets of Q i.e. C_Q(R) = {l_R(X), $\phi \neq$ X $\subset$ Q}. Indeed, if X $\subset$ Q then $l_R(X) = \bigcap_{x \in X} l_R(x)$ and for each x ∈ X, $l_R(x) \in C_Q(X)$ because R/l_R(x) $\subset$ Q and Q is $\mathcal{F}_Q$-torsion free by definition; so l_R(X) ∈ C_Q(R). Conversely let Y ∈ C_Q(R) and let X be the set {x ∈ Q, Yx = 0}. By definition Y $\subset$ l_R(X). Let a ∈ l_R(X). Define α: R/(Y : a) → R/Y, $\overline{\lambda} \mapsto \lambda\,\overline{a}$. Clearly α is monomorphic. For each f: R/(Y : a) → Q there exists a g: R/Y → Q such that g ∘ α = f by the injectivity of Q. But g($\overline{\lambda}$) = λx where x = g($\overline{1}$). We obtain: Yx = 0, thus x ∈ X and therefore ax = 0. C Consequently g ∘ α = 0, i.e. f = 0.

This entails that Hom_R (R/(Y : a), Q) = 0, i.e. (Y : a) ∈ $\mathcal{F}_Q$ or in other words a ∈ $Y^\sim$ = Y. □

Actually the foregoing example is a very general one. In fact, we prove next that each additive topology $\mathcal{F}$ may be cogenerated by an injective left R-module i.e. $\mathcal{F}$ is $\mathcal{F}_Q$ for some Q. This also establishes that $\mathcal{F}$-saturated ideals are nothing but left annihilator ideals of some injective module.

7.2.6. Proposition. Let $\mathcal{F}$ be an additive topology on R then $\mathcal{F} = \mathcal{F}_Q$ where Q $\cong$ E $(\bigoplus_{Y \in C_{\mathcal{F}}(R)} R/Y)$.

Proof. Let $\mathcal{T}_{\mathcal{F}}$ be the localizing class of $\mathcal{F}$ and $\mathcal{T}_Q = \{M \in$ R-mod, $\mathrm{Hom}_R(M,Q) = 0\}$. Obviously $\mathcal{T}_{\mathcal{F}} \subset \mathcal{T}_Q$. Take $M \in \mathcal{T}_Q$, then $\mathrm{Hom}_R(M/t_{\mathcal{F}}(M), Q) = 0$. If $M/t_{\mathcal{F}}(M) \neq 0$ pick $x \neq 0$ in $M/t_{\mathcal{F}}(M)$ and then $l_R(x) \in C_{\mathcal{F}}(R)$ yields $Rx \cong R/l_R(x) \subset Q$. Since Q is injective there exists a nonzero extension $M/t_{\mathcal{F}}(M) \to Q$, a contradiction. □

7.2.7. Definition. An R-module M is said to be *$\mathcal{F}$-Noetherian*, resp. *$\mathcal{F}$-Artinian* if for each ascending, resp. descending, chain of submodules of M, say $M_1 \subset M_2 \subset \ldots \subset M_n \subset \ldots$, resp. $M_1 \supset M_2 \supset \ldots \supset M_n \supset \ldots$, there is an index $m \in \mathbb{N}$ such that for each $n \geq m$ we have that $M_{n+1}/M_n \in \mathcal{T}_{\mathcal{F}}$, resp. $M_n/M_{n+1} \in \mathcal{T}_{\mathcal{F}}$. When $\mathcal{F} = \{R\}$ it follows that $\mathcal{T}_{\mathcal{F}} = \{0\}$ and then the above notions reduce to the usual notions of Noetherian, resp. Artinian, modules.
We say that a left R-module M is *$\mathcal{F}$-finitely generated*, resp. *cogenerated*, if there is a finitely generated R-submodule M' of M such that $M/M' \in \mathcal{T}_{\mathcal{F}}$, resp. if for every set $\{N_i,\ i \in J\}$ of submodules of M $\bigcap \{N_i,\ i \in J\} = 0$ implies $\bigcap_{j \in J'} N_j \in \mathcal{T}_{\mathcal{F}}$ for some finite subset $J' \subset J$.

7.2.8. Proposition. For a left R-module M the following assertions are equivalent:

1. M is $\mathcal{F}$-Noetherian ($\mathcal{F}$-Artinian).
2. Each submodule of M is $\mathcal{F}$-finitely generated (every factor module of M is $\mathcal{F}$-finitely cogenerated).
3. $C_{\mathcal{F}}(M)$ is Noetherian (Artinian).

Proof. The equivalence of 1. and 2. may be proved along the lines of the classical result for $\mathcal{F} = \{R\}$.

1.$\Rightarrow$ 3. Consider an ascending chain $N_1 \subset N_2 \subset \ldots \subset N_n \subset \ldots$ in $C_{\mathcal{F}}(M)$. By the assumption $N_{n+1}/N_n \in \mathcal{T}_{\mathcal{F}}$ for all $n \geq m$ for some $m \in \mathbb{N}$ but since N_m is $\mathcal{F}$-closed it follows that $N_n = N_m$ for all $n \geq m$.

3.$\Rightarrow$ 1. If $N_1 \subset \ldots \subset N_n \subset \ldots$ is an ascending chain of submodules of M then we obtain an ascending chain in $C_{\mathcal{F}}(M)$: $N_1^{\sim} \subset \ldots \subset N_n^{\sim} \subset \ldots$. By the assumption, there is an $m \in \mathbb{N}$ such that for each $n \geq m$ we have $N_{n+1}^{\sim} = N_n^{\sim} = \ldots = N_m^{\sim}$. Since

$N_n^{\sim}/N_n \in \mathcal{T}_{\mathcal{F}}$ it follows that $N_{n+1}^{\sim}/N_n \in \mathcal{T}_{\mathcal{F}}$ hence $N_{n+1}/N_n \in \mathcal{T}_{\mathcal{F}}$. In a similar way we may establish the equivalence in the Artinian case. □

7.2.9. Corollary. If N is a submodule of M then M is $\mathcal{F}$-Noetherian, resp. $\mathcal{F}$-Artinian, if and only if N and M/N are $\mathcal{F}$-Noetherian, resp. $\mathcal{F}$-Artinian.

7.2.10. Definition. An injective left R-module Q is Σ-*injective* resp. Δ-injective, if and only if R is $\mathcal{F}_Q$-Noetherian, resp. $\mathcal{F}_Q$-Artinian. We have the following characterization:

7.2.11. Proposition. Q is Σ-injective if and only if $Q^{(J)}$ is injective for any set J, if and only if $Q^{(\mathbb{N})}$ is injective.

Proof. Since Q is $\mathcal{F}_Q$-torsion free we may reproduce the proof of the classical Noetherian case to show that Q is Σ-injective implies that any direct sum of copies of Q is again injective.
For the converse, let us start from an ascending chain in $C_{\mathcal{F}_Q}(R)$, say $L_1 \subset L_2 \subset \dots \subset L_n \subset \dots$. According to Corollary 7.2.6. there is an $X_n \subset Q$ such that $L_n = l_R(X_n)$. If the chain is not stationary then we can choose an $x_n \in Q$ for each n, such that $L_n x_n = 0$ but $L_{n+1} x_n \neq 0$. Put $L = \bigcap_{n\geq 1} L_n$ and define f: $L \to Q^{(\mathbb{N})}$, $\lambda \mapsto (\lambda x_1, \lambda x_2, \dots)$. Since $Q^{(\mathbb{N})}$ is injective there exists an $y \in Q^{(\mathbb{N})}$ such that $f(\lambda) = \lambda y$ for all $\lambda \in L$. Put $y = (y_1, \dots, y_n, 0, \dots)$. Then it follows that $L_{n+1} x_n = 0$ for $n > m$, a contradiction. □

We say that a left R-module M has *$\mathcal{F}$-finite length* if the lattice $C_{\mathcal{F}}(M)$ has finite length and we denote this number by $l_{\mathcal{F}}(M)$. It is now clear that M has $\mathcal{F}$-finite length if and only if M is both $\mathcal{F}$-Noetherian and $\mathcal{F}$-Artinian. Moreover, if $N \subset M$ then M has $\mathcal{F}$-finite length if and only if N and M/N have $\mathcal{F}$-finite length and $l_{\mathcal{F}}(M) = l_{\mathcal{F}}(N) + l_{\mathcal{F}}(M/N)$.

7.2.12. Theorem. Suppose that ${}_RR$ is $\mathcal{F}$-Artinian. If a left R-module M is $\mathcal{F}$-Artinian then M is $\mathcal{F}$-Noetherian; in particular ${}_RR$ is $\mathcal{F}$-Noetherian.

Write L = $C_{\mathcal{F}}$(M) (note: 0 = $t_{\mathcal{F}}$(M), 1 = M). Let s(L) be the socle of the lattice L (cf. Chapter 1). Define inductively an ascending chain in L as follows: $0 = l_0 \subset l_1 \subset \dots \subset l_n \subset \dots$, where $l_1 = s(L), \dots, l_n = s([l_{n-1},1]), \dots$.
For x ∈ M let L_x be the sublattice $[0,(Rx)^{\sim}]$ in L, and we write $x^{\sim} = (Rx)^{\sim}$. In view of Proposition 7.2.4., L_x is isomorphic to $C_{\mathcal{F}}(Rx)$. If $x^{\sim} \in l_n$ then L_x has finite length (to prove this one may procede exactly as in Theorem 2.9.2.. Further, if $x^{\sim} \leq l_n$ and $x^{\sim} \not\leq l_{n-1}$ then we may prove by induction that $l(l_x) \geq n$. Indeed, when n = 1 there is nothing to prove.
Now, when $x^{\sim} \leq l_n$ and $x^{\sim} \not\leq l_{n-1}$ then we let $a = l_n \wedge (x^{\sim} \vee l_{n-1})$ and $b = l_n \wedge x^{\sim}$ and like in the proof of Theorem 2.9.2. we may derive that
$l_{n-1} \lneq a \leq l_n$ whilst $b \not\leq l_{n-1}$. Because $M = \bigvee_{x \in M} x^{\sim}$ there must exist an element y in M for which $y^{\sim} \leq b$ and $y^{\sim} \not\leq l_{n-1}$. Again we may now apply the same argumentation as in Theorem 2.9.2. □

7.2.13. Corollary. If Q is Δ-injective then it is Σ-injective. □

We now intend to provide a relative version of "Hilbert's Basis Theorem", to this end we need the following:

7.2.14. Lemma. Let φ: R → S be a morphism of rings. If $\mathcal{F}$ is an additive topology on R then $\varphi(\mathcal{F}) = \{H$ left ideal of S, $\varphi^{-1}(H : y) \in \mathcal{F}$ for all $y \in S\}$ is an additive topology on S.

Proof. As straightforward as can be. □

7.2.15. Theorem. Let $\mathcal{F}$ be an additive topology on R and let i be the canonical inclusion R → R[X] where X is an indeterminate commuting with R. If M is an $\mathcal{F}$-Noetherian R-module then M[X] is $i(\mathcal{F})$-Noetherian.

Proof. Let N be an R[X]-submodule of M[X]. For m ∈ ℕ let L_m(N) be the set $\{x \in M, X^m x + X^{m-1}x_{m-1} + \dots + x_0 \in N\}$. Clearly $\{L_m(N), m \in \mathbb{N}\}$ yields an increasing sequence of R-submodules of M. Let N ⊂ P ⊂ M[X] be R[X]-submodules such that: $L_m(P)/L_m(N) \in \mathcal{J}_{\mathcal{F}}$ for all m ≥ 0 then we claim that $P/N \in \mathcal{T}_{i(\mathcal{F})}$, i.e. $(N : p) \bigcap R \in \mathcal{F}$ for all p ∈ P, and we establish this by induction on the degree of p. If deg p = 0 then p ∈ M and $p \in L_0(P)$; since $L_0(P)/L_0(N) \in \mathcal{T}_{\mathcal{F}}$ it then follows

that $(L_0(N) : p) = (N: p) \bigcap R \in \mathcal{F}$. So we may suppose now that $d = \deg p \geq 1$ and that $(N : q) \bigcap R \in \mathcal{F}$ for each $q \in P$ with $\deg q \leq d - 1$.
Put $p = X^d y_d + X^{d-1} y_{d-1} + \ldots + y_0 \in P$, where $y_i \in M$, $i = 0,\ldots,d$. Since $y_d \in L_d(P)$ and $L_d(P)/L_d(N) \in \mathcal{T}_{\mathcal{F}}$ we get $(L_d(N) : y_d) \in \mathcal{F}$. Pick $a \in (L_d(N) : y_d)$. There is an $f_a \in N$ such that $f_a = X^d y_d a + \ldots + a_0$. Hence $p_1 = pa - f_a \in P$ and $\deg(p_1) \leq d - 1$ The induction hypothesis entails: $(N : p_1) \bigcap R = \{b \in R, p_1 b \in N\} \in \mathcal{F}$. Obviously, for all $a \in (L_d(N) : y_d)$ we have $\{b \in R, p_1 b \in N\} \subset \{b \in R, pab \in N\}$, thus $\{b \in R, pab \in N\} \in \mathcal{F}$. Put $I = \{b \in R, pb \in N\} = (N : p) \bigcap R$ and $J = (L_d(N) : y_d)$. Then $I \in \mathcal{F}$ and for each $a \in I$ we see that $(J : a) \in \mathcal{F}$, hence $J \in \mathcal{F}$.
Consider an ascending chain of submodules of M[X],
$N_0 \subset N_1 \subset \ldots \subset N_n \subset \ldots$. The associated double sequence $\{L_i(N_j), (i,j) \in \mathbb{N} \times \mathbb{N}\}$ of submodules of M yields an ascending chain of R-submodules whenever one fixes either i or j. Since M is $\mathcal{F}$-Noetherian, the set
$\{L_i(N_j), (i,j) \in \mathbb{N} \times \mathbb{N}\}$ has an $\mathcal{F}$-maximal element (i.e. an H such that for each $H \subsetneq P$ with P in the set we have $P/H \in \mathcal{T}_{\mathcal{F}}$), say $L_p(N_q)$. It follows that: $L_j(N_j)/L_p(N_q) \in \mathcal{T}_{\mathcal{F}}$ for all $i \geq p$, $j \geq q$ and therefore, $L_i(N_{j+1})/L_i(N_j) \in \mathcal{T}_{\mathcal{F}}$ for each $i \geq p$ and $j \geq q$. On the other hand, for each $0 \leq i \leq p-1$ there is an $n_i \in \mathbb{N}$ such that for all $j \geq n_i$ we have: $L_i(N_{j+1})/L_i(N_j) \in \mathcal{T}_{\mathcal{F}}$. Put $d = \max\{q, n_0, \ldots, n_{p-1}\}$. Then $L_i(N_{j+1})/L_i(N_j) \in \mathcal{T}_{\mathcal{F}}$ for all $i \geq 0$ and $j \geq d$.
Consequently $N_{j+1}/N_j \in \mathcal{T}_{i\mathcal{F}}$ for all $j \geq d$ and therefore M[X] is i($\mathcal{F}$)-Noetherian.□

We will eventually benefit from a graded version of the foregoing theorem and therefore we introduce relative finiteness conditions in R-gr.

7.2.16. Definition. Let R be a $\mathbb{Z}$-graded ring, $\mathcal{H}$ a graded filter on R and M a graded left R-module. We write $C^g_{\mathcal{H}}(M)$ for the modular lattice consisting of the graded submodules N of M such that M/N is gr-$\mathcal{H}$-torsion free. We say that M is *gr.$\mathcal{H}$-Noetherian*, resp. *gr.$\mathcal{H}$-Artinian*, whenever $C^g_{\mathcal{H}}(M)$ is a Noetherian, resp. Artinian, lattice. For $\mathcal{H} = \{R\}$ we obtain the usual notion of a gr-Noetherian, resp. gr-Artinian module. If φ: R → S is a graded morphism of graded rings (of degree zero) and $\mathcal{H}$ is a graded filter on R then $\varphi(\mathcal{H}) = \{L \in L_g(S), \varphi^{-1}(L : y) \in \mathcal{H}$ for all $y \in h(S)\}$ is a graded filter on S (easy!). □

At this point we need to recall some elementary facts concerning external homogenization, cf. also C. Năstăsescu, F. Van Oystaeyen [3].
If R is $\mathbb{Z}$-graded then the polynomial ring R[X] may be viewed as a $\mathbb{Z}$-graded ring

by putting $R[X]_n = \{\sum_{i+j=n} a_i X^j,\ a_i \in R_i\}$ and in a similar way we may define a $\mathbb{Z}$ - gradation on M[X] for any graded left R-module M. If x ∈ M we decompose it as $x = x_{-m} + ... + x_0 + ... + x_n$ where the x_i are homogeneous.
Put $x^* = x_{-m} X^{m+n} + ... + x_0 X^n + ... + x_n \in M[X]_n$. For an arbitrary R-submodule N of M we let N^* be the graded R[X]-submodule of M[X] generated by all the x^*, x ∈ N.
Conversely, for an u ∈ h(M[X]), say: $u = u_{-k} X^{k+j} + ... + u_0 X^j + ... + u_j$ with $u_i \in M_i$, we put $u_* = u_{-k} + ... + u_0 + ... + u_j \in M$. To any graded R[X]-submodule T of M[X] we associate an R-submodule of M defined by $T_* = \{u_*,\ u \in h(T)\}$.
The reader may verify without effort that the following properties hold:

1. if x ∈ M, $(x^*)_* = x$.
2. if u ∈ h(M[X]) then $u = (u_*)^* X^k$ where k = deg u - deg $(u_*)^*$.
3. $(N^*)_* = N$ for any R-submodule N of M.
4. if $L \subsetneq N$ then $L^* \subsetneq N^*$.
5. if x ∈ M then $(N : x)^* = (N^* : x^*)$.
6. if u ∈ h(M[X]) and L is a graded R[X]-submodule of M[X] then $(L : u)_* = (L_* : u_*)$.

The inclusion R → R[X] is a graded morphism (in terms of the gradation defined on R[X] as before) so a graded filter $\mathcal{H}$ on R leads to a graded filter i($\mathcal{H}$) on R[X].

7.2.17. Proposition. If a graded left R-module M is gr-$\mathcal{H}$-Noetherian then M[X] is gr-i($\mathcal{H}$)-Noetherian.

Proof. Similar to the proof of 7.2.15. because for a graded submodule N of M[X] it follows that L_m(N) is a graded R-submodule of M for all m ∈ $\mathbb{N}$.

In conclusion of this section we now investigate the relations between the graded and ungraded relative Noetherian properties.

7.2.18. Lemma. If $\mathcal{H}$ is a graded filter on R let $\mathcal{F}$ be the additive topology generated by $\mathcal{H}$ (i.e. the smallest additive topology containing $\mathcal{H}$). If M is a graded left R-module then $C^g_{\mathcal{H}}(M) \subset C_{\mathcal{F}}(M)$. Consequently if M is $\mathcal{F}$-Noetherian then M is also gr-$\mathcal{H}$-Noetherian.

Proof. It is clear that $\mathcal{F} = \{$L left ideal of R, L $\supset$ J for some J $\in \mathcal{H}\}$. If N $\in C^g_{\mathcal{H}}(M)$ and N $\notin C_{\mathcal{F}}(M)$ then there is an x $\in$ M, x $\notin$ N, such that (N : x) $\in \mathcal{F}$. If $(N : x)_g$ is the graded left ideal of R generated by (N : x) $\bigcap$ h(R) then there exists a J $\in \mathcal{H}$ such that J $\subset$ $(N : x)_g$.
Suppose that x_i is an homogeneous component of x such that $x_i \notin$ N. Since N is graded it is obvious that J $\subset$ $(N : x)_g \subset (N : x_i)$, hence $(N : x_i) \in \mathcal{H}$, but the latter is a contradiction thus N $\in C_{\mathcal{F}}(M)$. □

7.2.19. Theorem. If $\mathcal{F}$ is the additive topology on R generated by the graded filter $\mathcal{H}$ and M is a graded left R-module then M is gr-$\mathcal{H}$-Noetherian if and only if M is $\mathcal{F}$-Noetherian.

Proof. The above lemma takes care of the "if"-part, so let us suppose that M is gr-$\mathcal{H}$-Noetherian and let there be given a chain of R-submodules of M, $N_0 \subset N_1 \subset \ldots \subset N_n \subset \ldots$, with $N_k \in C_{\mathcal{F}}(M)$ for all k. We obtain a chain of graded R[X]-submodules of M[X]:

$$N_0^* \subset N_1^* \subset \ldots \subset N_n^* \subset \ldots \qquad (*)$$

We claim that N $\in C_{\mathcal{F}}(M)$ entails $N^* \in C^g_{i(\mathcal{H})}(M[X])$.
Suppose the claim is false i.e. assume that there exists an x $\in$ h(M[X]) such that x $\notin N^*$ and $(N^* : x) \bigcap R \in \mathcal{H}$. Clearly, $x_* \notin$ N.
We have: $(N^* : x) \bigcap R \subset (N^* : x)_*$. Indeed if a $\in$ R, a $= a_{-k} + \ldots + a_0 + \ldots + a_j$, is such that ax $\in N^*$ then $a_{-k}x \in N^*, \ldots, a_0x \in N^*, \ldots, a_jx \in N^*$ (because N^* is graded). Then $a_{-k}X^{k+j}x \in N^*, \ldots, a_jx \in N^*$ and hence:
$a^* = a_{-k}X^{k+j} + \ldots + a_0X^j + \ldots + a_j \in (N^* : x)$. But the latter yields a $= (a^*)_* \in (N^* : x)_*$. So we arrive at $(N^* : x)_*$ *in* $\mathcal{F}$ and $(N : x_*) = ((N^*)_* : x_*) = (N^* : x)_* \in \mathcal{F}$, a contradiction.
Thus (*) is a chain in $C^g_{i(\mathcal{H})}(M[X])$. Since M[X] is gr-i($\mathcal{H}$)-Noetherian by Proposition 7.2.17., (*) is stationary and so is the original chain. □

As a first application of the foregoing technicalities we relate gr-Σ-injectivity and Σ-injectivity.

Consider a graded left R-module Q which is gr-injective (i.e. injective in R-gr) and put $\mathcal{H}_Q = \{L \in L_g(R), \mathrm{Hom}_R(R/L, Q) = 0\}$. It is not hard to check that $\mathcal{H}_Q$ is a graded filter on R. We say that Q is *gr-Σ-injective*,, resp. *gr-Δ-injective*, if and only if R is gr-$\mathcal{H}_Q$-Noetherian, resp. gr-$\mathcal{H}_Q$-Artinian.

7.2.20. Proposition. $\mathcal{H}_Q = \mathcal{F}_{E(Q)} \bigcap L_g(R)$.

Proof. It is clear that $\mathcal{F}_{E(Q)} \bigcap L_g(R) \subset \mathcal{H}_Q$.
Conversely, if $J \in \mathcal{H}_Q$ consider $x \neq 0$ in Q and $a \in R$. We have to establish that $(J : a) \not\subset \mathrm{Ann}(x)$. Put $x = x_1 + ... + x_n$, $a = a_1 + ... + a_m$ homogeneous decompositions indexed in such a way that: $\deg x_1 < ... < \deg x_n$, $\deg a_1 < ... < \deg a_n$. Since $x_n \neq 0$ there is a $\lambda_1 \in h(R)$ such that $\lambda_1 a_1 \in J$ and $\lambda_1 x_n \neq 0$; there is a $\lambda_2 \in h(R)$ such that $\lambda_2\lambda_1 a_2 \in J$ and $\lambda_1\lambda_2 x_n \neq 0$, etc ..., $\lambda_m \in h(R)$ such that $\lambda_m . ... . \lambda_1 a_m \in J$ and $\lambda_m . ... . \lambda_1 x_n \neq 0$. Putting $\lambda = \lambda_m . ... . \lambda_1$ yields $\lambda x \neq 0$ and $\lambda a \in J$. □

7.2.21. Corollary. Let Q be a gr-injective graded left R-module, then Q is gr-Σ-injective if and only if E(Q) is Σ-injective.

Proof. Let $\mathcal{F}$ be the additive topology on R generated by $\mathcal{H}_Q$. From $\mathcal{H}_Q \subset \mathcal{F}_{E(Q)}$ it follows that $\mathcal{F} \subset \mathcal{F}_{E(Q)}$. But R is $\mathcal{F}$-Noetherian in view of Theorem 7.2.19. hence R is $\mathcal{F}_{E(Q)}$-Noetherian, i.e. E(Q) is σ-injective. The converse is clear as well. □

7.2.22. Remark. Let K be a gr-field, i.e. a commutative graded ring without proper graded ideals. It is very easy to check that $K \cong k[X, X^{-1}]$ where k is a field and X is an indeterminate such that deg X is the minimal positive really occuring in the gradation of K. It is obvious that K is gr-Artinian but not Artinian. However in general, if Q is a gr-injective R-module then it may be shown that Q is gr-injective if and only if E(Q) is Δ-injective (note: it is unnecessary to point out that k(X) is Artinian in the case where R = K is a gr-field).

§7.3 Relative Krull Dimension.

Let R be a ring, $\mathcal{F}$ a (left) additive topology on R.

7.3.1. Definition. For a left R-module M we define the *relative Krull dimension of*

M with respect to $\mathcal{F}$ to be the Krull dimension of the modular lattice $C_{\mathcal{F}}(M)$, $\mathrm{Kdim}_{\mathcal{F}}$ M = Kdim $(C_{\mathcal{F}}(M))$. For an ideal I of R and $\mathcal{F}_I = \mathcal{F}_{E(R/I)}$ we write Kdim_I for $\mathrm{Kdim}_{\mathcal{F}_I}$. Relative Krull dimension may also be defined recursively as follows: if M is $\mathcal{F}$-torsion put $\mu_{\mathcal{F}}(M) = -1$, if α is an ordinal such that $\mu_{\mathcal{F}}(M) \not< \alpha$ then we put $\mu_{\mathcal{F}}(M) = \alpha$ if for any descending chain $M \supset M_1 \supset M_2 \supset ... \supset M_n \supset ...$ of submodules there exists $n_0 \in \mathbb{N}$ such that $\mu_{\mathcal{F}}(M_n/M_{n+1}) < \alpha$ for all $n \geq N_0$.

7.3.2. Proposition. $\mathrm{Kdim}_{\mathcal{F}}(M) = \mu_{\mathcal{F}}(M)$ provided either side exists.

Proof. Put $\mathrm{Kdim}_{\mathcal{F}}(M) = \alpha$. If $\alpha = -1$ then $\mu_{\mathcal{F}}(M) \leq \alpha$ holds. In general $\mu_{\mathcal{F}}(M) > \alpha$ would entail the existence of a descending chain $M \supset M_1 \supset ... \supset M_n \supset ...$ of submodules of M such that $\mu_{\mathcal{F}}(M_n/M_{n+1}) \geq \alpha$ for all $n \geq n_0$, some $n_0 \in \mathbb{N}$ Consider $M \supset M_1^{\sim} \supset ... \supset M_n^{\sim} \supset ...$ in $C_{\mathcal{F}}(M)$. Since $\mathrm{Kdim}_F(M) = \alpha$ there is an $n_0' \in \mathbb{N}$ such that $\mathrm{Kdim}_{\mathcal{F}}(M_n^{\sim}/M_{1+n}^{\sim}) < \alpha$ for all $n \geq n_0'$. Moreover, since φ: $C_{\mathcal{F}}(M_n/M_{n+1}) \to C_{\mathcal{F}}(M_n^{\sim}/M_{1+n}^{\sim})$, $X/M_{n+1} \mapsto X^{\sim}/M_{1+n}^{\sim}$, is strictly increasing: $\mathrm{Kdim}_{\mathcal{F}}(M_n/M_{1+n}) < \alpha$ for all $n \geq n_0$, some n_0. The induction hypothesis entails that $\mu_{\mathcal{F}}(M_n/M_{1+n}) < \alpha$ for all $n \geq n_0$, a contradiction. Hence $\mu_{\mathcal{F}}(M) \leq \mathrm{Kdim}_{\mathcal{F}}(M)$.
In a similar way we may establish the converse inequality. □

7.3.3. Proposition. For any submodule N of M we have: $\mathrm{Kdim}_{\mathcal{F}}(M) = \sup\{\mathrm{Kdim}_{\mathcal{F}}(N), \mathrm{Kdim}_{\mathcal{F}}(M/N)\}$, if either side exists.

Proof. cf. Proposition 3.2.1. □

7.3.4. Corollary. If $\mathrm{Kdim}_{\mathcal{F}}(R)$ exists then for evry $\mathcal{F}$-finitely generated left R-module M we have: $\mathrm{Kdim}_{\mathcal{F}}(M) \leq \mathrm{Kdim}_{\mathcal{F}}(R)$.

Proof. Let $x_1, ..., x_n \in M$ be such that $M/Rx_1 + ... + Rx_n$ is $\mathcal{F}$-torsion, f: $R^n \to M$ the corresponding morphism with $\mathcal{F}$-torsion cokernel. By the proposition $\mathrm{Kdim}_{\mathcal{F}}(\mathrm{Im}(f)) \leq \mathrm{Kdim}_{\mathcal{F}}(R^n) = \mathrm{Kdim}_{\mathcal{F}}(R)$ and $\mathrm{Kdim}_{\mathcal{F}}(M) =$ $\sup\{\mathrm{Kdim}_{\mathcal{F}}(\mathrm{Im}(f)), \mathrm{Kdim}_{\mathcal{F}}(\mathrm{Coker}(f))\} = \mathrm{Kdim}_{\mathcal{F}}(\mathrm{Im}(f))$.
Consequently, $\mathrm{Kdim}_{\mathcal{F}}(M) \leq \mathrm{Kdim}_{\mathcal{F}}(R)$. □

If $C_{\mathcal{F}}$ (M) is α-critical for an ordinal $\alpha \geq 0$, then M is said to be an *$\mathcal{F}$-α-critical submodule.*

7.3.5. Corollary. Suppose that M is $\mathcal{F}$-torsion free and that $\mathrm{Kdim}_{\mathcal{F}}$ (M) exists, then:

1. $\mathrm{Kdim}_{\mathcal{F}}$ (M) $\leq$ sup {$\mathrm{Kdim}_{\mathcal{F}}$ (M/E) + 1, E essential in M} =
sup {$\mathrm{Kdim}_{\mathcal{F}}$ (M/E) + 1, E essential in M and M/E $\mathcal{F}$-torsion free}.
2. M has finite Goldie dimension.

Proof. 1. If E is an essential element in $C_{\mathcal{F}}$ (M) then E is essential in M, so it suffices to apply Proposition 3.2.3. an 7.2.4.
2. Direct from Proposition 3.2.2. and Corollary 7.2.5. □

7.3.6. Proposition. Let $\mathcal{F}_1, \mathcal{F}_2$ be additive topologies on R:

1. If $\mathcal{F}_1 \subset \mathcal{F}_2$ and $\mathrm{Kdim}_{\mathcal{F}_1}$ (M) exists then $\mathrm{Kdim}_{\mathcal{F}_2}$ (M) $\leq$ $\mathrm{Kdim}_{\mathcal{F}_1}$ (M).
2. Putting $\mathcal{F} = \mathcal{F}_1 \bigcap \mathcal{F}_2$ then $\mathrm{Kdim}_{\mathcal{F}}$ (M) = sup {$\mathrm{Kdim}_{\mathcal{F}_1}$ (M), $\mathrm{Kdim}_{\mathcal{F}_2}$ (M)}.

Proof. 1. Obvious from $C_{\mathcal{F}_2}$ (M) $\subset$ $C_{\mathcal{F}_1}$ (M).
2. By 1., $\mathrm{Kdim}_{\mathcal{F}}$ (M) $\leq$ sup {$\mathrm{Kdim}_{\mathcal{F}_1}$ (M), $\mathrm{Kdim}_{\mathcal{F}_2}$ (M)}. For the converse inequality we consider φ: $C_{\mathcal{F}}$ (M)$\to$ $C_{\mathcal{F}_1}$ (M) $\times$ $C_{\mathcal{F}_2}$ (M), x $\mapsto$ ($X^{\sim}$, $X^{\approx}$) where $\sim$ and $\approx$ denote the closure operations with respect to $\mathcal{F}_1$ and $\mathcal{F}_2$ resp. . It suffices to check that φ is strictly increasing. □

7.3.7 Corollary. Let I be a semiprime ideal of R such that R/I is a left Goldie ring. For any left R-module M we have: Kdim_I (M) = sup {Kdim_{P_i} (M), P_i a minimal prime ideal over I}.

Proof. By the proposition it is enough to establish that $\mathcal{F}_I = \bigcap_{i=1}^n \mathcal{F}_{P_i}$.
But R/I $\hookrightarrow \bigoplus_{i=1}^n$ R/P_i yields E(R/I) $\hookrightarrow \bigoplus_{i=1}^n$ E(R/P_i) and hence $\mathcal{F}_I \supset$
$\bigcap_{i=1}^n \mathcal{F}_{P_i}$.
Conversely if J $\in$ $\mathcal{F}_I$ then (J : x) $\cap$ $\mathcal{C}$(I) $\neq$ ϕ for all x $\in$ R. We claim that $\mathcal{C}$(I) $\subset$ $\mathcal{C}(P_i)$, i = 1,...,n. If c $\in$ $\mathcal{C}$(I) and λc $\in$ P_i then ($\bigcap_{j\neq i}$ P_j) λc $\subset$ I and therefore

$(\bigcap_{j\neq i} P_i)\ \lambda \subset I \subset P_i$. So $\lambda \in P_i$ follows and thus $c \in P_i$. Consequently: $(J : x) \cap C(P_i) \neq \phi$ for all $x \in R$, thus $J \in \mathcal{F}_{P_i}$ □

Whenever Kdim $(C_{\mathcal{F}}\ ({}_R R))$ exists we say that R has *$\mathcal{F}$-Krull dimension.* If I is an ideal of R and φ: R → R/I is the canonical epimorphism then we let $\varphi(\mathcal{F})$ be the direct image of $\mathcal{F}$ on the ring R/I, i.e. $\varphi(\mathcal{F}) = \{J/I,\ J \in \mathcal{F}\}$ and $C_{\varphi(\mathcal{F})}\ (R/I) = \{J/I,\ J \in C_{\mathcal{F}}(R),\ J$ a left ideal of $R\}$. In particular if $\mathcal{F} = \mathcal{F}_P$ where P is a prime ideal containing I, $\varphi(\mathcal{F}_P) = \mathcal{F}_{P/I}$.

7.3.8. Proposition. If R has $\mathcal{F}$-Krull dimension then:

1. If M ∈ R-mod has $\mathcal{F}$-Krull dimension then $\mathrm{Kdim}_{\mathcal{F}}$ (M) ≤ $\mathrm{Kdim}_{\mathcal{F}}$ (R).
2. If I is an ideal, φ: R → R/I the canonical epimorphism, then R/I has $\varphi(\mathcal{F})$-Krull dimension and $\mathrm{Kdim}_{\varphi(\mathcal{F})}$ (R/I) ≤ $\mathrm{Kdim}_{\mathcal{F}}$ (R).
3. If I is as in 2. and IM = 0 then $\mathrm{Kdim}_{\mathcal{F}}$ (M) = $\mathrm{Kdim}_{\varphi(\mathcal{F})}$ (M) if either side exists.

Proof. 1. Write $M = \sum_{i\in A} Rm_i$; then $M = \bigvee_{i\in A} (Rm_i)^{\sim}$. By Proposition 7.2.4. and n7.3.3. we have: $\mathrm{Kdim}_{\mathcal{F}}$ (M) = $\sup_{i\in A} \{\mathrm{Kdim}_{\mathcal{F}}\ (Rm_i)\}$ ≤ $\mathrm{Kdim}_{\mathcal{F}}$ (R).
2. This is clear because $C_{\varphi(\mathcal{F})}$ (R/I) → $C_{\mathcal{F}}$ (R), J/I ↦ J, is a strictly increasing map.
3. Obvious because $C_{\mathcal{F}}$ (M) and $C_{\varphi(\mathcal{F})}$ (M) are isomorphic. □

7.3.9. Proposition. Let R be a semiprime ring with $\mathcal{F}$-Krull dimension. If R is $\mathcal{F}$-torsion free then R is a left Goldie ring and $\mathrm{Kdim}_{\mathcal{F}}$ (R) = sup $\{\mathrm{Kdim}_{\mathcal{F}}$ (R/E) + 1, E essential and R/E $\mathcal{F}$-torsion free$\}$.

Proof. By Corollary 3.5., R has finite Goldie dimension. If K ≠ 0 is a left ideal then $C_{\mathcal{F}}$ (K) has Krull dimension, α say, thus $C_{\mathcal{F}}$ (K) contains an $\mathcal{F}$-critical element, A say (cf. Section 3.2.).
We check now that A is monoform (cf. Section 5.2.). If f ∈ Hom_R (B,A) for some submodule B of A, is nonzero such that Ker f ≠ 0, then $\mathrm{Kdim}_{\mathcal{F}}$ (Im (f)) = $\mathrm{Kdim}_{\mathcal{F}}$ (B/Ker (f)) < $\mathrm{Kdim}_{\mathcal{F}}$ (B) = α (since A is $\mathcal{F}-\alpha$-critical and B has the same property). But $\mathrm{Kdim}_{\mathcal{F}}$ (Im (f)) = $\mathrm{Kdim}_{\mathcal{F}}$ (A), contradiction. Hence Ker f = 0. Consequently, A is a monoform submodule of K. Theorem 5.2.3. entails that R is a left Goldie ring. Let L be an essential left ideal of R and let s be a regular element contained in

L. Since $Rs^n/Rs^{n+1} \simeq R/Rs$, $\text{Kdim}_{\mathcal{F}}(Rs^n/Rs^{n+1}) = \text{Kdim}_{\mathcal{F}}(R/Rs) \geq \text{Kdim}_{\mathcal{F}}(R/L)$. On the other hand, $Rs \supset Rs^2 \supset \ldots \supset Rs^n \supset \ldots$ is strictly descending hence $\text{Kdim}_{\mathcal{F}}(Rs^n/Rs^{1+n}) < \text{Kdim}_{\mathcal{F}}(R)$ for all $n \geq n_0$. In particular, $\text{Kdim}_{\mathcal{F}}(R/I) < \text{Kdim}_{\mathcal{F}}(R)$, hence $\text{Kdim}_{\mathcal{F}}(R/I) + 1 \leq \text{Kdim}_{\mathcal{F}}(R)$ and hence equality follows. □

7.3.10. Lemma. Let $\mathcal{F}$ be an additive topology on R and let P be a prime ideal of R such that R/P is a left Goldie ring then $P \in \mathcal{F}$ or $P \in C_{\mathcal{F}}(R)$.

Proof. If $P \notin C_{\mathcal{F}}(R)$ then $t_{\mathcal{F}}(R/P) \neq 0$, say $t_{\mathcal{F}}(R/P) = I/P$. Now I/P is an essential ideal of the left Goldie ring R/P, hence there is a $c \in I$ such that c mod P is regular in R/P. On the other hand there exists a $J \in \mathcal{F}$ such that $Jc \subset P$ thus $J \subset P$ or $P \in \mathcal{F}$. □

We write $\text{Spec}_{\mathcal{F}}(R)$ for the set of prime ideals P of R such that $P \notin \mathcal{F}$. The foregoing lemma entails that $\text{Spec}_{\mathcal{F}}(R) = \text{Spec}(R) \bigcap C_{\mathcal{F}}(R)$ for any ring with left Krull dimension. $\text{Spec}_{\mathcal{F}}(R)$ is a generically closed subset of Spec (R), cf. F. Van Oystaeyen, A. Verschoren [1], in particular if $Q \subset P$ are prime ideals such that $P \in \text{Spec}_{\mathcal{F}}(R)$ then $Q \in \text{Spec}_{\mathcal{F}}(R)$. Let $\text{Min}_{\mathcal{F}}(R)$ be the set of all minimal prime ideals P of R such that $P \notin \mathcal{F}$ i.e. $\text{Min}_{\mathcal{F}}(R)$ consists of the minimal elements in $\text{Spec}_{\mathcal{F}}(R)$.

7.3.11. Proposition. Let $\mathcal{F}$ be an additive topology on a ring R which has left Krull dimension, then: $\text{Kdim}_{\mathcal{F}}(R) = \sup\{\text{Kdim}_{\mathcal{F}}(R/P), P \in \text{Min}_{\mathcal{F}}(R)\}$.

Proof. $N = \bigcap_{i=1}^n P_i$ is the nilradical of R, then $N^m = 0$ for some $m \in \mathbb{N}$. Applying Proposition 7.3.3. repeatedly we arrive at: $\text{Kdim}_{\mathcal{F}}(R) = \sup\{\text{Kdim}_{\mathcal{F}}(N^{m-1}), \text{Kdim}_{\mathcal{F}}(N^{m-2}/N^{m-1}), \ldots, \text{Kdim}_{\mathcal{F}}(R/N)\}$. Each N^i/N^{i+1} is a finitly generated R/N-module, $1 \leq i \leq m-1$, so Corollary 7.3.4. yields $\text{Kdim}_{\mathcal{F}}(R) = \text{Kdim}_{\mathcal{F}}(R/N)$. But R/N embeds in $\bigoplus_{i=1}^n R/P_i$ and therefore: $\text{Kdim}_{\mathcal{F}}(R) \leq \sup\{\text{Kdim}_{\mathcal{F}}(R/P_i), i = 1,\ldots,n\} = \sup\{\text{Kdim}_{\mathcal{F}}(R/P), P \in \text{Min}_{\mathcal{F}}(R)\}$.
The converse inequality follows from Proposition 7.3.3. □

7.3.12. Corollary. If P is a minimal prime ideal of a ring R with left Krull dimension, then Kdim_P (R) = 0, i.e. E(R/P) is Δ-injective.

Proof. We have $\text{Min}_{\mathcal{F}_P}$ (R) = {P} and thus: Kdim_P (R) = Kdim_P (R/P) = $\text{Kdim}_{(0)}$ (R/P). Therefore we may assume that P = 0 and that R is a prime ring. But then $\mathcal{F}_0$ = {L, essential left ideal of R} = $\mathcal{F}_S$ where S is the set of regular elements of R. Now, if I ∈ $C_{\mathcal{F}_S}$ (R) then I = R ⋂ QI where Q is the simple Artinian ring of fractions (left) of R. It is clear that $C_{\mathcal{F}_S}$(R) is an Artinian lattice and therefore $\text{Kdim}_{\mathcal{F}_S}$ (R) = 0, hence Kdim_P (R) = 0. □

7.3.13. Proposition. Let R be a ring with $\mathcal{F}$-Krull dimension. If there is a strictly descending chain in $\text{Spec}_{\mathcal{F}}$ (R), say $R \supset P_0 \supset P_1 \supset \ldots \supset P_n$, then $\text{Kdim}_{\mathcal{F}}$ (R) ≥ n.

Proof. It is clear that $\text{Kdim}_{\mathcal{F}}$ $(R/P_o) \geq 0$. Assume that we have arrived at $\text{Kdim}_{\mathcal{F}}$ $(R/P_{n-1}) \geq$ n-1. Put $\overline{R} = R/P_n$ and $\overline{P} = P_{n-1}/P_n$. Since $\overline{P} \neq 0$, $\overline{P}$ is essential as a left ideal of $\overline{R}$ so by Goldie's theorem $\overline{P}$ contains a regular element $\overline{s}$ of $\overline{R}$. Consider, $\overline{R} \supset \overline{R}\overline{s} \supset \overline{R}\overline{s}^2 \supset \ldots \supset \overline{R}\overline{s}^m \supset \ldots$ .From $\overline{R}\overline{s}^m/\overline{R}\overline{s}^{m+1} \cong \overline{R}/\overline{R}\overline{s}$ and $\overline{R}\overline{s} \subset \overline{P}$ we derive that: $\text{Kdim}_{\mathcal{F}}$ $(\overline{R}\overline{s}^m/\overline{R}\overline{s}^{m+1}) \geq$ $\text{Kdim}_{\mathcal{F}}$ $(\overline{RP})$ = $\text{Kdim}_{\mathcal{F}}$ $(R/P_{n-1}) \geq$ n-1, fo all m. In view of Proposition 7.3.2. we obtain that $\text{Kdim}_{\mathcal{F}}$ $(\overline{R}) \geq$ n and consequently $\text{Kdim}_{\mathcal{F}}$ (R) ≥ n as desired. □

7.3.14. Corollary. If P is a prime ideal of the left Noetherian rinf R then ht(P) ≤ Kdim_P (R).

§7.4 Relative Krull Dimension Applied to the Principal Ideal Theorem.

Let R be a ring, $\mathcal{C}$ a class of left R-modules. The additive topology $\mathcal{F}$ is *cogenerated* by $\mathcal{C}$ if $\mathcal{F}$ = {L left ideal of R such that for all M ∈ $\mathcal{C}$, Hom_R (R/L, E(M)) = 0}. If $\mathcal{F}$ is cogenerated by {M} then we write $\mathcal{F} = \mathcal{F}_{E(M)}$ (or $\mathcal{F} = \mathcal{F}_M$). If $\mathcal{F}$ is cogenerated by $\mathcal{C}$ then $\mathcal{F} = \bigcap \{\mathcal{F}_{E(M)}, M \in \mathcal{C}\}$; it is clear that every M in $\mathcal{C}$ is torsion free for $\mathcal{F}$. For an ideal I of R we let G_I (R) = $\bigcap_{n\geq 0} I^n/I^{1+n}$ for the graded

ring associated to the I-adic filtration on R. (see Section 4.2.). We say that an ideal I has property (*) with respect to the additive topology $\mathcal{F}$ if $IK^{\sim} \subset (IK)^{\sim}$ for each left ideal K of R. Moreover, we define:
$G_{I,\mathcal{F}}(K) = \bigoplus_{n\geq 0} (I^n \cap (K + I^{n+1})^{\sim}/I^n)$.

7.4.1. Lemma. If I has property (*) with respect to $\mathcal{F}$ then $G_{I,\mathcal{F}}(K)$ is a graded left ideal of $G_I(R)$.

Proof. The lemma is a direct consequence of:
$I(I^n \cap (K + I^{n+1})^{\sim}) \subset I^{n+1} \cap I(K + I^{n+1})^{\sim} \subset I^{1+n} \cap (I(K + I^{n+1}))^{\sim} \subset I^{1+n} \cap (IK + I^{n+2})^{\sim} \subset I^{1+n} \cap (K + I^{n+2})^{\sim}$. □

Note that for commutative R property (*) holds for all ideals I of R; indeed if $a \in I$, $b \in K^{\sim}$ then there is a $J \in \mathcal{F}$ such that $Jb \subset K$ and thus $Jab = aJb \subset aK \subset IK$, meaning that $IK^{\sim}/IK$ is $\mathcal{F}$-torsion or $IK^{\sim} \subset (IK)^{\sim}$. A left R-module M is *$\mathcal{F}$-simple* ($\mathcal{F}$-cocritical) if M is $\mathcal{F}$-torsion free and $C_{\mathcal{F}}(M)$ consists of O,M only. Obviously M is $\mathcal{F}$-simple if and only if M is $\mathcal{F}$-torsion free and M/N is $\mathcal{F}$-torsion for each nonzero submodule N of M. If M is $\mathcal{F}$-simple then M is uniform and each nonzero submodule of M is again $\mathcal{F}$-simple. □

7.4.2. Proposition. Let R be a left Noetherian ring and let $\mathcal{F}$ be the additive topology cogenerated by $\mathcal{C}$ in R-mod. Let I be an ideal of R such that each $M \in \mathcal{C}$ is annihilated by some power of I and assume that I has the left Artin-Rees property, then:

1. If $K \in C_{\mathcal{F}}(R)$ is such that R/K is an essential extension of an $\mathcal{F}$-simple module then $I^n \subset K$ for some $n \in \mathbb{N}$.
2. If I has property (*) then for $K, L \in C_{\mathcal{F}}(R)$ such that $K \subset L$ and $G_{I,\mathcal{F}}(K) = G_{I,\mathcal{F}}(L)$ we have K = L.
3. Let $\mathcal{L}$ be the partially ordered set of graded left ideals $\bigoplus_{n\geq 0} K_n$ of $G_I(R)$ such that K_n is $\mathcal{F}$-closed in I^n/I^{n+1}, then $\mathrm{Kdim}_{\mathcal{F}}(R) \leq \mathrm{Kdim}(\mathcal{L})$ whenever $\mathcal{L}$ has Krull dimension.

Proof. 1. Let R/K be essential over the $\mathcal{F}$-simple module L/K. There is an $M \in \mathcal{C}$ such that $\mathrm{Hom}_R(L/K, E(M)) \neq 0$, hence there is a nonzero morphism

f: $L/K \to E(M)$ and $Im(f) \cap M \neq 0$. Put $L'/K = f^{-1}(M)$, $K \subsetneq L'$. By restricting f to L'/K we obtain a nonzero g: $L'/K \to M$ which has to be monomorphic because L'/K is $\mathcal{F}$-simple while M is $\mathcal{F}$-torsion free.
From $I^n M = 0$ for some $n \in \mathbb{N}$ we derive that $I^n (L'/K) = 0$ hence $I^n L' \subset K$. Select the left ideal H of R such that $IH \subset K \subsetneq H \subset L$. The Artin-Rees property for I entails that we may find an $m \in \mathbb{N}$ such that $I^m \cap H \subset IH$; hence,
$H \cap (K + I^m) = K + H \cap I^m \subset K + IH = K$ and therefore: $H \cap (K + I^m) = K$.
As an essential extension of a uniform module K has to be an irreducible left ideal of R, consequently $K + I^m = K$ and thus $I^m \subset K$.

2. From $G_{I,\mathcal{F}}(K) = G_{I,\mathcal{F}}(L)$ we deduce: $I^n \cap (K + I^{n+1}) \cong I^n \cap (L + I^{n+1})^{\sim}$ for all $n \in \mathbb{N}$. Assume that $K \neq L$. By the Noetherian assumption we may suppose that K is maximal amongst the $\mathcal{F}$-closed left ideals of R properly contained in L i.e. L/K is $\mathcal{F}$-simple.
Pick a left ideal H of R maximal with respect to $K \subset H$ and $L \cap H = K$. Then there is a canonical monomorphism φ: $L/K \to R/H$ such that R/H is an essential extension of Im φ. Thus R/H is $\mathcal{F}$-torsion free. From 1. we infer that $I^n \subset H$ for some $n \in \mathbb{N}$. Thus $(K + I^n)^{\sim} \subset H$ and from our hypohesis we obtain:
$I^{n-1} \cap L = I^{n-1} \cap (L + I^n)^{\sim} \cap L = I^{n-1} \cap (K + I^n)^{\sim} \cap L \subset I^{n-1} \cap H \cap L$ $= I^{n-1} \cap K$, or $I^{n-1} \cap L = I^{n-1} \cap K$. Now we may choose $m \in \mathbb{N}$ to be the least integer such that $I^m \cap K = I^m \cap L$. Consider the left ideal $L \cap (K + I^m)^{\sim}$. Since K is maximal in L we get: $L \cap (K + I^m)^{\sim} = L$ or $L \cap (K + I^m)^{\sim} = K$ First let us assume that $L = L \cap (K + I^m)^{\sim}$, then we obtain:
$L = L \cap (K + (I^m)^{\sim}) = K + (L \cap (I^m)^{\sim}) = K + (K \cap (I^m)^{\sim}) = K$,
a contradiction.
Hence we have to agree with $L \cap (K + I^m)^{\sim} = K$. By the hypothesis we now obtain:
$I^{m-1} \cap K = I^{m-1} \cap L \cap (K + I^m)^{\sim} = L \cap I^{m-1} \cap (L + I^m)^{\sim} = I^{m-1} \cap L$, contradicting the choice of m.

3. Follows easily from 2. □

Recall that an ideal I of R is invertible if there is an overring S of R and an R-subbimodule I^{-1} of S such that $II^{-1} = I^{-1}I = R$ (see Section 6.5.). An additive topology $\mathcal{F}$ on a left Noetherian ring R is I-*invariant*, for an invertible ideal I of R, if IL/IK and $I^{-1}L/I^{-1}K$ are $\mathcal{F}$-torsion free whenever $K \subset L$ are left ideals of R such that L/K is $\mathcal{F}$-torsion free.

7.4.3. Lemma. If I is an invertible ideal of R and $\mathcal{F}$ is I-invariant then I has property $(*)$ with respect to $\mathcal{F}$.

Proof. Let K be a left ideal of R. Since $R/(IK)^\sim$ is $\mathcal{F}$-torsion free it follows that $I^{-1}/I^{-1}(IK)^\sim$ is $\mathcal{F}$-torsion free. From $R \subset I^{-1}$ we obtain that $R/R \cap I^{-1}(IK)^\sim$ is $\mathcal{F}$-torsion free. Because $K \subset R \cap I^{-1}(IK)^\sim$ we obtain $K^\sim \subset R \cap I^{-1}(IK)^\sim$ and thus $K^\sim \subset I^{-1}(IK)^\sim$ or $IK^\sim \subset (IK)^\sim$. □

7.4.4. Theorem. (Jategaonkar, [3]). Let I be a proper invertible ideal in a left Noetherian ring R, let $\mathcal{F}$ be an I-invariant additive topology on R cogenerated by a class of R/I-modules, then: $\text{Kdim}_\mathcal{F}$ (R) = $\text{Kdim}_\mathcal{F}$ (R/I) + 1.

Proof. Since I has the Artin-Rees property (cf. Section 6.5.). From Proposition 7.4.2. we retain $\text{Kdim}_\mathcal{F}$ (R) $\leq$ Kdim ($\mathcal{L}$). For each $n \geq 0$ we let L_n be the left ideal of R such that $I^{n+1} \subset L_n \subset I^n$ and $K_n = L_n/I^{n+1}$. Then $I \subset I^{-n} L_n \subset R$. Since I^n/L_n is $\mathcal{F}$-torsion free and $\mathcal{F}$ is invariant, $R/I^{-n}L_n$ is $\mathcal{F}$-torsion free. The map sending $\bigoplus_{n\geq 0} K_n$ to the ascending (but possibly constant) chain $\{I^{-n}L_n/I, n \geq 0\}$ of closed left submodules of R/I, is strictly increasing. Hence, Kdim $\mathcal{L} \leq \text{Kdim}_\mathcal{F}$ (R/I) + 1. Proposition 7.4.2. then leads to $\text{Kdim}_\mathcal{F}$ (R) $\leq \text{Kdim}_\mathcal{F}$ (R/I) + 1. Since $\mathcal{F}$ is non-trivial, $I \notin \mathcal{F}$ and therefore $\text{Kdim}_\mathcal{F}$ (R/I) ≥ 0. Since $\mathcal{F}$ is I-invariant the lattices $C_\mathcal{F}$ (I^n/I^{n+1}) are isomorphic for all $n \geq 0$. Thus $\text{Kdim}_\mathcal{F}$ (R/I) $<$ Kdim$\mathcal{F}$ (R) and hence $\text{Kdim}_\mathcal{F} \geq 1 + \text{Kdim}_\mathcal{F}$ (R/I). □

7.4.5. Proposition. Let P be a prime ideal of a left Noetherian ring R and I an invertible ideal of R contained in P such that IP = PI, then $\mathcal{F}_P$ is I-invariant.

Proof. If $a \in I$ is such that $(IP : a) \in \mathcal{F}_P$ then $(IP : a) \cap \mathcal{C}(P) \neq \phi$ so $sa \in IP = PI$ for some $s \in \mathcal{C}(P)$, hence $saI^{-1} \subset P$. Since s is regular in R/P, $aI^{-1} \subset P$ and $a \in PI = IP$ follows. Thus I/IP is $\mathcal{F}_P$-torsion free.

Similarly one establishes that $I^{-1}/I^{-1}P$ is $\mathcal{F}_P$-torsion free. Consider left ideals $K \subset L$ of R such that L/K is $\mathcal{F}_P$-torsion free and suppose that IL/IK is not $\mathcal{F}_P$-torsion free. Then we may select left ideals, $K \subset K_1 \subsetneq L_1 \subset L$ such that there exists a monomorphism u: $L_1/K_1 \to R/P$ but Hom_R $(IL_1/IK_1, E(R/P)) = 0$. Since I is flat both as a left and right R-module (cf. Section 2.11.):

$1_I \otimes_R (L_1/K_1) \to I \otimes_R (R/P)$ is monomorphic, hence there exists a monomorphism $IL_1/IK_1 \to I/IP$. However, the fact that I/IP is $\mathcal{F}_P$-torsion free leads to $IL_1/IK_1 = 0$ or $IL_1 = IK_1$ and hence $L_1 = K_1$, a contradiction.

To establish that $I^{-1}L/I^{-1}K$ is $\mathcal{F}_P$-torsion free we procede in a similar way. □

7.4.6. Theorem. A prime ideal P in a left Noetherian ring R has height one when it is minimal over an invertible ideal I of R.

Proof. Let N be the prime radical of I, then $N^m \subset I \subset N$ for some $m \in \mathbb{N}$. Since $I^{-1}NI \subset I^{-1}I = R$ and $(I^{-1}NI)^m = I^{-1}N^mI \subset I$ it follows that $I^{-1}NI \subset N$. Similarly: $INI^{-1} \subset N$. Thus, $IN = NI$. Let $X = \{P_1 = P, ..., P_k\}$ be the set of prime ideals minimal over I, $N = \bigcap_{i=1}^k P_i$. The map $Q \to IQI^{-1}$ defines a permutation of X. Hence $I^nPI^{-n} = P$ or $I^nP = PI^n$. Since I^n is invertible, Proposition 7.4.5. and Theorem 7.4.4. yield: Kdimp (R) = 1 + Kdimp (R/I^n). Since Kdimp (R/I^n) = 0, cf. Corollary 7.3.11., and ht(P) < Kdim_P (R), cf. Corollary 7.3.13., we must have ht(P) ≤ 1. Since minimal prime ideals have nonzero annihilators they cannot contain an invertible ideal, hence ht(P) = 1. □

§7.5 Relative Gabriel Dimension.

There is some interest in introducing the notion of relative Gabriel dimension because it may be linked to properties of modules such that each quotient module has finite Goldie dimension.

7.5.1. Definition. Let $\mathcal{F}$ be an additive topology on R and let M be a left R-module. The *relative Gabriel dimension of* M with respect to $\mathcal{F}$ is defined to be the Gabriel dimension of the modular lattice $C_\mathcal{F}(M)$, we write: $\text{Gdim}_\mathcal{F}(M) = \text{Gdim}\, C_\mathcal{F}(M)$

7.5.2. Proposition. If N is a *a* submodule of M then: $\text{Gdim}_\mathcal{F}$ (M) = sup $\{\text{Gdim}_\mathcal{F}$ (N), $\text{Gdim}_\mathcal{F}$ (M/N)$\}$, if either side exists.

Proof. Direct from the results of Chapter 3, e.g. Proposition 3.4.3. □

7.5.3. Proposition. If $M = \bigoplus_{i\in \mathcal{A}} M_i$ then $\text{Gdim}_\mathcal{F}$ exists if and only if $\text{Gdim}_P\, M_i$ exists for each $i \in \mathcal{A}$ and in this case we have: $\text{Gdim}_\mathcal{F}$ (M) = sup $\{\text{Gdim}_\mathcal{F}$ (M_i), $i \in \mathcal{A}\}$.

Proof. Again a consequence of Proposition 3.4.3.. □

For a non-limit ordinal α, we say that M is *$\mathcal{F} - \alpha$-simple* if and only if $C_{\mathcal{F}}$ (M) is α-simple (cf. Section 3.4.) and we say that M is *Gabriel $\mathcal{F}$-simple* whenever M is $\mathcal{F} - \alpha$-simple for some ordinal α. It is clear from these definitions that M is $\mathcal{F} - \alpha$-simple exactly then when $\mathrm{Gdim}_{\mathcal{F}}$ (M) = $\mathrm{Gdim}_{\mathcal{F}}$ (N) = α and $\mathrm{Gdim}_{\mathcal{F}}$ (M/N) $< \alpha$ for every nonzero submodule N of M. In particular, if M is $\mathcal{F} - \alpha$-simple then M is $\mathcal{F}$-torsion free and any nonzero submodule of an $\mathcal{F} - \alpha$-simple module is again $\mathcal{F} - \alpha$-simple. In particular for $\alpha = 1$ we have that M is $\mathcal{F}$-1-simple if and only if M is $\mathcal{F}$-simple i.e. if and only if M is $\mathcal{F}$-torsion free and M/N is $\mathcal{F}$-torsion for each nonzero submodule N of M. As a direct consequence of Corollary 3.4.4. we may phrase:

7.5.4. Proposition. If $\mathrm{Gdim}_{\mathcal{F}}$ (M) exists then we have: $\mathrm{Gdim}_{\mathcal{F}}$ (M) = sup {$\mathrm{Gdim}_{\mathcal{F}}$ (S), S being Gabriel $\mathcal{F}$-simple in some factor module of M}.

7.5.5. Proposition. If $\mathrm{Kdim}_{\mathcal{F}}$ (M) exists then $\mathrm{Gdim}_{\mathcal{F}}$ exists and: $\mathrm{Kdim}_{\mathcal{F}}$ (M) $\leq$ $\mathrm{Gdim}_{\mathcal{F}}$ (M) $\leq$ $\mathrm{Kdim}_{\mathcal{F}}$ (M) + 1. If M is $\mathcal{F}$-Noetherian then $\mathrm{Gdim}_{\mathcal{F}}$ (M) = $\mathrm{Kdim}_{\mathcal{F}}$ (M).

Proof. Apply Theorem 3.5.1. and 3.5.2. □

Consider an ordinal α and let $(\text{R-mod})_\alpha$ be the class of all R-modules M with Gdim (M) $\leq \alpha$. Foregoing results entail that $(\text{R-mod})_\alpha$ is a localizing subclass of R-mod; let $\mathcal{F}_\alpha$ be the additive topology associated to $(\text{R-mod})_\alpha$, i.e. $\mathcal{F}_\alpha$ = {I left ideal of R, Gdim (R/I) $\leq \alpha$}.

7.5.6. Proposition. M is α-simple if and only if M is $\mathcal{F}_{\alpha-1}$-simple.

Proof. If M is α-simple then Gdim (N) = Gdim (M) = α and Gdim (M/N) $< \alpha$ for each nonzero submodule N of M. Thus M is $\mathcal{F}_{\alpha-1}$-torsion free and M/N is $\mathcal{F}_{\alpha-1}$-torsion for each nonzero submodule N of M, i.e. M is $\mathcal{F}_{\alpha-1}$-simple.
Conversely, if M is $\mathcal{F}_{\alpha-1}$-simple then M/N is $\mathcal{F}_{\alpha-1}$-torsion for each nonzero submodule

N of M whilst M is $\mathcal{F}_{\alpha-1}$-torsion free. Therefore, Gdim (M/N) $\leq$ α - 1 for each nonzero submodule N of M. In view of 4.1.6. we may conclude that M has Gabriel dimension and Gdim (M) $\leq$ α. Since M $\notin$ $(R\text{-mod})_{\alpha-1}$ it follows that Gdim (M) = α.

Since the nonzero submodules of M are also $\mathcal{F}_{\alpha-1}$-simple, it follows that Gdim (N) = Gdim (M) = α with Gdim (M/N) < α for such a submodule N, i.e. M is α-simple. □

7.5.7. Proposition. For a limit ordinal α, $\mathcal{F}_\alpha$ is the additive topology generated by $\mathcal{F}_\beta$, $\beta < \alpha$.

Proof. It is easy to establish that $(R\text{-mod})_\alpha$ is the least localizing subclass of R-mod containing $\bigcup_{\beta<\alpha} (R\text{-mod})_\beta$ e.g. this follows from the consideration of the Gabriel simple Loewy series (see Chapter 4). □

7.5.8. Proposition. Let α and β be given ordinals, $\mathcal{F}$ an additive topology on R. Assume that Gdim (X) $\leq$ α for all $\mathcal{F}$-torsion modules X. If M is a left R-module such that $\text{Gdim}_\mathcal{F}$ (M) exists, say $\text{Gdim}_\mathcal{F}$ (M) $\leq$ β, then M has Gabriel dimension and Gdim (M) $\leq$ $\alpha + \beta$.

Proof. By transfinite recursion on β. If $\beta = 0$ then $\text{Gdim}_\mathcal{F}$ (M) = 0 yields that M is $\mathcal{F}$-torsion and Gdim (M) $\leq$ α. Now we show that for a submodule M' of M, M/M' contains a γ-simple module for some $\gamma \leq \alpha + \beta$.

One easily reduces the problem to the case M' = 0. In this case M contains an $\mathcal{F} - \beta'$-simple module P for some $\beta' \leq \beta$ (if β is a limit ordinal then $\beta' < \beta$). Consider a nonzero submodule N of P, then $\text{Gdim}_\mathcal{F}$ (P/N) < β' and by the induction hypothesis we obtain that P/N has Gabriel dimension and Gdim (P/N) < $\alpha + \beta'$. Consequently P has Gabriel dimension and Gdim (P) $\leq \alpha + \beta' \leq \alpha + \beta$, so P must contain a γ-simple module with $\gamma \leq \alpha + \beta$ and thereore M also contains such a submodule. □

For an additive topology $\mathcal{F}$ on R we let $(R\text{-mod})_\alpha^\mathcal{F}$ be the class of all left R-modules satisfying $\text{Gdim}_\mathcal{F}$ (M) $\leq$ α. By Proposition 7.5.2. and 7.5.3., $(R\text{-mod})_\alpha^\mathcal{F}$ is a localizing subclass of R-mod and the corresponding additive topology is $\mathcal{F}_\alpha' = \{$I left ideal of R, $\text{Gdim}_\mathcal{F}$ (R/I) $\leq \alpha\}$. The proof of Proposition 7.5.8. may be modified in a straightforward way in order to prove:

7.5.9. Proposition. If $\mathrm{Gdim}_{\mathcal{F}'_\alpha}(M) \leq \beta$ then $\mathrm{Gdim}_{\mathcal{F}}(M) \leq \alpha + \beta$.

A left R-module M is called a *slim* module if each nonzero factor module of M contains a nonzero submodule with Krull dimension, (this notion has been called "D-module" in Lemonnier [4] but this is ambiguous terminology because D-modules are a topic of great interest in the theory of rings of differential operators).

7.5.10. Theorem (Lemonnier [4]). A slim module such that each factor module has finite Goldie dimension has Krull dimension.

Proof. We need a sublemma of a fairly general nature: let "$\mathcal{P}$" be a property such that if a left R-module M satisfies "$\mathcal{P}$" then there is a factor module of M containing a direct sum of two modules, one of those being nonzero and the other satisfying "$\mathcal{P}$", then every M satisfying "$\mathcal{P}$" has a factor module of infinite Goldie dimension. Indeed, for each $i \in \mathbb{N}$ we may construct submodules M_i, M_i', N_i, N_i' of M such that:
$M_{i-1}' \supset M_i \underset{\neq}{\supset} M_i' \underset{\neq}{\supset} N_i \supset N_{i-1}, M_i \supset N_i' \supset N_i$,
$M_i/N_i = N_i'/N_i \bigoplus M_i'/N_i$ with $N_{-1} = 0$ and $M_{-1}' = M$.
It follows that $N_i' \bigcap M_i' = N_i$ and it is easily checked that $N_i' \not\subset \bigcap_{k\geq 0} N_k$, $\sum_{i\geq 0} N_i'/N \cong \bigoplus_{i\geq 0} (N_i' + N)/N$, where $N = \bigcup_{k\geq 0} N_k$.
Now consider the property "$\mathcal{P}$" for a left R-module X:

a. X is slim,
b. factor modules of X have finite Goldie dimension,
c. X does not have Krull dimension.

Assume that M satisfies "$\mathcal{P}$". Applying Theorem 3.1.10. we may conclude that there is a strictly increasing Ψ: $\mathbb{D} \rightarrow L_R(M)$ where D is the set of diadic numbers and $L_R(M)$ is the lattice of left submodules of M. Since the latter is an upper continuous lattice we may extend Ψ to a strictly increasing map
$\overline{\Psi}$: $\mathbb{R} \rightarrow L_R(M)$, say $M_x = \overline{\Psi}(x)$, $x \in \mathbb{R}$. Up to replacing M by $M/\bigcap\{M_x, x \in \mathbb{R}\}$ we may assume that $\bigcap\{M_x, x \in \mathbb{R}\} = 0$.
By a. and b., M contains an essential submodule N which has Krull dimension. The map $x \mapsto M_x \bigcap N$ is not injective. From $\bigcap_{x\in\mathbb{R}} M_x = 0$ we derive the existence of $x < y$ such that $0 \neq M_x \bigcap N = M_y \bigcap N \underset{\neq}{\subset} N$. But in this case we have: M_y

$+ N/M_x \bigcap N \cong M_y/M_x \oplus N/M_x \bigcap N$ with $N/M_x \bigcap N$ and M_y/M_x a factor module of $M_y/M_x \bigcap N$.

On the other hand: $L_R\ (M_y/M_x) = [x,y]$ and therefore $L_R\ (M_y/M_x)$ contains a copy of $\mathbb{R}$.

By the sublemma M has a factor module with infinite Goldie dimension, a contradiction. □

7.5.11. Theorem. (R. Gordon J.C. Robson [2]). A left R-module M has Krull dimension if and only if M has Gabriel dimension and any factor module of M has finite Goldie dimension.

Proof. If Kdim (M) exists then the statements are equivalent. Conversely let Gdim (M) $= \alpha$, we prove that M has Krull dimension by transfinite recursion on α.

If $\alpha = 1$ then M is a semi-Artinian module. Let $M_1 \supset M_2 \supset ... \supset M_n \supset ...$, be a descending chain of submodules of M and put $N = \bigcap_{i\geq 0} M_i$. Up to replacing M by M/N we may suppose that N = 0. If s(M) is the socle of M then s(M) has finite length because of the hypothesis. Then there exists an $n_0 \in \mathbb{N}$ such that $M_n \bigcap s(M) = M_{n+1} \bigcap s(M)$ for all $n \geq n_0$. Since $0 = \bigcap_{i\geq 0} M_i$ we arrive at $M_n \bigcap s(M) = (\bigcap_{i\geq 0} M_i) \bigcap s(M) = 0$ and then the fact that s(M) is an essential submodule of M leads to $M_n = 0$. This proves that M is an Artinian module.

Next assume that α is a non-limit ordinal and consider the additive topology $\mathcal{F}_{\alpha-1}$ and the lattice $C_{\mathcal{F}_{\alpha-1}}$ (M) = L. For an N$\in C_{\mathcal{F}_{\alpha-1}}$ (M), M/N is $\mathcal{F}_{\alpha-1}$-torsion free while M/N contains an α-simple submodule. By Proposition 7.5.6., M/N contains an $\mathcal{F}_{\alpha-1}$-simple submodule so L is a semi-Artinian lattice.

On the other hand, our hypothesis implies that [a,1] has finite Goldie dimension for any $a \in L$. By an argument similar to the one above we may estblish that L is Artinian. Let $M_1 \supset M_2 \supset ... \supset M_n \supset ...$ be a descending chain of submodules of M. Since M is $\mathcal{F}_{\alpha-1}$-Artinian there is an $n_0 \in \mathbb{N}$ such that M_n/M_{n+1} is $\mathcal{F}_{\alpha-1}$-torsion for all $n \geq n_0$. Then Gdim $(M_n/M_{n+1}) < \alpha$ for all $n \geq n_0$. By the induction hypothesis: M_n/M_{n+1} has Kdim for all $n \geq n_0$. Application of Proposition 3.1.5. learns that M has Krull dimension.

Finally, assume that α is a limit ordinal. If N is an arbitrary submodule of M, Gdim $(M/N) \leq \alpha$. Since α is a limit-ordinal, M/N contains a β-simple module X for some $\beta < \alpha$.

The induction hypothesis, combined with the fact that X satisfies the hypothesis of Theorem 7.5.11., yields that X has Krull dimension. Consequently M is a slim module and so it has Krull dimension in view of Theorem 7.5.10. □

Note that a commutative domain D which has Gabriel dimension but not Krull dimension cannot be a slim module. Indeed, if D were a slim D-module then it has to contain a nonzero ideal I which has Krull dimension. For any $a \neq 0$ in I we would have an injective map φ_a: $D \to I$, $\lambda \to \lambda a$. Thus D must have Krull dimension, a contradiction.

This example establishes that Theorem 7.5.11. is stronger that 7.5.10.

§7.6 Relative Krull and Gabriel Dimensions of Graded Rings.

This section is based upon results of C. Nâstâsescu, S. Raianu, [1],[2]. The graded methods lead to a new proof for a result of R. Gordon, J.C. Robson [2] concerning the Gabriel dimension of polynomial rings (see Theorem 7.6.4.); the proof given here avoids the use of polynomial categories or quotient categories.

Throughout this section R will be a graded ring of type $\mathbb{Z}$.

7.6.1. Proposition. Let $\mathcal{H}$ be a graded filter on R, M a graded left R-module which is gr-$\mathcal{H}$-Noetherian and which has gr-Kdim$_{\mathcal{H}}$ (M). Let $\mathcal{F}$ be the additive topology on R generated by $\mathcal{H}$, then Kdim$_{\mathcal{F}}$ (M) exists and $\alpha \leq$ Kdim$_{\mathcal{F}}$ (M) $\leq \alpha + 1$.

Proof.

Step 1: The existence of gr-Kdim$_{i(\mathcal{H})}$ (M[X]). The proof is based on some results of Section 7.2., e.g. 7.2.15. and 7.2.17.. One constructs a strictly increasing map from $C^g_{i(\mathcal{H})}$ (M[X]) to the set of all ascending stationary chains of $C^g_{\mathcal{H}}$ (M) which we will denote by A_c ($C^g_{\mathcal{H}}$ (M)), by associating to an N $\in$ $C^g_{i(\mathcal{H})}$ (M[X]) the chain ($N_0^\sim$,...,$N_k^\sim$,...) where $N_k = \{x \in M, xX^k + ... + x_k \in N\}$ and $N_k^\sim = \{x \in M, (N_k : x) \in \mathcal{H}\}$. As in the proof of Theorem 7.2.15. it may be shown that, N $\subsetneq$ P $\in$ $C^g_{i(\mathcal{H})}$ (M[X]) yields $N_k^\sim \subset P_k^\sim$ for all k $\in \mathbb{N}$ and ($N_0^\sim, N_1^\sim$,...,$N_k^\sim$,...) $\neq$ ($P_0^\sim, P_1^\sim$,...,$P_k^\sim$,...). Thus we obtain gr-Kdim$_{i(\mathcal{H})}$ (M[X]) = Kdim ($C^g_{i(\mathcal{H})}$ (M[X])) $\leq$ Kdim (A_c($C^g_{\mathcal{H}}$ (M))) = Kdim ($C^g_{\mathcal{H}}$ (M)) + 1 = 1 + α.
Thus gr-Kdim$_{i(\mathcal{H})}$ (M[X]) exists and it is smaller than or equal to 1 + α.

Step 2: Kdim$_{\mathcal{F}}$ (M) $\leq$ gr-Kdim$_{i(\mathcal{H})}$ (M[X]). In fact this follows from the proof of Theorem 7.2.19. where it is shown that $N^* \in C^g_{i(\mathcal{H})}$ (M[X]) when N $\in$ $C_{\mathcal{F}}$ (M).

Indeed, since the map $N \mapsto N^*$ is strictly increasing it follows that: $\text{Kdim}_{\mathcal{F}}$ (M) $\leq$ gr-$\text{Kdim}_{i(\mathcal{H})}$ (M[X]) $\leq \alpha + 1$, where the first inequality is a consequence of Lemma 7.2.18. □

7.6.2. Theorem. Consider a graded filter $\mathcal{H}$ on R and let M be a graded left R-module such that gr-$\text{Gdim}_{\mathcal{H}}$ (M) $= \varepsilon + n$ where ε is either 0 or a limit ordinal and $n \in \mathbb{N}$. Let $\mathcal{F}$ be the additive topology generated by $\mathcal{H}$ then $\text{Gdim}_{\mathcal{F}}$ (M) exists and $\varepsilon + n \leq \text{Gdim}_{\mathcal{F}}$ (M) $\leq \varepsilon + 2n$.

Proof. First inequality may be derived from Lemma 7.2.18. For each ordinal we consider $\mathcal{F}_\alpha$ ($\mathcal{H}_\alpha$), the additive topology (graded filter) corresponding to the localizing class in R-mod, (R-gr) of all (graded) R-modules having relative (graded) Gabriel dimension at most α. By transfinite recursion on α we will establish: $\mathcal{H}_{\varepsilon+n} \subset \mathcal{F}_{\varepsilon+2n}$.

If $\alpha = 1$ then we have: gr-$\text{Gdim}_{\mathcal{H}}$ (M) = 1 and we may assume that M is gr-$\mathcal{H}$-simple. Then gr-$\text{Kdim}_{\mathcal{H}}$ (M) = 0 and $\text{Kdim}_{\mathcal{F}} \leq 1$ in view of 7.4.1..

Thus $\text{Gdim}_{\mathcal{F}}$ (M) ≤ 2 because of Proposition 7.5.5.. Assume that the assertion holds for all ordinals smaller than $\alpha = \varepsilon + n$. If $n \neq 0$ then we reduce the problem to the case where M is gr-α-$\mathcal{H}$-simple i.e. we may assume that gr-$\text{Kdim}_{\mathcal{H}_{2+n+1}}$ (M) = 0. Hence $\text{Kdim}_{\mathcal{F}_{\varepsilon+2n-2}}$ (M) ≤ 1 by Proposition 7.6.1., and by the induction hypothesis. But then $\text{Gdim}_{\mathcal{F}_{\varepsilon+2n-2}}$ (M) ≤ 2, cf. 7.5.5., and $\text{Gdim}_{\mathcal{F}}$ (M) $\leq \varepsilon + 2n - 2 + 2 = \varepsilon + 2n$ follows from Proposition 7.5.9.. If $\alpha = \varepsilon$ is a limit ordinal then $M = \bigcup_{\lambda<\varepsilon} M_\lambda$ where M_λ is a graded left R-module such that gr-$\text{Gdim}_{\mathcal{H}}$ (M_λ) $\leq \lambda$ for each $\lambda < \varepsilon$. The induction hypothesis now implies that $\text{Gdim}_{\mathcal{F}}(M_\lambda) < \varepsilon$ and thus $\text{Gdim}_{\mathcal{F}}$ (M) $\leq \varepsilon$. □

7.6.3. Remark. The foregoing may be generalized for rings graded by a finitely generated abelian group.

7.6.4. Theorem (Gordon-Robson, [2]). Let R be a ring, M a left R-module having Gdim (M) $= \varepsilon + n$ where ε is either zero or a limit ordinal and $n \in \mathbb{N}$, then we have:

a. if $n \neq 0$, $2 + n + \leq$ Gdim (M[X]) $\leq \varepsilon + 2n$.

b. if $n = 0$, Gdim (M[X]) $= \varepsilon$.

Proof. A modification of the proof of Theorem 7.4.2. up to using the following result (instead of Proposition 7.4.1.). □

7.6.5. Proposition. Let $\mathcal{F}$ be an additive topology on R and let M be an $\mathcal{F}$-Noetherian left R-module such that $\mathrm{Kdim}_{\mathcal{F}}$ (M) $= \alpha$ then $\mathrm{Kdim}_{j(\mathcal{F})}$ (M[X]) $= \alpha + 1$, where j: R $\rightarrow$ R[X] is the natural inclusion.

Proof. We claim that $\mathrm{Kdim}_{j(\mathcal{F})}$ (M[X]) = Kdim $(A_c(C_{\mathcal{F}}$ (M))) $= \alpha + 1$.
If $N_0 \subset N_1 \subset ... \subset N_k \subset ...$ is a stationary chain in $C_{\mathcal{F}}$ (M) then
N $= N_0 \oplus XN_1 \oplus ... \oplus X^kN_k \oplus ... \in C_{j(\mathcal{F})}$ (M[X]) because otherwise there would exist an x $\notin$ N such that $(N : x)_{R[X]} \in j(\mathcal{F})$. But then $(N : x)_{R[X]} \bigcap R \in \mathcal{F}$ would lead to $(N_i : x_i)_R \in \mathcal{F}$, where x $= x_0 + Xx_1 + ... + X^kx_k$ and i is such that $x_i \notin N_i$.
The latter contradicts $N_i \in C_{\mathcal{F}}$ (M). It follows directly from this that $\alpha + 1 =$ Kdim $(A_c$ $(C_{\mathcal{F}}$ (M))) $\leq$ $\mathrm{Kdim}_{j(\mathcal{F})}$ (M[X]).
The converse inequality is an adaptation of the proof of step 1 in Proposition 7.4.2..□

§7.7 Exercises.

(154) We say that additive topologies $\mathcal{F}, \mathcal{G}$, are compatible if $t_{\mathcal{F}}\ t_{\mathcal{G}} = t_{\mathcal{G}}\ t_{\mathcal{F}}$. If I is an ideal of a left Noetherian ring R and $\mathcal{F}$ and $\mathcal{F}_I$ are compatible then I has property (*) (cf. beginning of Section 7.4.) with respect to $\mathcal{F}$.

(155) Let R be a left Noetherian ring. Show that it is equivalent to establish the principal ideal theorem for a prime ideal P minimal over a principal ideal generated by a normalizing element or for a prime ideal P minimal over an invertible ideal I, by constructing the generalized Rees ring $\check{R} \cong \bigoplus_{n\in\mathbb{Z}} I^{n-m}$ where m is chosen such that $I^mP = PI^m$ (noting that $\check{R}I^m = \check{R}X^{-1}$ if we let $\check{R}$ be given as $\sum_{n\in\mathbb{Z}} I^nX^n$)
Hint: Prove that $\check{P} = \check{R}P$ has height one and deduce that P has height one too, using the fact that for prime ideals Q not containing I we have QI = IQ = I $\bigcap$ Q.

Bibliographical Comments to Chapter 7.

In this chapter we aimed to include a presentation of relative Krull and Gabriel dimension without referring to quotient categories.

The first two sections contain the basic notions and some elementary results. We point out the relative version of the Hopkins-Levitzki theorem (Miller-Teply theorem's [1]). The proof of this theorem follows the proof given by C. Nâstâsescu in [7], [8]. (for details see also the book of T. Albu and C. Nâstâsescu [1]).

The relation between the graded and ungraded relative Noetherian property follows the papers of C. Nâstâsescu and S. Raianu [2],[4].

After the general results on relative Krull dimension in Section 7.3. we apply these to obtain proof of Jategaonkar's principal theorem.

In Section 7.5. we apply a result of Lemonnier [4] and provide a new approach to a result of R. Gordon, J.C. Robson [2] (here: Theorem 7.5.11.).

In Section 7.6. we use some technical tricks making use of relative Gabriel dimension of graded modules to derive a proof of another result of R. Gordon. J.C. Robson [2] (here Theorem 7.6.4.) without making use of polynomial categories. The nature of the Chapter 7 is somewhat more technical, as suggested in the introduction it may be skipped at first reading without destroying the continuity of the book.

Chapter 8

Homological Dimensions.

§8.1 The Projective Dimension of a Module.

Let R be a ring; the class of left R-modules will be denoted by R-mod as before while mod-R will stand for the class of right R-modules.

8.1.1. Proposition. An exact sequence of R-modules,

$$0 \longrightarrow M' \xrightarrow{f} M \xrightarrow{g} M'' \longrightarrow 0,$$

yields a commutative diagram with exact rows and columns:

(1)

$$\begin{array}{ccccccccc}
 & & 0 & & 0 & & 0 & & \\
 & & \downarrow & & \downarrow & & \downarrow & & \\
0 & \longrightarrow & N' & \longrightarrow & N & \longrightarrow & N'' & \longrightarrow & 0 \\
 & & \downarrow & & \downarrow & & \downarrow & & \\
0 & \longrightarrow & P' & \longrightarrow & P & \longrightarrow & P'' & \longrightarrow & 0 \\
 & & \downarrow & & \downarrow & & \downarrow & & \\
0 & \longrightarrow & M' & \xrightarrow{f} & M & \xrightarrow{g} & M'' & \longrightarrow & 0 \\
 & & \downarrow & & \downarrow & & \downarrow & & \\
 & & 0 & & 0 & & 0 & &
\end{array}$$

where P',P,P" are projective left R-modules. Moreover, up to picking the projective left R-modules P' and P" one may choose the exact sequences $0 \to N' \to P' \to M' \to 0$ and $0 \to N'' \to P'' \to M'' \to 0$ arbitrarily.

Proof. Let P' and P" be projective modules and ε': P' $\to$ M', ε": P" $\to$ M" surjective morphisms, N' = Ker (ε') and N" = Ker (ε"). Put P = P' $\oplus$ P". Projectivity of P" yields the existence of h: P" $\to$ M such that gh = ε". Define ε: P $\to$ M, (x',x") $\mapsto$ (f $\circ$ ε') (x') + h(x"). For y $\in$ M, there is an x" $\in$ P" such that g(y) = ε"(x") = g(h(x")), hence y - h(x") $\in$ Ker g = Im f. Then y - h(x") = f(ε'(x')) for some x' $\in$ P', thus ε(x',x") = y or ε is onto.
Now consider the morphism α: P' $\to$ P, $\mapsto$ (x',0) and β: P $\to$ P", (x',x") $\mapsto$ x", then the following diagram is commutative and its rows are exact:

(2)

$$\begin{array}{ccccccccc} 0 & \longrightarrow & P' & \xrightarrow{\alpha} & P & \xrightarrow{\beta} & P'' & \longrightarrow & 0 \\ & & \downarrow{\scriptstyle \varepsilon'} & & \downarrow{\scriptstyle \varepsilon} & & \downarrow{\scriptstyle \varepsilon''} & & \\ 0 & \longrightarrow & M' & \xrightarrow[f]{} & M & \xrightarrow[g]{} & M'' & \longrightarrow & 0 \end{array}$$

Put N = Ker (ε) and let N' $\to$ N, N $\to$ N" be the maps induced by α and β respectively. It is clear that 0 $\to$ N' $\to$ N $\to$ N" $\to$ 0 is exact and (2) leads to (1) in the obvious way. □

8.1.2. Corollary. An exact sequence

$$0 \longrightarrow M' \xrightarrow[f]{} M \xrightarrow[g]{} M'' \longrightarrow 0$$

gives rise to a commutative diagram with exact rows and columns:

(3)

$$
\begin{array}{ccccccccc}
 & & 0 & & 0 & & 0 & & \\
 & & \downarrow & & \downarrow & & \downarrow & & \\
0 & \longrightarrow & N' & \xrightarrow{u} & N & \xrightarrow{v} & N'' & \longrightarrow & 0 \\
 & & \downarrow & & \downarrow & & \downarrow & & \\
0 & \longrightarrow & P'_{n-1} & \longrightarrow & P_{n-1} & \longrightarrow & P''_{n-1} & \longrightarrow & 0 \\
 & & \downarrow & & \downarrow & & \downarrow & & \\
\vdots & & \vdots & & \vdots & & \vdots & & \vdots \\
 & & \downarrow & & \downarrow & & \downarrow & & \\
0 & \longrightarrow & P'_0 & \longrightarrow & P_o & \longrightarrow & P''_0 & \longrightarrow & 0 \\
 & & \downarrow & & \downarrow & & \downarrow & & \\
0 & \longrightarrow & M' & \xrightarrow[f]{} & M & \xrightarrow[g]{} & M'' & \longrightarrow & 0 \\
 & & \downarrow & & \downarrow & & \downarrow & & \\
 & & 0 & & 0 & & 0 & &
\end{array}
$$

where P'_i, P_i, P''_i are projective left R-modules, $0 \leq i \leq n\text{-}1$.

Proof. From the proposition by induction on $n \in \mathbb{N}$. □

8.1.3. Proposition (Schanuel). Given the exact sequences:

$$0 \longrightarrow N \xrightarrow{i} P \xrightarrow{j} M \longrightarrow 0,$$
$$0 \longrightarrow N' \xrightarrow{i'} P' \xrightarrow{j'} M' \longrightarrow 0,$$

where P and P' are projective left R-modules.
If $M \cong M'$ then $N \oplus P' \cong N' \oplus P$.

Proof. Let $h: M \to M'$ be an isomorphism. There exists an $f: P \to P'$ such that $h \circ j = j' \circ f$. Since $j' \circ f \circ i = h \circ j \circ i = 0$ there is a unique $g: N \to N'$ such that $f \circ i = i' \circ g$. Consider the sequence

$$0 \longrightarrow N \xrightarrow{\alpha} P \oplus N' \xrightarrow{\beta} P' \longrightarrow 0,$$

where $\alpha(x) = (i(x), g(x))$ and $\beta(y,z) = i'(z) - f(y)$. Obviously the latter sequence is exact. Since P' is projective the sequence splits, hence $N \oplus P' \cong N' \oplus P$. □

8.1.4. Corollary. Given $M \cong M'$ and the exact sequences:

$$0 \to N \to P_n \to P_{n-1} \to \dots \to P_0 \to M \to 0$$
$$0 \to N' \to P'_n \to P'_{n-1} \to \dots \to P'_0 \to M' \to 0,$$

where all P_i, P'_i are projective left R-modules, then we obtain:
$N \oplus P'_n \oplus P_{n-1} \oplus \dots \oplus \dots \cong N' \oplus P_n \oplus P'_{n-1} \oplus \dots \oplus \dots$. In particular N is projective if and only if N' is projective.

8.1.5. Definition. A *projective resolution* of a left R-module M is an exact sequence

$$\dots \to P_n \to P_{n-1} \to \dots \to P_1 \to P_0 \to M \to 0 \tag{4}$$

where each P_i is a projective left R-module for all $i \geq o$. If $P_k \neq 0$ and $P_n = 0$ for all $n > k$ then we say that the resolution (4) has length k. If this does not happen then (4) is said to have infinite length.

8.1.6. Proposition. Each left R-module M has a projective resolution.

Proof. (we freely use results of Section 2.1. and consequent sections). Let P_0 be a free left R-module such that we have an epimorphism $P_o \to M$. Suppose that an exact sequence:

$$P_n \xrightarrow{f} P_{n-1} \longrightarrow \dots \longrightarrow P_0 \longrightarrow M \longrightarrow 0,$$

with $P_0, \dots, P_n$ being projective left R-modules, has been constructed. If f is injective then M has a projective resolution with $P_k = 0$ for all $k > n$; otherwise we let P_{n+1} be a projective module such that there is an epimorphism h: $P_{n+1} \to \text{Ker}(f)$. Put $g = i \circ h$ where i: $\text{Ker}(f) \to P_n$ is the canonical inclusion. Obviously the newly obtained sequence:

$$P_{n+1} \xrightarrow{g} P_n \longrightarrow \dots \longrightarrow P_0 \longrightarrow M \longrightarrow 0$$

is again exact and this finishes the proof of the induction step. □

We may now phrase:

8.1.7. Definition. The *projective dimension* of a left module M, denoted by pd_R (M), or simply pd (M), is the least n for which M has a projective resolution of length n. If such an n does not exist then we put pd (M) = ∞. In particular pd (M) = 0 if and only if M is projective.

8.1.8. Corollary. Given an exact sequence

$$0 \to N \to P_{n-1} \to \ldots \to P_1 \to P_0 \to M \to 0$$

where the left R-modules $P_0, \ldots, P_{n-1}$ are projective, then pd (M) $\leq$ n if and only if N is projective.

Proof. If N is projective then pd (M) $\leq$ n. Suppose now that pd (M) $\leq$ n. Then there exists a projective resolution of M of type:

$$0 \to P'_n \to P'_{n-1} \to \ldots \to P'_1 \to P'_0 \to M \to 0,$$

application of Corollary 8.1.4. finishes the proof. □

8.1.9. Proposition. Let 0 → M' → M → M" → 0 be an exact sequence.

1. If pd (M') $\leq$ n and pd (M") $\leq$ n then pd (M) $\leq$ n.
2. If pd (M) $\leq$ n and pd (M") $\leq$ n then pd (M') $\leq$ n.
3. If pd (M') $\leq$ n and pd (M) $\leq$ n then pd (M") $\leq$ n + 1.

Proof. We will use the diagram (3). If pd (M') $\leq$ n and pd (M") $\leq$ n then the above corollary implies that N' and N" are projective. Therefore,
0 → N' → N → N" → 0 is a split sequence and hence N $\cong$ N' $\bigoplus$ N" is projective too. By the above corollary we obtain that pd (M) $\leq$ n and so 1. follows.
One may prove 2. in a similar way.
For 3. suppose that pd (M') $\leq$ n and pd (M) $\leq$ n. It follows that N' and N are projective. Exactness of the sequence:

$$0 \longrightarrow N' \xrightarrow{u} N \xrightarrow{i'' \circ v} P''_{n-1} \longrightarrow \ldots \longrightarrow P''_0 \longrightarrow M" \longrightarrow 0$$

yields that pd (M") $\leq$ n + 1. □

8.1.10. Corollary. Let N be a submodule of the left R-module M.

1. If pd (M) > pd (N) then pd (M/N) = pd (M).
2. If pd (M) < pd (N) then pd (M/N) = pd (N) + 1.
3. If pd (M) = pd (N) then pd (M/N) $\leq$ pd (M) + 1.

Proof. 3. follows from Proposition 8.1.9.(3.).
1. We may suppose that pd (M) = n < ∞. Let M = P/K, where P is a projective left R-module. Then N = Q/K where Q $\subset$ P and M/N $\cong$ P/Q. Projectivity of P and Corollary 8.1.8. lead to: pd (K) = pd (M) - 1 and pd (Q) = pd (M/N) - 1. By the foregoing proposition: pd (M) $\leq$ pd (M/N) $\leq$ pd (M) + 1. Assume that pd (M/N) = pd (M) + 1 = n + 1. From the exactness of the sequence
$0 \to K \to Q \to N \to 0$ and envoking Proposition 8.1.9. we obtain:
n = pd (Q) $\leq$ max {pd (K), pd(N)} = n - 1, a contradiction.
Consequently: pd (M/N) = pd (M).
2. If pd (N) = ∞ then pd (M/N) = ∞ because of Proposition 8.1.9. and hence we may assume henceforth that pd (N) = n < ∞. Then pd (M) $\leq$ n - 1. If pd (M/N) $\neq$ pd (N) + 1, then pd (M/N) = pd (N) = n. With notation as in 1. we obtain: pd (Q) = n - 1 and pd (K) $\leq$ n - 2. Applying 1. it follows that pd (N) = pd (Q) = n - 1, a contradiction. So the conclusion must be that pd (M/N) = n + 1, as claimed. □

8.1.11. Proposition. For a family of left R-modules $\{M_i, i \in \Lambda\}$ we have: pd $(\bigoplus_{i\in\Lambda} M_i) = \sup\{\text{pd}(M_i), i \in \Lambda\}$.

Proof. For each $i \in \Lambda$ and $k \geq 0$ we consider the exact sequence:

$$0 \to N_i \to P_i^k \to P_i^{k-1} \;..\to P_i^0 \to M_i \to 0 \tag{5}$$

where P_i^k, $o \leq l \leq k$, is projective. Summing (5) over $i \in \Lambda$ yields:

$$0 \to \bigoplus_{i\in\Lambda} N_i \to \bigoplus_{i\in\Lambda} P_i^k \to \dots \to \bigoplus_{i\in\Lambda} P_i^0 \to \bigoplus_{i\in\Lambda} M_i \to o \tag{6}$$

where the $\bigoplus_{i\in\Lambda} P_i^l$ ar projective for each $l \in [0,k]$. Corollary 8.1.8. and the sequences (5) and (6) yield the desired equality.

8.1.12. Definition. The *left global (or homological) dimension* of R is defined to be sup {pd (M), M ∈ R-mod} = l.gldim (R). The *right global dimension of R* is the left global dimension of the opposite ring R^0.

8.1.13. Proposition. Let S ≠ ϕ be a multiplicatively closed set of R such that R satisfies the left Ore conditions with respect to S. If M∈ R-mod then $pd_{S^{-1}R}$ (S^{-1}M) ≤ pd_R (M) and in particular, l.gldim (S^{-1}M) ≤ l.gldim (R).

Proof. Put n = pd_R (M). If n = ∞ then the inequality is clear. When n < ∞ there exists a projective resolution:

$$0 \to P_n \to P_{n-1} \to \dots \to P_1 \to P_0 \to M \to 0.$$

Exactness of the localization at S (see Section 2.8.) yields exaxctness of:

$$0 \to S^{-1}P_n \to \dots \to S^{-1}P_1 \to S^{-1}P_0 \to S^{-1}M \to 0,$$

and each $S^{-1}P_i$ is a projective S^{-1} R-module. Therefore:
$\text{pd}_{S^{-1}R}$ ($S^{-1}M$) ≤ pd_R (M). Furthermore, since every S^{-1} R-module is isomorphic to some S^{-1} M for some R-module M, the second statement follows easily. □

8.1.14. Theorem (M. Auslander, [1]). Let $\{M_\alpha, \alpha \in \mathcal{A}\}$ be a family of submodules of the left R-module M, indexed by a set $\mathcal{A}$ of ordinal numbers. Assume that $M_\alpha \subset M_\beta$ if $\alpha \leq \beta$ and that $M = \bigcup_{\alpha\in\mathcal{A}} M_\alpha$. If pd $(M_\alpha/\bigcup_{\beta<\alpha} M_\beta) \leq$ n for each $\alpha \in \mathcal{A}$ then pd (M) ≤ n.

Proof. If n = 0 then $M_\alpha/\bigcup_{\beta<\alpha} M_\beta$ is projective and therefore
$M_\alpha = \bigcup_{\beta<\alpha} M_\beta \oplus M'_\alpha$ where $M'_\alpha = M_\alpha/\bigcup_{\beta<\alpha} M_\beta$ is also projective. On the other hand, it is equally obvious that $M \cong \bigoplus_{\alpha\in\mathcal{A}} M'_\alpha$. Thus M is projective and pd (M) = 0.

We use induction on n assuming now that n > 0.

As before, set $M'_\alpha = M_\alpha/\bigcup_{\beta<\alpha} M_\beta$ and let F'_α be a free R-module projecting onto M'_α with kernel K'_α, say. Put $F_\alpha = \bigoplus_{\beta\leq\alpha} F'_\beta$. Then F_α is a free R-module and $F_\beta \subset F_\alpha$ for $\beta \leq \alpha$. By transfinite recursion we may construct the epimorphisms f_α: $F_\alpha \to M_\alpha$ with kernel K_α = Ker (f_α) such that the restriction of f_α to F_β is f_β for $\beta \leq \alpha$ and such that $K'_\alpha \cong K_\alpha/\bigcup_{\beta<\alpha} K_\beta$. Indeed, let us consider the following diagram:

$$\begin{array}{ccccccccc}
 & & 0 & & 0 & & 0 & & \\
 & & \downarrow & & \downarrow & & \downarrow & & \\
0 & \to & \bigcup_{\beta<\alpha} K_\beta & \to & K_\alpha & \to & K'_\alpha & \to & 0 \\
 & & \downarrow & & \downarrow & & \downarrow & & \\
0 & \to & \bigcup_{\beta<\alpha} F_\beta & \to & F_\alpha & \to & F'_\alpha & \to & 0 \\
 & & \downarrow & & \downarrow & & \downarrow & & \\
0 & \to & \bigcup_{\beta<\alpha} M_\beta & \to & M_\alpha & \to & M'_\alpha & \to & 0 \\
 & & \downarrow & & \downarrow & & \downarrow & & \\
 & & 0 & & 0 & & 0 & &
\end{array}$$

We suppose that all f_β: $F_\beta \to M_\beta$ with the required properties have been constructed for the ordinals $\beta \in \mathcal{A}$, $\beta < \alpha$. Obviously, $\bigcup_{\beta<\alpha} F_\beta = \bigoplus_{\beta<\alpha} F'_\beta$. But $F_\alpha = \bigcup_{\beta<\alpha} F_\beta \bigoplus F'_\alpha$, so we may use Proposition 8.1.. to complete the diagram and then it is easy enough to verify that f_α has the desired properties.
We put F = $\bigoplus_{\alpha\in\mathcal{A}} F'_\alpha$. Then F=$\bigcup_{\alpha\in\mathcal{A}} F_\alpha$ and there exists a unique morphism f: F $\to$ M such that f| $F_\alpha = f_\alpha$ for all $\alpha \in \mathcal{A}$. It is clear that f is an epimorphism and K = Ker (f) = $\bigcup_{\alpha\in\mathcal{A}} K_\alpha$. From Corollary 8.1.4. we retain that pd $(K'_\alpha) \leq$ n - 1 for all $\alpha \in \mathcal{A}$, hence pd (K) $\leq$ n - 1 because of the induction hypothesis.
Finally, Corollary 8.1.10. yields pd (M) $\leq$ n. □

8.1.15. Corollary (M. Auslander, [1]). The (left) global dimension of a ring R equals the supremum of the projective dimensions of all cyclic left R-modules i.e., d = l.gldim (R) = sup {pd (R/I), I left ideal of R} = d'.

Proof. It is clear that d' $\leq$ d. If M is a left R-module then we order M by a set $\mathcal{A}$ of ordinals i.e. M = $\{x_\alpha, \alpha \in \mathcal{A}\}$. Let M_α be the submodule of M generated by all x_β, $\beta \leq \alpha$. Then the factor module $M_\alpha / \bigcup_{\beta<\alpha} M_\beta$ is generated by the class of the element x_α and hence it is cyclic. Then d $\leq$ d' because of Theorem 8.1.4. □

8.1.16. Theorem (First Change of Rings Theorem). Let R be a ring and let I be a proper invertible ideal of R. If M is a nonzero left R/I-module such that $\text{pd}_{R/I}$

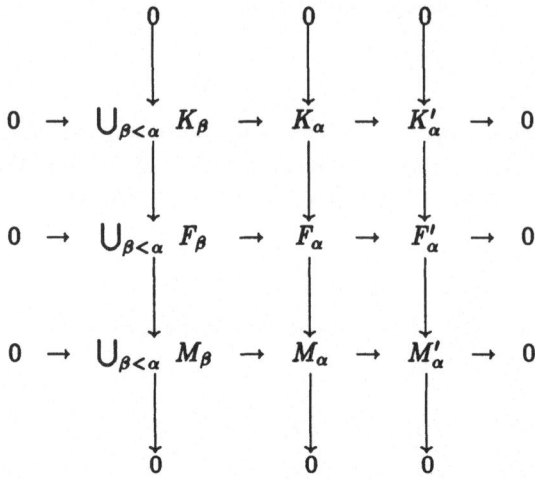

(M) = n < ∞ then pd_R (M) = n + 1. In particular, if l.gldim (R/I) < ∞ then we have: l.gldim (R) $\geq$ 1 + l.gldim (R/I).

Proof. If n = 0 then M is a projective R/I-module and as such it is a direct summand in a free R/I-module L. Since I is invertible, R/I cannot be projective as an R-module (otherwise R = I $\bigoplus$ J for some left ideal J of R would yield IJ = 0 and hence J = 0), hence pd (R/I) = 1. Proposition 8.1.11. then entails that
pd_R (M) $\leq$ 1. From IM = 0 we may conclude that M is not a projective left R-module and thus pd_R (M) = 1.
Now the proof continues by induction on n. Let n > 0 and construct the exact sequence of R/I-modules: 0 $\to$ K $\to$ L $\to$ M $\to$ 0 where L is a free R/I-module. Since $pd_{R/I}$ (M) = n we find that $pd_{R/I}$ (K) = n - 1 and the induction hypothesis then yields that pd_R (K) = n (using Corollary 8.1.10.(2)). From pd_R (L) = 1 (and again using the same corollary) we obtain that pd_R (M) = n + 1 if n > 1.
Only the case n = 1 remains now. In this case (3) of Corollary 8.1.10. entails that pd_R (M) $\leq$ 2. First consider the case where pd_R (M) $\leq$ 1 and let
0 $\to$ N $\to$ P $\to$ M $\to$ 0 be an exact sequence of left R-modules, where P is projective. Because IM = 0, IP $\subset$ N and we obtain an exact sequence of R/I-modules:
0 $\to$ N/IP $\to$ P/IP $\to$ M $\to$ 0. The splitting of the exact sequence
0$\to$ IP/IN $\to$ N/IN $\to$ N/IP $\to$ 0 entails that IP/IN is a projective R/I-module. But we have IP/IN $\simeq$ I $\otimes_R$ P/N and
P/N $\cong$ I^{-1} $\otimes_R$ (I $\otimes_R$ P/N) $\cong$ (I^{-1} $\otimes_R$ R/I) $\otimes_{R/I}$ (IP/IN) $\cong$ (I^{-1}/R) $\otimes_{R/I}$ IP/IN. The projectivity of I^{-1} as a left R-module entails that I^{-1}/R is a projective left R/I-module and the Hom-tensor relation: $Hom_{R/I}$ ((I^{-1}/R) $\otimes_{R/I}$ (IP/IN),Y) $\cong$ $Hom_{R/I}$ (IP/IN, $Hom_{R/I}$ (I^{-1}/R,Y)) we obtain that $Hom_{R/I}$ ((I^{-1}/R) $\otimes_{R/I}$ (IP/IN),–) is an exact functor. Therefore the "domain"-module is a projective R/I-module and consequently M = P/N is also a projective R/I-module, a contradiction. The conclusion: pd_R (M) = 2, is exactly what we need to complete the proof. □

8.1.17. Theorem (Second Change of Rings Theorem). Let a be a normalizing element of R (i.e. aR = Ra). Let M be a left R-module and assume a has no annihilator in R nor in M, then we have: $pd_{R/aR}$ (M/aM) $\leq$ pd_R (M).

Proof. If pd_R (M) = ∞ then we have nothing to prove. If pd_R (M) = n is finite we continue by induction on n. If n = 0 then M is projective and M/aM is projective as an R/aR-module. Suppose now that n > 0. Let F be a free R-module fitting in an exact sequence 0 $\to$ K $\to$ F $\to$ M $\to$ 0. Then pd_R (K) = n - 1 and

$pd_{R/aR}$ (K/aK) $\leq$ n - 1, by the induction hypothesis. The exactness of R/aR $\otimes_R$ K $\to$ R/aR $\otimes_R$ F $\to$ R/aR $\otimes_R$ M $\to$ 0 leads to an exact sequence:

$$K/aK \xrightarrow[\alpha]{} F/aF \xrightarrow[\beta]{} M/aM \longrightarrow 0.$$

By definition Ker (β) = Im (α) = K + aF/aF $\cong$ K/K $\bigcap$ aF. Since a does not annihilate any element of M it follows that K $\bigcap$ aF = aK and therefore we obtain an exact sequence:

$$0 \to K/aK \to F/aF \to M/aM \to 0.$$

Since $pd_{R/aR}$ (K/aK) $\leq$ n - 1 we may apply Proposition 8.1.9.(3) in order to derive that $pd_{R/aR}$ (M/aM) $\leq$ n. □

8.1.18. Theorem (Third Change of Rings Theorem). Let R be a left Noetherian ring and let a be a normalizing element of R contained in the Jacobson radical J(R) of R. Let M be a finitely generated left R-module and assume that a is a non-zero-divisor on both R and M, then $pd_{R/aR}$ (M/aM) = pd_R (M).

Proof. By the foregoing theorem, $pd_{R/aR}$ (M/aM) $\leq$ pd_R (M). So we may assume that $pd_{R/aR}$ (M/aM) = n < ∞ and we continue by induction on n.
If n = 0 then M/aM is a projective R/aR-module. If M/aM were free as an R/aR-module then M would be a free R-module. To establish this claim let $\hat{x}_1,...,\hat{x}_n$ be a basis for M/aM as an R/aR-module. Consider N = $(x_1, ..., x_n)$, the R-submodule of M generated by representatives x_i for $\hat{x}_i$. We have M = aM + N and Nakayama's lemma yields M = N. A relation $\sum_{i=1}^n \lambda_i x_i = 0$ leads to $\lambda_i \in$ aR = Ra, say $\lambda_i = a\mu_i$ with $\mu_i \in$ R, i = 1.....n. But a $\sum_{i=1}^n \mu_i x_i = 0$ yields $\sum_{i=1}^n \mu_i x_i = 0$ by the assumptions an a. Repetition of this argument leads to $\lambda_{i,k} \in$ R, uniquely determined by k $\in \mathbb{N}$, such that $\lambda_i = a^k \lambda_{i,k}$. From $\lambda_{i,k} = a\, \lambda_{i,k+1}$ for all k $\geq$ 0 it follows that the chain of left ideals of R: $R\lambda_{i,1}$, $R\lambda_{i,2}$, ..., $R\lambda_{i,k}$, ..., is an ascending chain. The left Noetherian hypothesis on R yields that $R\lambda_{i,n} = R\lambda_{i,n+1}$ for all n $\geq n_0$, some $n_0 \in \mathbb{N}$. Hence $\lambda_{i.n+1} = b\lambda_{i,n}$ for some b $\in$ R. Consequently:
(1 - ba) $\lambda_{i,n+1} = 0$, but 1 - ba is invertible in R because a $\in$ J(R), hence $\lambda_{i,n+1} = 0$ and $\lambda_i = 0$ follows. So we have proved that $\{x_1, ..., x_n\}$ is an R-basis for M.
Next assume that M/aM is projective as an R/aR-module. Let F be a free R-module fitting in an exact sequence of left R-modules 0 $\to$ K $\to$ F $\to$ M $\to$ 0. As in the proof of the foregoing theorem we arrive at an exact sequence:
0 $\to$ K/aK $\to$ F/aF $\to$ M/aM $\to$ 0. By the projectivity of M/aM we obtain:
F/aF $\cong$ M/aM $\oplus$ K/aK $\cong$ M $\oplus$ K/a (M $\oplus$ K). By the "free"-case it is evident that M $\oplus$ K is a free R-module and thus M is a projective R-module, i.e. pd_R (M)

= 0.

Next assume that n > 0. Let F be a free left R-module fitting in as exact sequence o → K → F → M → 0. Again we obtain the exact sequence
0 → K/aK → F/aF → M/aM → 0, as before. Combining $pd_{R/aR}$ (M/aM) = n > 0 and Corollary 8.1.10.(2), we obtain that R/aR (K/aK) = n - 1 and the induction hypothesis in turn implies that pd_R (K) = n - 1. Proposition 8.1.9.(3) then allows to conclude that pd_R (M) ≤ n, as desired. □

8.1.19. Corollary. With notations, conventions and assumptions as in the foregoing theorem we also obtain: pd_R (M/aM) = 1 + pd_R (M) =
1 + $pd_{R/aR}$ (M/aM), whenever M ≠ 0 and $pd_{R/aR}$ (M/aM) < ∞.

§8.2 Homological Dimension of Polynomial Rings and Rings of Formal Power Series.

The R[X]-module M[X] = R[X] $\otimes_R$ M can be obtained from the following left R-module $M^{(\mathbb{N})}$ by introducing the following scalar multiplication: if p(X) = $a_0 + a_1X + ... + a_nX^n$ ∈ R[X], and f = $(m_0, m_1, ..., m_k, ...)$ ∈ $M^{(\mathbb{N})}$, the we put p(X) f = $(n_0, n_1, ..., n_k, ...)$ ∈ $M^{(\mathbb{N})}$, where $n_k = \sum_{c+j=k} a_i m_j$.

8.2.1. Lemma. For any left R[X]-module M there is an exact sequence of R[X]-modules:

$$0 \longrightarrow M[X] \underset{i}{\longrightarrow} M[X] \underset{j}{\longrightarrow} M \longrightarrow o,$$

where: i$(m_0, m_1, ..., m_k, ...) = (Xm_0, Xm_1 - m_0, XM_2 - m_1, ..., Xm_k - m_{k-1}, ...)$
j$(m_0, m_1, ..., m_k, ...) = \sum_{i \geq o} X^i m_i$.

Proof. It is obvious that i is injective, j is surjective and Ker (j) ⊃ Im (i). On the other hand, if f = $(m_0, m_1, ..., m_k, 0, ..., 0, ...)$ ∈ Ker (j) then we may look at g = $(n_0, n_1, ..., n_{k-1}, 0, ..., 0, ...)$ where $n_s = -m_{s+1} - Xm_{s+2} - ... - X^{k-s-1}m_k$,
0 ≤ s ≤ k - 1, then i(g) = f and so we have verified that Ker (j) = Im (i).

8.2.2. Lemma. For any left R-module M we have $\text{pd}_{R[X]}$ (M[X]) $\leq$ pd_R (M).

Proof. If M is a free R-module then M[X] is a free R[X]-module and it also follows that M[X] is a projective R[X]-module when M is a projective R-module. It is not restrictive to assume that pd_R (M) $= n < \infty$. Let $o \to P_n \to P_{n-1} \to ... \to P_0 \to M$ be a projective resolution of M. Since R[X] is free (hence flat) both as a left and right R-module we obtain an exact sequence: $0 \to P_n[X] \to$ pn $P_{n-1}[X] \to ... \to P_0[X] \to M[X] \to 0$, hence $\text{pd}_{R[X]}$ (M[X]) $\leq$ n. □

Remark. Lemma 2.2. also follows from Theorem 8.1.17. because X is a normalizing element of R[X] and R[X]/(X) $\cong$ R.

8.2.3. Theorem (Hilbert's Syzygy Theorem). With notations as before: l.gldim (R[X]) = 1 + l.gldim (R).

Proof Put n = l.gldim (R). If $n = \infty$ then l.gldim (R[X]) $= \infty$. Indeed if l.gldim (R[X]) $= s < \infty$ and M is a left R-module then we may consider the exact sequence: $0 \to N \to P_{s-1} \to ... \to P_0 \to M \to 0$ where P_i, $0 \leq i \leq s - 1$, is a projective left R-module. Exactness of $0 \to N[X] \to P_{s-1}[X] \to ... \to P_0[X] \to M[X] \to 0$, combined with $\text{pd}_{R[X]}$ (M[X]) $\leq$ s, yields that N[X] is a projective left R[X]-module. Then N is a projective left R-module and thus pd_R (M) $\leq$ s. Since M is arbitrary, l.gldim (R) $\leq$ s, a contradiction.
So we may reduce the problem to the case where $n < \infty$. Since X is not a zero-divisor on R[X], Theorem 8.1.16. implies l.gldim (R[X]) $\geq$ 1 + l.gldim (R). The converse inequality follows from the foregoing lemmas. □

8.2.4. Corollary. l.gldim (R[$X_1, ..., X_n$]) = n + l.gldim (R).

8.2.5. Corollary. If K is a division ring, l.gldim (K[$X_1, ..., X_n$]) = n.

For a left R-module M we let M [[X]] be the *module of formal power series* consisting of all elements of the form: $m_0 + m_1 X + ... + m_i X^i + ...$, $m_i \in$ M, and scalar multiplication defined by $(aX^p)(mX^q)$ = am X^{p+q} for a ∈ R, m ∈ M. It is clear that M [[X]] is a left R[[x]]-module, which is generally not isomorphic to R[[x]] $\otimes_R$ M. For an u ∈ M[X], say u = $m_0 + m_1X + ... + m_kX^k + ...$, we let the *order of u*, ord(u), as the least natural number k such that $m_k \neq 0$.

8.2.6. Proposition. If M is a left Noetherain R-module then M[[X]] is a left Noetherian R[[X]]-module.

Proof. Let N be an R[[X]]-module of M[[X]]. We denote by L_i = {m ∈ M | there exists a formal series f ∈ N which is of the form f = mX^i + terms of higher degree}. Clearly L_i is a R-submodule of M. Since N is an R[[X]]-submodule of M[[X]] then $L_i \subseteq L_{i+1}$ for any i ≥ 0.
We have the ascending chain of R-submodules of M: $L_0 \subset L_1 \subset ... \subset L_i \subset L_{i+1} \subset ...$ Since M is a Noetherian R-module, there exists an r ≥ 0 such that $L_r = L_{r+1}$ = ... Since M is Noetherian the submodules $L_0, L_1, ..., L_r$ are R-finitely generated. We denote by $\{m_{ij}\}_{j=1,...,n_i}$ a set of generators for L_i (0 ≤ i ≤ r). There exist f_{ij} ∈ L such that f_{ij} is of the form $f_{ij} = m_{ij}X^i$ + terms of higher degree.
Proceeding inductively we establish that the set $\bigcup_{i=0}^r \{f_{ij} \mid j = 1,...,n_i\}$ is a set of generators for L as R[[X]]-module. The result may be derived from Theorem 4.2.6.4. if we observe that M[[X]] is exhaustively filtered by putting F_n (M[[X]]) = {u ∈ M[[X]], deg (u) ≥ n} and that the associated graded module is M[X] over R[X] = G (R[[X]]). □

8.2.7. Theorem. If R is left Noetherian then l.gldim (R[[X]]) = 1 + l.gldim (R).

Proof. It is easily seen that X ∈ J(R[[X]]). Assume that l.gldim (R) = n < ∞. Theorem 8.1.16. implies that l.gldim (R[[X]]) ≥ 1 + n.
Conversely, let M be a finitely generated left R[[X]]-module with $pd_{R[[X]]}$ (M) = k. If k = 0 then k ≤ n + 1 is obvious. If k ≠ 0 we consider a free R[[X]]-module L of finite rank fitting in an exact sequence 0 → K → L → M → 0. Since R[[X]] is left Noetherian, K is finitely generated. Then Corollary 8.1.10.(2) leads to $pd_{R[[X]]}$ (K) = k - 1. Theorem 8.1.18. may be applied because X is not a zerodivisor on K, it follows that: $pd_{R[[X]]}$ (K) = pd_R (K/XK) ≤ n, thus k - 1 ≤ n and k ≤ n + 1. Now l.gldim (R[[X]]) ≤ n + 1 in view of Corollary 8.1.15.. when n = ∞ then, for each t ∈ ℕ there exists a left ideal I of R such that pd_R(I) ≥ t. The left ideal J = I[[X]] of R[[X]] satisfies $pd_{R[[X]]}$ (J) ≥ t in view of Theorem 8.1.18. and therefore we also

obtain l.gldim (R[[X]]) $= \infty$. □

Some Remarks.

1. If φ is an automorphism of R then we can construct the ring of twisted polynomials R[X,φ] and the ring of twisted formal power series R[[X,φ]], in both rings multiplication is fully described by the rule: $Xa = \varphi(a)X$ for all $a \in R$. Since φ is an automorphism, X is a normalizing element in R[X,φ] and also in R[[X,φ]]. Using the "change of rings" theorems and proceeding like in the proofs of Theorem 8.2.3. and Theorem 8.2.7. it is straightforward to prove the following results:

a. l.gldim (R[x,φ]) = 1 + l.gldim (R).
b. If R is left Noetherian then R[[X,φ]] is left Noetherian (this also may be derived by using the natural filtration on R[[X,φ]] for which G(R[[X,φ]]) = R[X,φ], as in 8.2.6.) and in this case: l.gldim (R[[X,φ]]) = 1 + l.gldim(R).

2. Fields has proved in [1] that: l.gldim (R[X,φ] $\leq$ 1 + l.gldim (R), is valid for an injective endomorphism of R. On the other hand there exist examples (e.g. an unpublished one due to Kaplansky) of a ring R (non-Noetherian) such that l.gldim (R[[X]])) > 1 + l.gldim (R).

§8.3 Injective Dimension of a Module.

The notion of injectivity is dual to that of projectivity and this duality allows to transpose several results of Section 8.1.. Let us start by listing some results which may be obtained by obvious dualization of similar results in Section 8.1., the proofs have been ommitted.

8.3.1. Proposition. Every exact sequence of left R-modules,

$$0 \longrightarrow M' \xrightarrow{f} M \xrightarrow{g} M" \longrightarrow 0,$$

fits in a commutative diagram:

$$
\begin{array}{ccccccccc}
 & & 0 & & 0 & & 0 & & \\
 & & \downarrow & & \downarrow & & \downarrow & & \\
0 & \longrightarrow & M' & \xrightarrow{f} & M & \xrightarrow{g} & M'' & \longrightarrow & 0 \\
 & & \downarrow & & \downarrow & & \downarrow & & \\
0 & \longrightarrow & Q' & \longrightarrow & Q & \longrightarrow & Q'' & \longrightarrow & 0 \\
 & & \downarrow & & \downarrow & & \downarrow & & \\
0 & \longrightarrow & N' & \longrightarrow & N & \longrightarrow & N'' & \longrightarrow & 0 \\
 & & \downarrow & & \downarrow & & \downarrow & & \\
 & & 0 & & 0 & & 0 & &
\end{array}
$$

where Q', Q, Q" are injective left R-modules and rows and columns are exact. Moreover, the exact sequences $0 \to M' \to Q' \to N' \to 0$, $0 \to M'' \to Q'' \to N'' \to 0$, may be chosen arbitrary with Q', Q" injective.

8.3.2. Corollary. An exact sequence of left R-modules

$$0 \longrightarrow M' \xrightarrow{f} M \xrightarrow{g} M'' \longrightarrow 0$$

may be introduced in a commutative diagram with exact rows and columns:

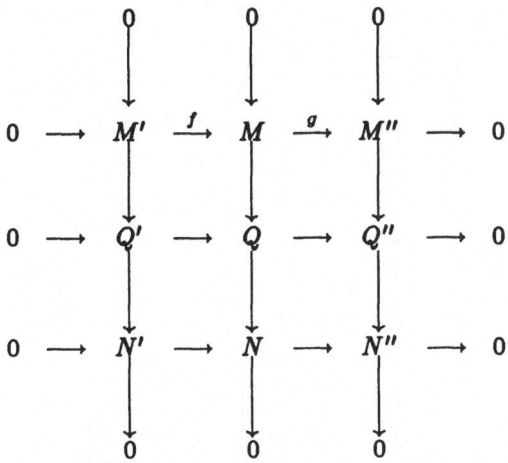

$$\begin{array}{ccccccccc}
 & & 0 & & 0 & & 0 & & \\
 & & \downarrow & & \downarrow & & \downarrow & & \\
0 & \longrightarrow & M' & \xrightarrow{f} & M & \xrightarrow{g} & M'' & \longrightarrow & 0 \\
 & & \downarrow & & \downarrow & & \downarrow & & \\
0 & \longrightarrow & Q' & \longrightarrow & Q & \longrightarrow & Q'' & \longrightarrow & 0 \\
 & & \downarrow & & \downarrow & & \downarrow & & \\
\vdots & & \vdots & & \vdots & & \vdots & & \vdots \\
 & & \downarrow & & \downarrow & & \downarrow & & \\
0 & \longrightarrow & Q'_{n-1} & \longrightarrow & Q_{n-1} & \longrightarrow & Q''_{n-1} & \longrightarrow & 0 \\
 & & \downarrow & & \downarrow & & \downarrow & & \\
0 & \longrightarrow & N' & \longrightarrow & N & \longrightarrow & N'' & \longrightarrow & 0 \\
 & & \downarrow & & \downarrow & & \downarrow & & \\
 & & 0 & & 0 & & 0 & &
\end{array}$$

where Q'_i, Q_i, Q''_i, $0 \leq i \leq n - 1$, are injective left R-modules.

8.3.3. Proposition (Schanuel). Given the exact sequences:
$0 \to M \to Q \to N \to 0$ and $0 \to M' \to Q' \to N' \to 0$, where Q and Q' are injective left R-modules. If $M \cong M'$ then $N \oplus Q' \cong N' \oplus Q$.

8.3.4. Corollary. Given the exact sequences:

$$0 \to M \to Q_0 \to \dots \to Q_{n-1} \to Q_n \; Y \; N \to 0,$$
$$0 \to M' \to Q'_0 \to \dots \to Q'_{n-1} \to Q'_n \to N' \to 0,$$

where Q_i, Q'_i, $o \leq i \leq n$, are injective left R-modules. If $M \cong M'$ then:
$N \oplus Q'_n \oplus Q_{n-1} \oplus \dots \cong N' \oplus Q_n \oplus Q'_{n-1} \oplus \dots$. In particular, N is injective if and only if N' is injective.

8.3.5. Definition. An *injective resolution*, resp. *minimal injective resolution*, of a left R-module M is an exact sequence of left R-modules:

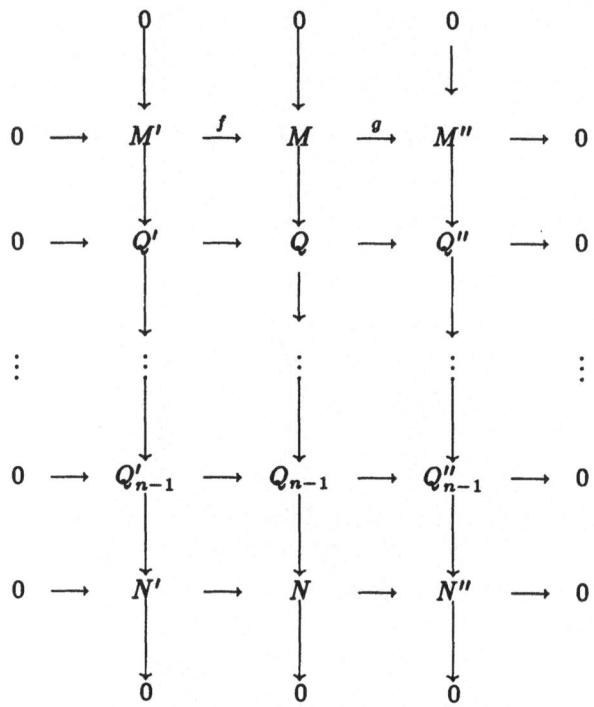

$$0 \longrightarrow M \xrightarrow{\varepsilon} Q_0 \xrightarrow{d_0} Q_1 \xrightarrow{d_1} \dots \longrightarrow Q_n \longrightarrow \dots$$

where each Q_n is an injective left R-module, resp. Q_n is the injective hull of Im d_{n-1} and Q_0 is the injective hull of Im ε, for all $n \geq 0$. If $Q_k \neq 0$ but $Q_n = 0$ for all n $>$ k then the injective resolution, resp. minimal injective resolution, is said to have *length k*. If no such k exists we say that it has infinite length.

8.3.6. Proposition. Any left R-module M has a minimal injective resolution.

Proof. Let $Q_0 = E(M)$ be the injective hull of M, ε: M $\rightarrow$ Q_0 the canonical inclusion. Consider $Q_1 = E_R\ (Q_0/\text{Im}\ \varepsilon)$ and d_0: $Q_0 \rightarrow Q_1$ the map induced by $Q_0 \rightarrow Q_0/\text{Im}\ \varepsilon$, $Q_2 = E_R\ (Q_1/\text{Im}\ d_0)$, ... The exact sequence

$$0 \longrightarrow M \xrightarrow{\varepsilon} Q_0 \xrightarrow{d_0} Q_1 \xrightarrow{d_1} Q_2 \longrightarrow \dots$$

is clearly a minimal injective resolution. □

8.3.7. Definition. The *injective dimension* of a left R-module M, denoted by inj.dim$_R$ (M) or simply inj.dim (M), is the smallest natural number n for which there exists an injective resolution of M of length n. If such an n does not exist we write inj.dim (M) $= \infty$.

8.3.8. Corollary. If there is an exact sequence of left R-modules,

$$0 \rightarrow M \rightarrow Q_0 \rightarrow Q_1 \rightarrow \dots \rightarrow Q_{n-1} \rightarrow N \rightarrow 0,$$

where all the Q_i are injective left R-modules then inj.dim (M) $\leq$ n if and only if N is an injective left R-module.

8.3.9. Proposition. Given an exact sequence of left R-modules,
$0 \rightarrow M' \rightarrow M \rightarrow M" \rightarrow 0$, then we have:

1. If inj.dim (M') $\leq$ n and inj.dim (M") $\leq$ n then inj.dim (M) $\leq$ n.
2. If inj.dim (M') $\leq$ n and inj.dim (M) $\leq$ n then inj.dim (M") $\leq$ n.
3. If inj.dim (M") $\leq$ n and inj.dim (M) $\leq$ n then inj.dim (M') $\leq$ n+1.

8.3.10. Corollary. Let N be a left R-submodule of $M \in$ R-mod.

1. If inj.dim (M) > inj.dim (M/N) then inj.dim (N) = inj.dim (M).
2. If inj.dim (M) < inj.dim (M/N) then inj.dim (N) = 1+inj.dim (M/N).
3. If inj/dim (M) = inj.dim (M/N) then inj.dim (N) $\leq$ 1+inj.dim (M/N).

8.3.11. Theorem. Let there be given injective resolutions of the left R-module M, say:

$$0 \longrightarrow M \xrightarrow{\varepsilon} E_0 \xrightarrow{d_0} E_1 \xrightarrow{d_1} E_2 \longrightarrow \ldots \text{ and}$$

$$0 \longrightarrow M \xrightarrow{\varepsilon'} E'_0 \xrightarrow{\delta_0} E'_1 \xrightarrow{\delta_1} E'_2 \longrightarrow \ldots$$

for any $n \geq 0$ there exists a morphism f_n: $E_n \to E'_n$ such that all squares in the following diagram are commutative:

(*)

$$\begin{array}{ccccccccccc} 0 & \longrightarrow & M & \xrightarrow{\varepsilon} & E_0 & \xrightarrow{d_0} & E_1 & \xrightarrow{d_1} & E_2 & \xrightarrow{d_2} & \ldots \\ & & \downarrow 1_M & & \downarrow f_0 & & \downarrow f_1 & & \downarrow f_2 & & \\ 0 & \longrightarrow & M & \xrightarrow{\varepsilon'} & E'_0 & \xrightarrow{\delta_0} & E'_1 & \xrightarrow{\delta_1} & E'_2 & \xrightarrow{\delta_2} & \ldots \end{array}$$

Proof. Since E'_0 is injective there exists f_0: $E_0 \to E'_o$ such that $f_0 \circ \varepsilon =$ $\varepsilon' \circ$ f. Since E_0 is an essential extension of Im ε, f_0 must be monomorphic. Hence $E'_0 =$ Im $f_0 \oplus Q'_0$ and Im $\delta_0 = \delta_0$ ((Im f_0) $\oplus$ δ_0 (Q'_0). For the latter equality it suffices to note that y $\in \delta_0$ (Im f_0) $\bigcap$ δ_0 (Q'_0) entails y $= \delta_0$ $(x_0) = \delta_0$ (x'_0) for some $x_0 \in$ Im f_0, $x'_0 \in Q'_0$, i.e. $x_0 - x'_0 \in$ Ker $\delta_0 =$ Im $\varepsilon' \subset$ Im f_0, therefore $x'_0 = 0$ and y = 0.
Let E(δ_0(Im f_0)) be the injective hull of δ_0 (Im f_0), then e obtain: E(δ_0(Im f_0)) $\subset E'_1$ and E(δ_0(Im f_0)) $\bigcap$ $\delta_0(Q'_0) = 0$. The restriction $\delta_0 \mid Q'_0$ is injective, hence $\delta_0(Q'_0)$ is an injective left R-module. Thus $Q_1 =$ E(δ_0(Im f_0)) $\oplus$ $\delta_0(Q_0\prime)$ $\oplus$ Q'_1.
On the other hand, f_0 induces an isomorphism g_0: Im $d_0 \to \delta_0$ (Im f_0) such that $g_0 \circ d_0 = \delta_0 \circ f_0$. From this we derive the existence of an isomorphism f_1: $E_1 \to$ E(δ_0(Imf_0)) such that $f_1 \circ d_0 = \delta_0 \circ f_0$. We have:
Im $d_1 = d_1$(E(δ_0(Im f_0)))+ $d_1(\delta_0(Q'_0)$ + $\delta_1(Q'_1) = \delta_1$(Im f_1) $\oplus$ $\delta_1(Q'_1)$. Indeed, δ_1(Im f_1) $\bigcap$ $\delta_1(Q'_1) = 0$ because if $y_1 \in \delta_1$(Im f_1) $\bigcap$ $\delta_1(Q'_1)$ then $y_1 = \delta_1(x_1) =$ $\delta_1(x'_1)$ for some $x_1 \in$ Im f_1 and $x'_1 \in Q'_1$; from $x_1 - x'_1 \in$ Ker $\delta_1 =$ Im δ_0 and Im $\delta_0 \bigcap Q'_1 = 0$ it follows that $x'_1 = 0$ or $y_1 = 0$.
Now consider E(δ_1(Im f_1)). We have $\delta(Q'_1) \bigcap$ E(δ_1(Im f_1)) = 0 and therefore $E'_2 =$ E(δ_1(Im f_1)) $\oplus$ $\delta_1(Q'_1)$ $\oplus$ Q'_2 (where $\delta_1(Q'_1)$ is injective). Continuing the argument

this way we arrive at an isomorphism f_2: $E_2 \rightarrow E(\delta_1(\text{Im } f_1))$ such that $f_2 \circ d_1 = \delta_1 \circ f_1$, and so on. □

8.3.12. Corollary. Two minimal injective rsolutions of a left R-module M have the same length.

Proof. Consider two minimal injective resolutions of M connected by the morphisms f_n as in the above theorem (see diagram (*)).
Since Im f_0 is essential in E_0' and we now assume that the injective resolutions are minimal it follows that Im $f_0 = E_0'$ and f_0 is therefore an isomorphism. Since $f_0 \circ d_0 = \delta_0 \circ f_0$ and f_0 is an isomorphism we arrive at Im $f_1 \supset$ Im δ_0 and therefore Im f_1 is essential in E_1'.
From Im $f_1 \cong E_1$ it follows that Im $f_1 = E_1'$ and hence f_1 is an isomorphism, and so on... □

8.3.13. Corollary. The injective dimension of a left R-module M equals the length of a minimal injective resolution for M.

8.3.14. Definition. The *global left injective dimension* of R, denoted by l.gl.inj.dim (R), is defined to be the supremum of the injective dimensions of arbitrary left R-modules.

Philosophically speaking, it is in the nature of the notion of injective dimension that it should be closely related to the notion of projective dimension. In fact we shall prove that both dimensions coincide but only after having made it clear that this is not a complete triviality.

8.3.15. Definition. Consider left R-modules M and N. We say that Ext (M,N) = 0 if and only if any left R-module T containing N such that $T/N \cong M$ necessarily contains N as a direct summand. It is not hard to verify that the following statements hold:

a. A left R-module Q is injective if and only if Ext (M,Q) = 0 for all left R-modules M.

b. A left R-module P is projective if and only if Ext (P,N) = 0 for all left R-modules N.

Of course Ext(-,-) may be defined in a more elegant way by making use of derived functors and some homological algebra but we avoid the unnecessary "complications" at this moment by only using the nullity of Ext (-,-).

8.3.16. Proposition (Kaplansky, [2]). Let M,N be left R-modules. The following statements are equivalent:

1. Ext (M,N) = 0.
2. For any given exact sequence $0 \longrightarrow K \xrightarrow{f} L \xrightarrow{g} M \longrightarrow 0$,
$0 \longrightarrow \mathrm{Hom}_R(M,N) \xrightarrow{g^*} \mathrm{Hom}_R(L,N) \xrightarrow{f^*} \mathrm{Hom}(K,N) \longrightarrow o$ is also exact.
3. If $0 \longrightarrow K \xrightarrow{f} L \xrightarrow{g} M \longrightarrow 0$ is a fixed exact sequence where L is a projective left R-module, then the following sequence is exact
$0 \longrightarrow \mathrm{Hom}_R(M,N) \xrightarrow{g^*} \mathrm{Hom}_R(L,N) \xrightarrow{f^*} \mathrm{Hom}(K,N) \longrightarrow 0$ is exact.

Proof. It will be sufficient to establish surjectivity of f^* in order to prove the implication 1 ⇒ 2. Let h: K → N be left R-linear. Put $A = N \oplus L$, $B = \{(-h(x), f(x)), x \in K\}$ is a left R-submodule of A and put C = A/B. Define α: N → C, $x \mapsto (x,0)$ mod B, and β: C → M, (x,y) mod B $\mapsto$ g(y). Clearly, these morphisms are well-defined and they fit into an exact sequence:

$$0 \longrightarrow N \xrightarrow{\alpha} C \xrightarrow{\beta} M \longrightarrow 0. \qquad (*)$$

By 1., (∗) splits. So there exists an α': C → N such that $\alpha' \circ \alpha = 1_N$.
If i: L → A is the canonical injection defined by $y \mapsto (0,y)$, and π: A → L being the canonical surjection, then we let h': L → N be given by $h' = \alpha' \circ \pi \circ i$. If $x \in K$ then $h'(f(x)) = \alpha'((o,f(x)) \bmod B) = \alpha'((h(x,0)) \bmod B) = (\alpha' \circ \alpha)(h(x)) = h(x)$. Hence $h' \circ f = h$ and this establishes surjectivity of f^*.

2 ⇒ 3. Obvious.

3 $\Rightarrow$ 1. Let T be a left R-module containing N such that $T/N \cong M$. Since L is projective there exist u: $L \rightarrow T$, v: $K \rightarrow N$, making the following diagram commutative:

$$\begin{array}{ccccccccc}
0 & \longrightarrow & K & \xrightarrow{f} & L & \xrightarrow{g} & M & \longrightarrow & 0 \\
 & & \big\downarrow v & & \big\downarrow u & & \big\downarrow 1_M & & \\
0 & \longrightarrow & N & \longrightarrow & T & \longrightarrow & M & \longrightarrow & 0
\end{array}$$

By the hypothesis, there exists a morphism w: $L \rightarrow N$ such that $w \circ f = v$. For an x $\in$ T there exists an x' $\in$ L such that $\beta(x) = g(x')$. Put $y = x + w(x') - u(x')$. Then we calculate: $\beta(y) = \beta(x) + \beta(w(x')) - \beta(u(x')) = \beta(x) - g(x') = \beta(x) - \beta(x) = 0$. Consequently $y \in N$.
In order to check that y does not depend on the choice of x' let $x_1' \in L$ be such that $\beta(x) = g(x_1')$. Then, $g(x' - x_1') = 0$ implies $x' - x_1' = f(z)$ for some $z \in K$ and thus $w(x' - x_1') = w(f(z)) = v(z)$, $u(x' - x_1') = u(f(z)) = v(z)$, and finally:
$x + w(x') - u(x') = x + w(x_1') - u(x_1')$ as desired. This enables us to define
α: $T \rightarrow N$, $x \mapsto x + w(x) - u(x')$, which is clearly an R-linear map. If $x \in N$ then $\beta(x) = 0$ and hence $\beta(x) = g(0)$ or $\alpha(x) = x + w(0) - u(0) = x$, i.e. $\alpha \mid N = 1_N$. This establishes a splitting for the exact sequence

$$0 \longrightarrow N \longrightarrow T \xrightarrow{\beta} M \longrightarrow 0$$

proving that N is a direct summand of T. □

Combining Baer's criterion with the above proposition we obtain:

8.3.17. Corollary. A left R-module Q is injective if and only if Ext $(R/I,Q) = 0$ for every left ideal I of R.

By dualization we prove:

8.3.18. Proposition (Kaplansky, [2]). Let M and N be left R-modules. The following statements are equivalent:

1. Ext $(M,N) = 0$.
2. For any eact sequence $0 \rightarrow N \rightarrow L \rightarrow K \rightarrow 0$,

$$0 \longrightarrow \mathrm{Hom}_R\,(M,N) \longrightarrow \mathrm{Hom}_R\,(M,L) \longrightarrow \mathrm{Hom}_R\,(M,K) \longrightarrow 0 \qquad (**)$$

is again exact.
3. If $o \to N \to L \to K \to 0$ is a fixed exact sequence where L is injective then (**) is exact.

8.3.19. Corollary. Given the exact sequences of left R-modules:

$$0 \longrightarrow K \xrightarrow{u} P \xrightarrow{v} M \longrightarrow 0 \text{ and } 0 \longrightarrow N \xrightarrow{u'} Q \xrightarrow{v'} L \longrightarrow 0$$

where P is projective and Q is injective, then Ext (M,L) = 0 if and only if Ext (K,N) = 0.

Proof. Suppose Ext (M,L) = 0. Let f: $K \to L$ be a morphism. By Proposition 8.3.16. there exists a morphism g: $P \to L$ such that $g \circ u = f$. Since P is projective there exists h: $P \to Q$ suuch that $v' \circ h = g$. Then $v' \circ (h \circ u) = f$, so we may apply Proposition 8.3.18. to conclude that Ext (K,N) = 0.
Conversely, when Ext (K,N) = 0 then one proves that Ext (M,L) = 0 in a similar way. □

8.3.20. Theorem. The global projective dimension of R equals the global injective dimension of R.

Proof. Let us show how to establish that l.gl.inj.dim (P) $\leq$ l.gldim (R) (the converse may be proved in a similar way).
It is not restrictive to assume that l.gldim (R) = n is finite. Let there be given an exact sequence of left R-modules:

$$0 \to M \to Q_0 \to \dots \to Q_{n-1} \to N \to 0 \qquad (*)$$

where each Q_i, $0 \leq i \leq n - 1$, is an injective left R-module. Since pd (A) $\leq$ n for any left R-module A, there exists a projective resolution of A having length at most n, say:

$$0 \to P_n \to P_{n-1} \to .. \to P_0 \to A \to 0 \qquad (**)$$

Projectivity of P_n leads to Ext $(P_n,M) = 0$; so in view of (*) and (**) as well as Corollary 8.3.19. we obtain that Ext (A,N) = 0. The fact that A is arbitrary entails that N is injective and hence inj.dim (M) $\leq$ n.
The latter entails that gl.inj.dim (R) $\leq$ n. □

8.3.21. Corollary. Let R be a ring, $\mathcal{G}$ a class of left R-modules and suppose that any left R-module M contains a nonzero submodule which is in $\mathcal{G}$. Then:

l.gldim (R) = sup $\{pd_R(A), A \in \mathcal{G}\}$.

Proof. That sup $\{pd_R(A), A \in \mathcal{G}\} \leq$ l.gldim R is obvious.

In order to establish the converse inequality we may assume that sup $\{pd_R(A), A \in \mathcal{G}\} = n < \infty$. Consider a left R-module M and an exact sequence:

$0 \to M \to Q_0 \to \dots \to Q_{n-1} \to N \to 0$. If $A \in \mathcal{G}$ then Ext (A,N) = 0 (exactly as in the proof of 8.3.20.). Let X' be a submodule of X and f: X' $\to$ N a module morphism. Consider the set $\mathcal{F} = \{(Y,g), X' \subset Y \subset X, g: Y \to N \text{ and } g \mid X' = f\}$ and order it by putting $(Y_1, g_1) \leq (Y_2, g_2)$ if and only if $Y_1 \subset Y_2$ and $g_2/Y_1 = g_1$. The set $\mathcal{F}$ is inductively ordered. By Zorn's lemma we may select a maximal element (X_0, f_0) in $\mathcal{F}$. If $X_0 \neq X$ then X/X_0 contains a nonzero submodule X_0'/X_0 isomorphic to a module of $\mathcal{G}$.

Since Ext $(X_0'/X_0, N) = 0$, it follows from 8.3.16. that there exists a morphism f_0': $X_0' \to N$ such that $f_0' \mid X_0 = f_0$. Then $(X_0', f_0') \in \mathcal{F}$ and $(X_0, f_0) < (X_0', f_0')$ leads to a contradiction unless $X_0' = X$ but this yields the injectivity of N. It follows that inj.dim (M) $\leq$ n and by Theorem 8.3.20. we may conclude that l.gldim (R) $\leq$ n. □

8.3.22. Corollary. (Auslander). For a ring R, l.gldim (R) = sup {pd (R/I), I a left ideal of R}.

Proof. Put $\mathcal{G}$ = {R/I, I left ideal in R} and apply the foregoing. □

The right global dimension of R, denoted by r.gldim (R), is defined to be l.gldim (R^o). An example due to L. Small will show that the left and right dimensions may be different, i.e. for $R = \begin{pmatrix} \mathbb{Z} & \mathbb{Q} \\ 0 & \mathbb{Q} \end{pmatrix}$ we have l.gldim (R) = 2, r.gldim (R) = 1. For Noetherian rings the behaviour of both dimensions is much nicer.

8.3.23. Theorem (Boratynski [1]). Let J(R) be the Jacobson radical of a left Noetherian ring R, then l.gldim (R) = sup {inj.dim (R/I), I a left ideal of R containing J(R)} = n.

Proof. If n is infinite then l.gldim (R) $\leq$ n is obvious (and this is the only inequality we really have to prove), so assume that n < ∞. Let M be a finitely generated left R-module such that J(R) M = 0. If M is generated by $x_1, x_2, ..., x_s$ then we will prove by induction on s that inj.dim (M) $\leq$ n.

If s = 1 then the assertion is clear.

If s > 1 then we consider the submodule M' generated by $\{x_1, ..., x_{s-1}\}$. The induction hypothesis yields that inj.dim (M') $\leq$ n. Because M/M' is a cyclic module annihilated by J(R) it follows that inj.dim (M/M') $\leq$ r. Proposition 8.1.9. entails inj.dim (M) $\leq$ n. In view of Auslander's theorem we must show that

pd_R (M) $\leq$ n for every finitely generated left R-module. Consider an exact sequence $0 \to K \to P_{n-1} \to ... \to P_0 \to M \to 0$, where each P_i is a finitely generated left R-module. Let now N be finitely generated left R-module with the property J(R)N = 0. Since inj.dim (N) $\leq$ n there exists an injective resolution:

$0 \to N \to Q_0 \to Q_1 \to ... \to Q_n \to 0$. From Ext $(M,Q_n) = 0$ we may derive that Ext (K,N) = 0 by a repeated application of Corollary 8.3.19.. The left Noetherian hypothesis entails that K is finitely generated hence there exists an exact sequence: $0 \longrightarrow L \xrightarrow{i} P \longrightarrow K \longrightarrow 0$, where P is a finitely generated projective module and also L is finitely generated. Since Ext (K,N) = 0 we arrive at an exact sequence:

$0 \to$ Hom (K,N) $\to$ Hom (P,N) $\to$ Hom (L,N) $\to 0$ for every N with the property J(R) N = 0. In particular we may consider N = L/ J(R)L, π: L $\to$ N being the canonical morphism. The exactness of the foregoing sequence yields that there is an f $\in$ Hom (P,N) mapping to π i.e. $f \circ i = \pi$. Projectivity of P yields the existence of a morphism $\pi \circ g \circ i = f \circ i = \pi$, hence $\pi \circ (g \circ i - 1_L) = 0$. Put $h = g \circ i - 1_L$. From $\pi \circ h = 0$, Im (h) $\subset$ J(R) L and Im $(1_L + h)$ + J(R) L = L follows from $x = (1_L + h)(x) - h(x)$ for every x $\in$ L. Nakayama's lemma entails that

Im $(1_L + h)$ = L and thus $1_L + h$ is surjective.

Sublemma. If M is a left Noetherian R-module and u: M $\to$ M is an epimorphism then u is an isomorphism.

Proof. Consider the ascending chain: Ker u $\subset$ Ker u^2 $\subset$ ... $\subset$ Ker u^n $\subset$ ... . For some n $\in \mathbb{N}$ we have Ker u^n = Ker u^{n+1}. If x $\in$ Ker u then u(x) = 0 but since u^n is surjective there is an y $\in$ M such that $x = u^n(y)$, thus $u^{n+1}(y) = 0$ or y $\in$ Ker u^{n+1} = Ker u^n i.e. x = 0.

Now we return to the proof of the theorem. Since $1_L + h = g \circ i$ is an isomorphism there exists a morphism g': P $\to$ L such that $g' \circ i = 1_L$, i.e.

$$0 \longrightarrow L \xrightarrow{i} P \longrightarrow K \longrightarrow 0$$

is split. Then L is a direct summand of P hence projective too; similar for K. It follows from exactness of $0 \to K \to P_{n-1} \to ... \to P_0 \to M \to 0$ that pd_R (M) $\leq$ n as claimed. □

We say that R is *semi-local* whenever R/J(R) is a semisimple ring. As an immediate consequence of the preceding theorem we have:

8.3.24. Corollary. If R is a left Noetherian semi-local ring then l.gldim (R) = inj.dim (R/J(R)).

8.3.25. Theorem (M. Auslander). If R is a left and right Noetherian ring then r.gldim (R) = l.gldim (R).

Proof. Put r.gldim (R) = n. If n = ∞ then l.gldim (R) $\leq$ n holds, so we take n to be finite. For a finitely generated left R-module M we may consider the following exact sequence (since R is left Noetherian):

$$0 \to N \to P_{n-1} \to \dots \to P_1 \to P_0 \to M \to 0,$$

where each P_i is a finitely generated projective left R-module and N is finitely generated. Let $C_{\mathbb{Z}}$ be an injective cogenerator over $\mathbb{Z}$, then we obtain the exact sequence:

$$0 \to \mathrm{Hom}_{\mathbb{Z}}(M,C) \to \mathrm{Hom}_{\mathbb{Z}}(P_0,C) \to \dots \to \mathrm{Hom}_{\mathbb{Z}}(P_{n-1},C) \to \mathrm{Hom}_{\mathbb{Z}}(N,C) \to 0.$$

Envoking Proposition 2.11.9. we may conclude that each $\mathrm{Hom}_{\mathbb{Z}}(P_i,C)$ is an injective right R-module. Since r.gldim (R) = n it follows that inj.dim ($\mathrm{Hom}_{\mathbb{Z}}$ (M,C)) $\leq$ n, consequently $\mathrm{Hom}_{\mathbb{Z}}$ (N,C) is an injective right module. Hence N is a flat left R-module (use Proposition 2.11.9. again) and then Proposition 2.11.13. entails that N is projective. Therefore pd_R (M) $\leq$ n and we may use Auslander's theorem and conclude that l.gldim (R) $\leq$ n.
The converse may be established by interchanging left and right. □

In conclusion of this section we provide a characterization of rings R with l.gldim (R) = 0 or l.gldim (R) = 1.

8.3.26. Theorem. The following assertions are equivalent for a ring R,

1. l.gldim (R) = 0,

2. r.gldim (R) = 0,
3. R is semisimple Artinian.

Proof. 3. $\Rightarrow$ 1. If I is a left ideal of R then I is a direct summand of $_RR$ and therefore R/I is a projective left R-module. By Auslander's theorem l.gldim (R) = 0 follows. The implication 3 $\Rightarrow$ 2 follows in exactly the same way.
1. $\Rightarrow$ 3. (similarly for 2. $\Rightarrow$ 3.). If I is a left ideal of R then pd_R (R/I) = 0 yields that R/I is a projective R-module i.e. I is a direct summand of R and therefore R is semisimple Artiian. □

8.3.27. Theorem. The following assertions are equivalent for a ring R:

1. l.gldim (R) $\leq$ 1.
2. Every left ideal of R is a projective R-module.
3. Every submodule of a left R-module is projective.
4. Every factor module of an injective module is injective.

Proof. 1. $\Leftrightarrow$ 2. Follows from Auslander's Theorem. The equivalences 1 $\Leftrightarrow$ 3. and 1. $\Leftrightarrow$ 4. are completely trivial.

A ring R satisfying the conditions of the foregoing theorem is called a *left hereditary ring. Right hereditary rings* may be defined in a similar way. In the absence of the Noetherian conditions both concepts are different.

§8.4 The Flat Dimension of a Module.

The use of projective, or dually injective, modules in defining the global dimension is intrinsically connected to properties of the Hom functor (and its derived functors). If one considers the tensor product then the class of modules obtained by demanding certain exactness properties of the functor M $\otimes$ - is the class of flat modules. The use of flat resolutions instead of projective ones leads to some new concepts that will be related however, even in a very strict way, to the dimensions introduced in foregoing sections.

8.4.1. Definition. A *flat resolution* of a left R-module M is an (infinite) exact sequence

$$... \to P_n \to P_{n-1} \to .. \to P_1 \to P_0 \to M \to 0 \qquad (*)$$

where each P_i is a flat left R-module. If $P_k \neq 0$ and $P_n = 0$, $n \geq k$, then the resolution (*) is said to have length k, otherwise we say that it has infinite length. The *flat dimension* of M, sometimes refered to as the *weak (projective) dimension* of M, denoted by fd_R (M) or wd_R (M) or simply by fd (M), is the least number n such that M has a flat resolution of length n. If no such n exists then we put fd (M) = 0.

In particular fd (M) = 0 if and only if M is flat. Since a projective resolution is clearly also a flat resolution we have that fd (M) $\leq$ pd (M) for each left R-module M. Put $M^* = \text{Hom}_{\mathbb{Z}}(M, \mathbb{Q}/\mathbb{Z})$, the character module of M, which is a right R-module.

8.4.2. Proposition. For a left R-module M, fd (M) = r.inj.dim (M^*).

Proof. That inj.dim (M^*) $\leq$ fd (M) is clear. Now assume inj.dim (M^*) = n. When $n = \infty$, fd (M) $\leq$ n is clear, so we may assume that inj.dim (M^*) = n and $n < \infty$. Consider an exact sequence:

$$0 \to N \to P_{n-1} \to ... \to P_1 \to P_0 \to M \to 0$$

where each P_i is a flat module. We obtain an exact sequence of right modules

$$0 \to M^* \to P_0^* \to P_1^* \to ... \to P_{n-1}^* \to N^* \to 0.$$

Since each P_i^* is an injective right R-module (cf. Proposition 2.11.9.) and inj.dim (M^*) = n it follows that N^* is an injective right R-module.
Thus N is flat and fd (M) $\leq$ n follows. □

Remark. As an application of Proposition 8.4.2. and Proposition 8.3.9. we obtain the following Corollaries. If $0 \to M' \to M \to M" \to 0$ is an exact sequence of left R-modules then:

1. If fd (M') $\leq$ n, fd (M") $\leq$ n, then fd (M) $\leq$ n.
2. If fd (M") $\leq$ n and fd (M) $\leq$ n then fd (M') $\leq$ n.
3. If fd (M') $\leq$ n and fd (M) $\leq$ n then fd (M") $\leq$ n + 1.

8.4.3. Corollary. For a left R-module M the following assertions are equivalent:

1. We have fd (M) $\leq$ n.
2. Every projective resolution of M has a flat (n-1)-th kernel.

Proof. The implication 2. $\Rightarrow$ 1. is obvious. In order to establish the converse let us consider an exact sequence of left R-modules:

$$0 \to N \to P_{n-1} \to \ldots \to P_1 \to P_0 \to M \to o$$

where each P_i is projective. From the exact sequence:
$0 \to M^* \to P_0^* \to \ldots \to P_{n-1}^* \to N \to o$, where each P_i^* is an injective right module, and Proposition 8.4.2. we may conclude that inj.dim $(M^*) \leq$ n and hence N^* is injective or N is flat. □

8.4.4. Definition. Let ${}_RN$ be a left R-module, M_R a right R-module. We write r.Tor (M,N) = 0, resp. l.Tor (M,N) = 0, if for every exact sequence of right, resp. left, R-modules of the form $0 \to K \to L \to M \to 0$, resp. $0 \to Q \to P \to N \to 0$, we have the exact sequence $0 \to K \otimes N \to L \otimes N \to M \otimes N \to 0$, resp.
$0 \to M \otimes Q \to M \otimes P \to M \otimes N \to 0$. It is clear from this that ${}_RN$ is flat, resp. M_R is flat, if and only if l.Tor (M,N) = 0, resp. r.Tor (M,N) = 0, for each right, resp. left, R-module M, resp. N.

Here we point out that we have chosen to define the property Tor (M,N) = 0 without going into detail concerning the theory of derived functions of M $\otimes$ – (just like for Ext in the foregoing section).

8.4.5. Proposition. Let M_R and ${}_RN$ be as before, then the following assertions are equivalent:

1. l.Tor (M,N) = 0.
2. There exists a projective right R-module L and an exact sequence
$0 \to K \to M \to 0$ such that $0 \to K \otimes_R N \to L \otimes_R N \to M \otimes_R N \to 0$ is exact.
3. Ext $(M, N^*) = 0$.

Proof 1. $\Rightarrow$ 2. is obvious and 3. $\Rightarrow$ 1. follows from Proposition 8.3.16. and

Proposition 2.11.5., so we only have to prove the implication 2. $\Rightarrow$ 3.
Since $\mathbb{Q}/\mathbb{Z}$ is an injective cogenerator of $\mathbb{Z}$ mod, the sequence
$0 \to K \otimes N \to L \otimes N \to M \otimes N \to 0$ is exact if and only if the sequence
$0 \to \mathrm{Hom}_{\mathbb{Z}}(M \otimes N, \mathbb{Q}/\mathbb{Z}) \to \mathrm{Hom}_{\mathbb{Z}}(L \otimes N, \mathbb{Q}/\mathbb{Z}) \to \mathrm{Hom}_{\mathbb{Z}}(K \otimes N, \mathbb{Q}/\mathbb{Z})$
$\to 0$ is exact. From Proposition 8.3.16. we now retain: Ext $(M,N^*) = 0$. □

8.4.6. Proposition. If M_R and ${}_RN$ are as above then the following assertions are equivalent:

1. r.Tor $(M,N) = 0$.
2. There exists a projective left R-module P and an exact sequence of left R-modules $0 \to Q \to P \to N \to o$ such that the sequence $0 \to M \otimes Q \to M \otimes P \to M \otimes N \to 0$ is exact.
3. Ext $(N,M^*) = 0$.

8.4.7. Theorem. If M_R, ${}_RN$ are as above then the following statements are equivalent:

1. l.Tor $(M,N) = 0$.
2. r.Tor $(M,N) = 0$

Proof. 1. $\Rightarrow$ 2. Consider the exact sequence of left R-modules:
$0 \to Q \to P \to N \to 0$ where P is projective, hence we have an exact sequence
$0 \to N^* \to P^* \to Q^* \to 0$ where P^* is an injective right R-module. Proposition 8.3.18. yields the exactness of:
$0 \to \mathrm{Hom}_R(M,N^*) \to \mathrm{Hom}_R(M,P^*) \to \mathrm{Hom}_R(M,Q^*) \to 0$ and so we obtain the exact sequence: $0 \to \mathrm{Hom}_{\mathbb{Z}}(M \otimes_R N, \mathbb{Q}/\mathbb{Z}) \to \mathrm{Hom}_{\mathbb{Z}}(M \otimes_R P, \mathbb{Q}/\mathbb{Z}) \to \mathrm{Hom}_{\mathbb{Z}}(M \otimes_R Q, \mathbb{Q}/\mathbb{Z}) \to 0$.
Then $0 \to M \otimes_R Q \to M \otimes_R P \to M \otimes_R N \to 0$ is exact and Proposition 8.4.6. then yields that r.Tor $(M,N) = 0$.

2. $\Rightarrow$ 1. A symmetric version of 1. $\Rightarrow$ 2. □

Remark. The theorem allows to write Tor $(M,N) = 0$ without specifying "left" or "right".

8.4.8. Theorem. Consider the exact sequences:

$$0 \to M_R \to P_R \to N_R \to 0,$$
$$0 \to {}_RK \to {}_RQ \to {}_RL,$$

where P_R and ${}_RQ$ are projective. Then Tor (M,L) = 0 if and only if Tor (N,K) = 0.

Proof. The first exact sequence yields an exact sequence of left R-modules $0 \to N^* \to P^* \to M^* \to 0$ where P^* is injective. By Corollary 8.3.19. we have Ext (K,N^*) = 0 if and only if Ext (L,M^*) = 0. Propositioon 8.4.6. yields Tor (N,K) = 0 if and only if Tor (M,L) = 0. □

8.4.9. Definition. The *left weak dimension* of R is defined by: l.wdim (R) = sup {fd (M), M a left R-module}. Similarly, r.wdim (R) = sup {fd (M), M a right R-module}.

8.4.10. Theorem. For any ring R, l.wdim (R) = r.wdim (R).

Proof. Put n = l.wdim (R). If n = ∞ then r.wdim (R) ≤ n is clear, so assume that n < ∞. Let M_R be a right R-module, ${}_RN$ a left R-module and consider the exact sequence: $0 \to P_n \to \dots \to P_1 \to P_0 \to N \to 0$, where each P_i is flat. The sequence $0 \to N^* \to P_0^* \to \dots \to P_n^* \to 0$ is exact and P_i^* is an injective right R-module for each i = 0,..,n.
Now let there be given an exact sequence $0 \to K \to Q_{n-1} \to \dots \to Q_0 \to M \to 0$ where each Q_i is flat. Since Ext (M,P_n^*) = 0, Corollary 8.3.11. entails Ext (K,N^*) = 0, thus Tor (K,N) = 0 in view of Proposition 8.4.6.. Since N is arbitrary K is flat. Therefore fd (M) ≤ n in view of Corollary 8.4.3. Consequently r.wdim (R) ≤ n.
In formally the same way one may establish the converse inequality and the theorem as well. □

In view of the theorem we just write wdim (R) for the *weak dimension* of R.

8.4.11. Corollary. For any ring R, wdim (R) ≤ inf {l.gldim (R), r.gldim (R)}.

To compute wdim (R) it will be sufficient to look at the flat dimension of cyclic modules.

8.4.12. Theorem., For any ring R we have the following equalities: wdim (R) = sup {fd (R/I), I a finitely generated left ideal of R} = sup {fd (J), J a finitely generated right ideal of R}.

Proof. In view of Proposition 2.11.11. the right R-module M_R is flat if and only if Tor (M,R/I) = 0 for every finitely generated left ideal I of R. Continue the proof as in the proof of Theorem 8.4.10., then we arrive at wdim (R) = r.wdim (R) $\leq$ sup {fd (R/I), I is a finitely generated left ideal of R}.

8.4.13. Corollary. For a left Noetherian ring R: l.gldim (R) = wdim (R). For a right Noetherian ring R: r.gldim (R) = wdim (R).

Proof. Apply the theorem taking into account that the finitely presented flat left modules are exactly the finitely generated projective ones, (cf. Chapter 2, Proposition 2.11.14.). □

We characterize rings with wdim (R) = 0 in the following:

8.4.14. Theorem. For any ring R, the following assertions are equivalent:

1. wdim (R) = 0.
2. Every left (right) R-module is flat.
3. For any a $\in$ R there exists a b $\in$ R such that a = aba.
4. Every left (right) finitely generated ideal of R is a direct summand of R.

Proof. 1. $\Rightarrow$ 2. By definition.
2. $\Rightarrow$ 3. Apply Corollary 2.11.12. for M = R, K = Ra, I = aR, so we obtain that aRa = aR $\wedge$ Ra, hence a $\in$ aRa and there is a b $\in$ R for which a = aba.
3. $\Rightarrow$ 4. Consider a left ideal I = $Ra_1 + \ldots + Ra_n$ of R. By induction on n we

establish that I is generated by an idempotent, e say. If I = Ra then a = aba for some b ∈ R and then ba = (ba)(ba) yields Rba = Ra with ba = $(ba)^2$.
Suppose next that I = $Re_1 + Re_2$ where $e_i^2 = e_i$, i = 1,2. Clearly I = Re_1 + $R(e_2 - e_2e_1)$. But $R(e_2 - e_2e_1)$ = Rf for some idempotent f. Since $(e_2 - e_2e_1)e_1 = 0$ it follows that $fe_1 = 0$. We may calculate $(e_1 + f - e_1f)^2 = e_1+f-e_1f$ and also $Re_1 + Re_2 = R(e_1+f-e_1f)$.
So I is generated by some idempotent element e of R.
4. ⇒ 1. Envoke Theorem 8.4.12. □

A ring R satisfying the equivalent conditions of the foregoing theorem is said to be a *Von Neumann regular* ring. As an example: every direct product of a family of central simple algebras (or even simple Artinian rings) is Von Neumann regular.

A ring R is said to be *left semi-hereditary* if and only if every finitely generated left ideal of R is projective. In view of Theorem 8.4.12. we obtain that wdim (R) ≤ 1 for a left semi-hereditary ring R.

8.4.15. L. Small's Example (cf [3]). Put R = $\begin{pmatrix} \mathbb{Z} & \mathbb{Q} \\ 0 & \mathbb{Q} \end{pmatrix}$. Then l.gldim (R) = 2 and r.gldim (R) = 1.

Proof (and remarks). We make the following general observations:

o.1. If I is an ideal of R, M a left R-module such that IM = 0 then we have:

a. pd_R (M) ≤ pd_R (R/I) + l.gldim (R/I)
b. fd_R (M) ≤ fd_R (R/I) + wdim (R/I).

If $\mathrm{pd}_{R/I}$ (M) = ∞ then l.gldim (R/I) = ∞ and a. is obvious, so assume that $\mathrm{pd}_{R/I}$ (M) = n < ∞ hence l.gldim (R/I) ≥ n. There is an exact sequence of R/I-modules: $0 \to P_n \to \dots \to P_1 \to P_0 \to M \to 0$ where each P_i is a projective R/I-module. From $\mathrm{pd}_R(P_i) = \mathrm{pd}_R$ (R/I), o ≤ i ≤ n, again a. follows. To prove claim b. one proceeds in a similar way.

o.2. If $I_1, \dots, I_n$ are ideals of R such that $I_1 \cdot \dots \cdot I_n = 0$ then

a. l.gldim (R) ≤ max {l.gldim (R/I_k) + pd_R (R/I_k), k = 1,...,n}
b. wdim (R) ≤ max {wdim (R/I_k) + fd_R (R/I_k), k = 1,...,n}.

Indeed, for a left R-module M we consider the filtration:

$0 = (I_1 \cdot .. \cdot I_n)M \subset (I_2 \cdot ... \cdot I_n)M \subset ... \subset I_nM \subset M$. Proposition 8.1.9. yields: pd_R (M) $\leq$ max $\{\mathrm{pd}_R((I_{k+1} \cdot ... I_n)M/(I_k \cdot ... \cdot I_n)M), k = 1,...,n\}$. Now $I_k((I_{k+1} \cdot ... \cdot I_n)M/(I_k \cdot ... \cdot I_n)M) = 0$ and so we may apply o.1. and obtain a. (a similar result is valid for the r.gldim (R)).
The assertion concerning wdim (R) may be established in a similar way.

o.3. For $R = \begin{pmatrix} \mathbb{Z} & \mathbb{Q} \\ 0 & \mathbb{Q} \end{pmatrix}$ now, put $e = \begin{pmatrix} 1 & 0 \\ 0 & 0 \end{pmatrix}$, $f = \begin{pmatrix} 0 & 0 \\ 0 & 1 \end{pmatrix}$ and I = eR, J = Rf. The latter one-sized ideals are in fact two-sided and R/I ≅ Q, R/J ≅ $\mathbb{Z}$, IJ = 0.
In view of o.2. we obtain: l.gldim (R) $\leq$ max {l.gldim (R/I) + pd_R (R/I), l.gldim (R/J) + pd_R (R/J)} = max {1, pd_R (R/I)}.
However I ≅ Re $\oplus$ $_RX$ as a left R-module, where $X = \begin{pmatrix} 0 & \mathbb{Q} \\ 0 & 0 \end{pmatrix}$. Since JX = 0, $_RX$ is a left R/J-module. From $\mathrm{pd}_{\mathbb{Z}}$ (Q) = 1 and R = R/J $\oplus$ J (i.e. R/J is a projective left R-module) we obtain that pd_R (X) = 1 and hence pd_R (R/I) = 2 or l.gldim (R) $\geq$ 2. But then we have l.gldim (R) = 2.
Now wdim (R) $\leq$ max {wdim (R/I) + fd_R (R/J), wdim(R/J) + fd_R (R/J)} = max {1,fd_R (R/I)}. Since $\mathrm{fd}_{\mathbb{Z}}$ (Q)= 0 we find that fd_R ($_RX$) = 0 and therefore fd_R (R/I) = 1. Hence: wdim (R) $\leq$ 1.
Since R is right Noetherian but not left Noetherian (cf. Section 2.15.) we may conclude that r.gldim (R) = wdim (R) $\leq$ 1. As R is not semisimple, r.gldim (R) = 1 follows. □

§8.5 The Artin-Rees Property and Homological Dimensions.

Let I be an ideal of R. If R is a left Noetherian ring then we put $\mathcal{F}_I$ = {J left ideal of R such that $I^n \subset$ J for some n ∈ $\mathbb{N}$ }

8.5.1. Lemma. For a left Noetherian ring R, $\mathcal{F}_I$ is an additive topology on R.

Proof. Very easy. □

A left R-module M is $\mathcal{F}_I$-torsion if and only if for every m ∈ M, I^km = 0 for some k ∈ $\mathbb{N}$

8.5.2. Definition. An additive topology $\mathcal{F}$ on R is stable if the torsion class is stable under taking injective hulls.

8.5.3. Proposition. Let I be an ideal of a left Noetherian ring R. The folowing assertions are equivalent:

1. I has the Artin-Rees property.
2. I has the Artin-Rees property for left ideals, i.e. for every left ideal K of R and any $n \in \mathbb{N}$ there exists an $h(n) \in \mathbb{N}$ such that $I^{h(n)} \bigcap K \subset I^n K$.
3. For each left R-module M which is $\mathcal{F}_I$-torsion we have $E(M) = \bigcup_{k \geq 0} (0 : I^k)$, where $(0 : I^k) = \{x \in E(M), I^k x = 0\}$.
4. $\mathcal{F}_I$ is a stable additive topology.

Proof. 1.⇒2. and 3.⇒4. are obvious.
2.⇒3. Put E_1 equal to the union of the $(0 : I^k)$. Since $M \subset E_1$ it will be sufficient to prove that E_1 is injective. Consider a morphism f: $K \to E_1$ for some left ideal K of R. Since K is finitely generated, $f(K) \subset (0 : I^k)$ for some $k \in \mathbb{N}$. Pick n such that $I^n \bigcap K \subset I^k K$. We have $f(I^k K) \subset I^k f(K) = 0$ and therefore $I^n \bigcap K \subset \mathrm{Ker}\, f$. The morphism $\bar{f}$ induced by f fits in a diagram of R/I^n-modules (we assume $n > k$):

$$\begin{array}{ccccc} 0 & \longrightarrow & K/I^n \bigcap K & \longrightarrow & R/I^n \\ & & \downarrow \bar{f} & \swarrow \bar{g} & \\ & & (0 : I^n) & & \end{array}$$

Since $(0 : I^n)$ is an injective (R/I^n)-module there exists a morphism $\bar{g}$: $R/I^n \to (0 : I^n)$ extending $\bar{f}$. The composition $R \to R/I^n \to (0 : I^n)$ yields the desired extension of f to R.

4.⇒1. Let N be a submodule of a finitely generated left R-module M. Since $N/I^n N$ is $\mathcal{F}_I$-torsion, $E_R\,(N/I^n N)$ is $\mathcal{F}_I$-torsion too. Consider the diagram:

$$\begin{array}{ccccc} 0 & \longrightarrow & N & \longrightarrow & M \\ & & \pi \downarrow & & \swarrow \overline{f} \\ & & N/I^nN & & \\ & & i \downarrow & \swarrow & \\ & & E_R(N/I^nN) & & \end{array}$$

where π,i are the canonical morphisms, and $\overline{f}$ exists by definition of E_R (N/I^n N). Thus Ker $(\overline{f}) \bigcap$ N = Ker f, (f = i $\circ$ π), i.e. Ker $(\overline{f}) \bigcap$ N = I^nN. Since M is finitely generated, f(M) is a finitely generated submodule of E_R (N/I^nN). Therefore there exists an h(n) $\in \mathbb{N}$ such that $I^{h(n)}$ f(M) = 0, i.e. $I^{h(n)}$M $\subset$ Ker $(\overline{f})$ and I^nM $\supset$ $I^{h(n)}$M $\bigcap$ N follows. □

8.5.4. Proposition. Let $\mathcal{F}$ be a stable additive topology on a left Noetherian ring R. If M is an $\mathcal{F}$-torsion left R-module then inj.dim (M) $\leq$ sup {pd(R/I), I $\in$ $\mathcal{F}$}.

Proof. Put n = sup {pd (R/I), I $\in$ $\mathcal{F}$}, as usual we may assume that n $<$ ∞. Consider an exact sequence

$$0 \to M \to Q_0 \to Q_1 \to \dots \to Q_{n-1} \to N \to 0$$

where each Q_i is an injective left R-module. Since $\mathcal{F}$ is stable we may assume that each Q_i is also $\mathcal{F}$-torsion and therefore N is $\mathcal{F}$-torsion. Exactly as in the proof of Corollary 8.3.21. we may establish that Ext (R/I,N) = 0 whenever I $\in$ $\mathcal{F}$ and also that every morphism f: X' $\to$ N defined on a submodule X' of an $\mathcal{F}$-torsion module X may be extended to X. Now in general let X' be a submodule of an arbitrary left R-module X and let f: X' $\to$ N be a given morphism. There exists a morphism g: X $\to$ E(N) such that g | X' = f, hence Ker (f) equals X' $\bigcap$ Ker (g) and we obtain a commutative diagram:

$$\begin{array}{ccccc} 0 & \longrightarrow & X' & \longrightarrow & X \\ & & \downarrow \pi' & & \downarrow \subset \\ 0 & \longrightarrow & X'/Ker\ f & \longrightarrow & X/Ker\ g \\ & & \downarrow \overline{f} & \swarrow & \downarrow \overline{g} \\ & & N & & E(N) \end{array}$$

where π and π' are canonical epimorphisms and $f = \overline{f} \circ \pi'$, $g = \overline{g} \circ \pi$. Since $\overline{f}, \overline{g}$ are monomorphic it follows that X'/Ker f and X/Ker g are $\mathcal{F}$-torsion. Therefore there is an h: X/Ker g $\to$ N such that $\overline{f} = h \mid (X'/\text{Ker } f)$. Now it is clear that f is the restriction of $h \circ \pi$ and consequently N is injective, proving the proposition. □

8.5.5. Theorem. Let I be an ideal contained in the Jacobson radical J(R) of a left Noetherian ring R. If I has the AR-property then l.gldim (R) $\leq$ pd_R (R/I) + l.gldim (R/I).

Proof. In view of Theorem 8.3.23. we have: l.gldim (R) = sup {inj.dim (M), M a left R-module such that IM = 0}. The AR-property yields that $\mathcal{F}_I$ is stable so the proposition yields: l.gldim (R) $\leq$ sup {pd_R (R/K), K $\in$ $\mathcal{F}_I$}. For K $\in$ $\mathcal{F}_I$, there exists an $n \geq 0$ such that K $\supset$ I^n. If M is a left R-module such that I^mM = 0 for some $m \geq 0$ then it is clear that pd_R (M) = sup {pd_R ($I^k M/I^{k+1}M$), k = 1,...,n}. Hence, l.gldim (R) $\leq$ sup {pd_R (M), M is finitely generated and IM = 0}.
By the observation o.1. of L. Small's example (Section 8.4.) we have: pd_R (M) $\leq$ pd_R (R/I) + l.gldim (R/I). Therefore, l.gldim (R) $\leq$ pd_R (R/I) + l.gldim (R/I).
□

8.5.6. Definition. The ring R is said to be *(semi-)local* if R/J(R) is (semi-)simple.

8.5.7. Corollary. Let R be a left Noetherian semi-local ring. If J(R) has the AR-property then l.gldim (R) = pd_R (R/J(R)).

Proof. Follows from Theorem 8.5.5. and the fact that the left global dimension of R/J(R) is zero.

8.5.8. Corollary. Let R be a left Noetherian local ring such that J(R) has the left AR-property. If M is a left R-module with nonzero socle such that l.gldim (R) = n $< \infty$, then pd_R (M) = n.

Proof. Corollary 8.5.7. implies l.gldim (R) = pd_R (S), where S is a simple R-module.

Since the socle s(M) $\neq 0$ there exists an exact sequence
$0 \to S \to M \to M/S \to 0$. The result follows by applying Corollary 8.1.10. □

§8.6 Regular Local Rings.

An element a of a ring R is said to be normalizing if aR = Ra. We say that $a_1,...,a_n \in R$ form a *normalizing set* if: a_1 is a normalizing element of R, $\overline{a_i}$ is a normalizing element of $R/(a_1,...,a_{i-1})$, i = 2,...,n, where $\overline{a_i}$ is the image of a_i in $R/(a_1,...,a_{i-1})$. A normalizing set $\{a_1,...,a_n\}$ is said to be a *regular normalizing set* if a_1 is regular and $\overline{a_i}$ is regular in $R/(a_1,...,a_{i-1})$. A set of generators for an ideal I of R is said to be a *normalizing set of generators* if it is also a normalizing set of R.

8.6.1. Proposition. Let I be an ideal of the left Noetherian ring which is contained in the Jacobson radical J(R) of R. If I has a regular normalizing set of generators consisting of n elements, then:

1. Kdim (R) = n + Kdim (R/I).
2. If l.gldim (R/I) $< \infty$ then l.gldim (R) = n + l.gldim (R/I).
3. pd_R (R/I) = n.

Proof. If n = 1 then I = aR = Ra for some regular normalizing a in R. In this case $S = \{1, a, a^2, ..., a^m, ...\}$ satisfies the Ore conditions hence there is a left and right ring of fractions $T = S^{-1}R = RS^{-1}$. Therefore I is an invertible ideal of R and we may apply Theorem 6.5.6. to obtain Kdim (R) = 1 + Kdim (R/I).
The assertion 1. now follows by an easy induction on n.
2. By induction on n and Theorem 8.1.18.
3. If n = 1 then pd_R (R/aR) = 1 + $\mathrm{pd}_{R/aR}$ (R/aR) = 1 (see Theorem 8.1.16.). We continue to obtain the proof by induction on n.
Assume n > 1, denote: $R_1 = R/a_1R$. Then $\overline{a_2},..,\overline{a_n}$ is a regular normalizing set of generators for $I_1 = I/a_1R$. The induction hypothesis yields: $\mathrm{pd}_{R_1}(R_1/I_1) = n - 1$. Apply Theorem 8.1.16., then we obtain: $\mathrm{pd}_R(R_1/I_1) = 1 + \mathrm{pd}_{R_1}(R_1/I_1) = n$. Since $R_1/I_1 = R/I$ we conclude that pd_R (R/I) = n. □

A ring is said to be *completely local* if the non-units of R form an ideal or equivalently if R/J(R) is a division ring. We say that R is a (left) *n-dimensional regular local* ring if it is a comletely local left Noetherian ring such that J(R) has a regular normalizing set of generators consisting of n elements. An ideal P of R is said to be *completely prime* whenever R/P is a domain; obviously a completely prime ideal is a prime ideal.

8.6.2. Lemma. Let a be a regular normalizing element of R contained in J(R), where R is a left Noetherian ring. Put P = aR = Ra.

1. If P is a prime ideal then R is a prime ring.
2. If P is completely prime then R is a domain.

Proof. Since P is an invertible ideal it has the Artin-Rees property and thus $\bigcap_{n\geq 0} P^n = 0$ (cf. Section 6.5.).
1. If P is a prime ideal and xRy = 0 then x ∈ P or y ∈ P. If x ∈ P, x ≠ 0 then there exists an n such that $x \in P^n$, $x \notin P^{n+1}$. Hence there is a $z \notin P$ such that $x = a^n z$ but since a^n is regular this leads to zRy = 0. Hence zRy ⊂ P and thus y ∈ P. If y ≠ 0 then $y \in P^m$, $y \notin P^{m+1}$ for some m ∈ ℕ and hence there is an $u \notin P$ such that $y = ua^m$. Consequently zRu = 0 since a^m is regular. But then zRu ⊂ P with $z, u \notin P$ is a contradiction unless y = 0, i.e. R is a prime ring.

2. Assume xy = 0 with x ≠ 0 and run along the lines of proof of 1. □

8.6.3. Theorem. Let R be a left Noetherian local ring such that J(R) has a regular normalizing set of generators consisting of exactly n elements, then l.gldim (R) = pd_R (R/J(R)) = Kdim (R) = cl.Kdim (R) = n. Moreover R is a prime ring. If R is completely local (i.e. R is also regular there) then R is a domain.

Proof. The fact that R is prime (resp. a domain) follows from the lemma using induction on n. Envoking Proposition 8.6.1. we obtain: l.gldim (R) = pd_R (R/J(R)) = Kdim (R) = n. We now show that cl.Kdim (R) = n, also by induction on n. If n = 1 then R is prime and by the Principal Ideal Theorem we may conclude that cl.Kdim (R) ≥ 1. Since cl.Kdim (R) ≤ Kdim (R) = 1 it follows that cl.Kdim (R) = 1.
Now assume that n > 1, and consider the ring $R_1 = R/a_1R$. By the induction hypothesis we obtain cl.Kdim (R_1) = n - 1 and R_1 is a prime ring. Since R is a prime ring, it follows that cl.Kdim R ≥ n and then finally this implies that cl.Kdim

(R) = n. □

8.6.4. Examples.

1. Commutative regular local rings play an important part in algebraic geometry.

2. Let K be a division ring and let R = K$[[X_1, ..., X_n]]$ be the ring of formal power series in n commuting indeterminates. Then R is an n-dimensional regular local ring.

3. If R is an n-dimensional regular local ring and φ is an automorphism of R, we put S = R$[[X,\varphi]]$ where multiplication of formal series is defined by the rule Xa = φ(a)X. Then S is an (n+1)-dimensional regular local ring.

4. (Walker, [1]). Let g be a finite dimensional Lie algebra over a field k and let $\mathcal{L}$ be an ideal of g. Let R be the universal enveloping algebra of g over k and let P be the ideal of R generated by $\mathcal{L}$. In case g is nilpotent, the localization R_P exists and it is a regular local ring.

§8.7 Exercises.

(156) For any prime p and any integer n > 1, gldim $\mathbb{Z}/p^n\mathbb{Z} = \infty$.

(157) Let $R_1, ..., R_n$ be rings and R = $\prod_{i=1}^n R_i$. Prove that l.gldim (R) = sup {l.gldim (R_i), i = 1,...,n}.

(158) For any ring R and any n $\in \mathbb{N}$: l.gldim (R) = l.gldim (M_n(R)).

(159) Let φ: R $\rightarrow$ S be a ring morphism, M a left S-module. We write φ_*(M) for the left R-module M with scalar multiplication a.x = φ(a)x for a $\in$ R, x $\in$ M. Prove the following properties:

1.pd_R (φ_*(M)) $\leq$ pd_S (M) + pd_R (S).
2. If S is a flat R-module then inj.dim_R (φ_*(M)) $\leq$ inj.dim_S (M).
3. If N is a left R-module and S is a flat right R-module then pd_S (S $\otimes_R$ N) $\leq$ pd_R (N).
4. If S is a projective right R-module then: inj.dim_S (S $\otimes_R$ N) $\leq$ inj.dim_R (N).

(160) Let there be given a sequence of R-modules:

$$M_0 \xrightarrow{f_0} M_1 \xrightarrow{f_1} M_2 \longrightarrow ... \longrightarrow M_n \xrightarrow{f_n} M_{n+1} \longrightarrow ...$$

and put M = $\underset{n}{\xrightarrow{\lim}} M_n$. Prove the following statements:
1. There exists an exact sequence of the form:

$$0 \longrightarrow \bigoplus_{n=0}^{\infty} M_n \xrightarrow{i} \bigoplus_{n=0}^{\infty} M_n \xrightarrow{j} M \longrightarrow 0,$$

where i$(m_0, m_1, ...) = \sum_{i\geq o} u_i(m_i)$ where u_i: M_i $\rightarrow$ M are the canonical morphisms.
2. If pd_R (M_i) $\leq$ n for all i $\geq$ 0, then pd_R (M) $\leq$ n + 1.

(161) The ring T_n (R) of lower triangular matrices over a ring R satisfies: l.glim (T_n(R)) = 1 + l.gldim (R). Thus T_n(R) is left hereditary if and only if R is semi-simple.

(162) (A. Goldie). A commutative semi-hereditary ring has no nilpotent elements.

(163) If R is a semiprime ring with principal left ideals then R is left hereditary.

(164) (A. Goldie). If R is a prime ring such that left ideals are principal then R $\cong$ M_n (D) where D is a left hereditary domain.

(165) (Kaplansky). If R is a left hereditary ring then a projective left R-module is isomorphic to a direct sum of left ideals of R.

(166) A commutative domain R is said to be a Dedekind domain if every ideal of R is invertible. Prove that the following assertions are equivalent:

1. R is a Dedekind domain.
2. R is hereditary.
3. An R-module Q is injective if and only if it is divisible.

(167) Let R be a subring of a ring S such that there exists a sub-bimodule ${}_RX_R$ of ${}_RS_R$ such that S = R $\bigoplus$ X. Prove that l.gldim R $\leq$ l.gldim S + pd_R (S).

(168) If R is a (left) non-singular ring (the singular radical is zero) then we let Q_{max}(R), the maximal quotient ring of R, be given by Hom_R (E(R),E(R)); it is a Von Neumann regular ring.

1. Prove that the following statements are equivalent:
a. Q_{max}(R) is a semisimple ring.
b. Every essential left ideal of R contains a finitely generated essential left ideal of R.
c. Q_{max}(R)I = Q_{max}(R), for every essential left ideal I of R.
2. Prove the following statements:
a. If Q_{max}(R) is semisimple then every left Q_{max} (R)-module is injective as an R-module.
b. If Q_{max}(R) is semisimple then every right Q_{max}(R)-module is flat as an R-module.
c. If Q_{max}(R) is semisimple and M is a non-singular left R-module then Q_{max} (R) $\bigotimes_R$ M is an injective envelope of M.

(169) A is said to be a *Baer ring* if every left annihilator ideal is generated by an idempotent. Such a ring is evidently both left and right non-singular. If R is a ring such that every left ideal contains a nonzero idempotent then A = B $\bigoplus$ C where B is a reduced ring and every nonzero ideal of C contains a nonzero nilpotent element (Utumi).

(170) For a non-singular ring R the following statements are equivalent:

1. Q_{max}(R) is flat as a right R-module.
2. Q_{max}(R) $\bigotimes_R$ M is a projective left Q_{max}(R)-module for every finitely generated submodule M of a free left R-module.
3. Q_{max}(R) $\bigotimes_R$ I is projective as a left Q_{max}(R)-module for every finitely generated left ideal I of R.

(171) Let R be left non-singular, M a left R-submodule of Q_{max}(R) then Z(Q_{max} (R) $\bigotimes_R$ M) is the kernel of the canonical map Q_{max}(R) $\bigotimes_R$ M $\to$ Q_{max}(R). In particular if we assume that Q_{max}(R) is flat as a right R-module then every finitely generated submodule of a free left R-module is projective.

(172) (Sandomierski). The following properties of the ring R are equivalent:

1. **R is left semi-hereditary.**
2. **R is left non-singular, every left ideal of R is a flat module and $Q_{max}(R)$ is flat as a left R-module.**

Bibliographical Comments for Chapter 8.

In this chapter we aimed to present the basic theory concerning projective and injective dimension (as well as global dimensions of rings) without referring to the functors Ext^n (-,-), Tor_n (-,-). The study of these dimensions is based upon Schanuel's lemma (Proposition 8.1.3. and 8.3.3.).

The presentation in the first three sections follows the lines of argumentation of Kaplansky, [2]. The "Change of Rings" theorems are given here in a slightly more general form when compared to Kaplansky, [2], mainly because of the extended applicability in the next chapter. These theorems also allow to compute the global dimension of polynomial rings (Hilberts' Syzygy Theorem) and of formal power series rings. In addition the first three paragraphs contain Auslander's theorem (8.1.14.) and the study of the injective dimension of a module using minimal injective resolutions. We use Corollary 8.3.19. (Kaplansky [2]) in the study of the global dimension of a left Noetherian ring e.g. Theorem 8.3.23. (Boratinsky [1]) and Theorem 8.3.25. (Auslander [1]).

A characterisation of rings with left global dimension at most one is given at the end of Section 8.3.

In our treatment of the flat dimension of a module the basic notion involved is that of the character module of a module. The main rsults in Section 8.4. are 8.4.7., 8.4.10., 8.4.12., and Corollary 8.4.12 which provides a new proof of a result of M. Auslander's. Similar to the final part of Section 8.3., the final results of Section 8.4. characterize rings with weak dimension zero, i.e. Von Neumann regular rings.

In the exercises we hinted at some properties of the maximal quotient ring which provides at several occasions examples of Von Neumann regular rings.

In Section 8.5. we follow Boratinsky and present results concerning the homological dimension of rings R such that J(R) has the AR-property e.g. Theorem 8.5.5. and Corollary 8.5.7.

The final section deals with local regular rings in the sense of Walker, [1]. The main result here is probably 8.6.3. which extends to the non-commutative case the well-known characterization of commutative local rings.:

Theorem Let R be a commutative local regular ring with maximal ideal W, then the following assertions are equivalent:

1. $\text{Dim}_{R/W}\ (W/W^2) = \text{Kdim } R = n.$

2. gldim (R) = n.
3. pd_R (R/W) = n.
4. W is generated by an R-sequence $a_1, ..., a_n$.

The following chapter will be devoted to the study of rings having finite global dimension, aiming to provide an extension of the above mentioned theorem to some classes of Noetherian non-commutative rings.

Some typical references for the contents of chapter 8 are:

M. Auslander [1]; I,N. Bernstein [2]; M. Boratynsky [1]; H. Cartan and S. Eilenberg [1]; C. Faith [1]; K.L. Fields [1],[2]; I. Kaplansky [1],[2]; C. Năstăsescu [9]; L.W. Small [2],[3]; R. Walker [1].

Chapter 9

Rings of Finite Global Dimension.

In this chapter we study the connection between the Krull dimension and the homological dimension of a left Noetherian ring. Main results in this chapter will be Theorem 9.3.10. and 9.4.10. However the result of Theorem 9.3.10. may be strenghtened considerably (as we point out in the bibliographical comments at the end) but we did not include these more extended versions here because they require more techniques than what we are willing to introduce here.

In this chapter we will take for granted some properties of the functors $\mathrm{Ext}^n(-,-)$; the reader who is not familiar with these should first consult Cartan, Eilenberg [1].

For the local characterization of projective dimension of a module we follow the approach used by H. Bass in [1].

§9.1 The Zariski Topology.

Let R be a ring with identity. X = Spec (R) is the set of all proper prime ideals of R (see also Section 5.3.) and X_m = Max (R) is the set of all proper maximal ideals of R. It is obvious that Max (R) $\neq \phi$ and that Max (R) $\subset$ Spec (R). For an ideal I of R we let: $V(I) = \{P \in \mathrm{Spec}(R), I \subset P\}$,
$V_m(I) = \{\omega \in \mathrm{Max}(R), W \supset I\}$.

These sets satisfy a number of properties:

1. $V(R) = \phi$, $V(0) = X$.
2. $V(\sum_{i \in A} I_i) = \bigcap_{i \in A} V(I_i)$ for any family $\{I_i, i \in A\}$ of ideals of R.
3. $V(I) \bigcup V(J) = V(I \bigcap J)$ for I,J, ideals of R.

In view of these properties these subsets define a topology on X = Spec (R) and this topology is called the *Zariski topology*. The open sets are given by X(I) = X - V(I), I an ideal of R. The Zariski topology induces on X_m a topology with closed sets $V_m(I)$. One easily checks that X(I) = X(J) if and only if I and J have the same radical i.e. the open sets are in bijective correspondence with the semiprime ideals of R.

9.1.1. Proposition. Both X and X_m are quasi-compact (i.e. like compact but without the Hausdorff property, in fact X is satisfying the weaker separation condition T_1 but almost never the Hausdorff condition).

Proof. Suppose that $X = \bigcup_{\alpha \in A} X(I_\alpha) = X(\sum_{\alpha \in A} I_\alpha)$. Then $\sum_{\alpha \in A} I_\alpha = R$ and using the fact that 1 is a finite sum of elements of the I_α it follows that $X = \bigcup_{i=1} X(I_{\alpha_i})$. A similar proof works for X_m. □

If f: R → R' is a ring morphism and P' is a prime ideal of R' then $f^{-1}(P')$ need not be a prime ideal of R as one can see by looking at the morphism $\begin{pmatrix} K & K \\ 0 & K \end{pmatrix} \to \begin{pmatrix} K & K \\ K & K \end{pmatrix}$, where K is a field. For commutative rings R' and R it is well-known that f: R → R' defines a map $\tilde{f}$: Spec (R') → Spec (R), $Q \mapsto f^{-1}(Q)$. From $\tilde{f}^{-1}(V(I)) = V(J)$ where J = R'f(I) it follows that $\tilde{f}$ is continuous.

Now let R be a commutative ring and A an R-algebra with centre Z(A) containing R. If S is a multiplicative set of R with 1 ∈ S then $S^{-1}A$ is an $S^{-1}R$-algebra and we let φ_S: A → $S^{-1}A$ be the canonical ring morphism. In this particular situation, it is true that for Q ∈ Spec ($S^{-1}A$), $\varphi_S^{-1}(Q)$ ∈ Spec (A).

9.1.2. Proposition. The map $\tilde{\varphi}_S$: Spec ($S^{-1}A$) → Spec (A) induces a homeomorphism between Spec ($S^{-1}A$) and $Y = \{P \in \text{Spec}(A), P \bigcap S = \phi\}$, the inverse map being given by: $p \mapsto S^{-1}p$.

Proof. Easy. □

Let A be an R-algebra, S a multiplicative system of R. If M,N are left A-modules then we may consider th following commutative diagram:

(*)

$$\begin{array}{ccc} Hom_A(M,N) & \xrightarrow{\alpha_{M,N}} & S^{-1}Hom_A(M,N) \\ & \searrow^{\gamma_{M,N}} & \downarrow \beta_{M,N} \\ Hom_A(S^{-1}M, S^{-1}N) & \underset{\delta_{M,N}}{\longleftarrow} & Hom_{S^{-1}A}(S^{-1}M, S^{-1}N) \end{array}$$

where α is the canonical morphism, $\gamma(f) = S^{-1}f$ for all f $\in$ Hom_A (M,N) and δ is the inclusion morphism and for β we have β $(s^{-1}f) = s^{-1}$ $(S^{-1}f)$.

9.1.3. Proposition. In the diagram (*), δ is an isomorphism and β is an isomorphism if M is a finitely generated A-module.

Proof. That δ is an isomorphism is clear. That β is an isomorphism whenever M $= A^n$, n $\in$ N, is evident. Th general statement now follows by applying the "Five Lemma" to the exact sequence $A^m \to A^n \to$ M $\to$ 0 which exists because of the finite presentation assumption.

§9.2 The Local Study of Homological Dimension.

Let A be an R-algebra. If p is a prime ideal of R, let S_p = R - p and write $A_p = S_p^{-1}$A, $R_p = S_p^{-1}$R. If M is a left A-module we write $M_p = S_p^{-1}$M. Clearly M_p is a left A_p-module. Moreover, if N is a projective left A-module then N_p is a projective left A_p-module.

9.2.1. Proposition. Let A be an R-algebra and let M be a left A-module. Define U_n(M) = {p $\in$ Spec (R), pd_{A_p} $(M_p) \leq$ n}. If there exists an exact sequence:

$$P_{n+1} \longrightarrow P_n \longrightarrow \dots \longrightarrow P_0 \underset{\epsilon}{\longrightarrow} M \longrightarrow 0$$

where each P_i, $0 \leq i \leq n+1$, is a finitely generated projective left A-module, then U_n (M) is open in Spec (R).

In particular pd_R (M) $\leq$ n if and only if U_n (M) = Spec (R).

Proof. By induction on n. Consider the case n = 0.

There exists an exact sequence: $P_1 \longrightarrow P_0 \xrightarrow{\varepsilon} M \longrightarrow o$ where each P_i is a finitely generated projective A-module. Consider the map φ: Hom_A (M,P_0) $\rightarrow$ Hom_A (M,M) induced by ε and let α be the image of 1_M in coker (φ). Since φ is an R-homomorphism it follows that M is projective if and only if $1_M \in$ Jm (φ), if and only if $\alpha = 0$.

Since both P_o and M are finitely presented A-modules then for each

p $\in$ Spec (R) we may identify the corresponding mapping Hom_A (M_p, $(P_o)_p$) $\rightarrow$ Hom_{A_p} (M_p, M_p) with the localization φ_p of φ. Therefore M_p is A_p-projective if and only if $\alpha_p = \frac{\alpha}{1}$ is zero in coker $(\varphi)_p$ = coker (φ_p). But $\alpha_p = 0$ if and only if there exists an s $\notin$ p such that $s\alpha = 0$, i.e. if and only if p $\not\supset$ I where I = $l_R(\alpha)$. Therefore U_0 (M) = X(I). Moreover M is projective if and only if $\alpha = 0$, if and only if I = R, if and only if V(I) = ϕ or U_0 (M) = Spec (R). In case n > 0 we consider the exact sequence $0 \longrightarrow K \longrightarrow P_0 \xrightarrow{\varepsilon} M \longrightarrow 0$

where K = Ker (ε). Then pd_R (M) $\leq$ n if and only if pd_R (K) $\leq$ n - 1. Therefore U_n (M) = U_{n-1} (K). Since there is an exact sequence :

$$0 \longrightarrow P_{n+1} \longrightarrow P_n \longrightarrow .. \longrightarrow P_1 \longrightarrow K \longrightarrow 0$$

we may conclude (using induction) that U_{n-1} (K) is open and pd_R (K) $\leq$ n-1 if and only if U_{n-1} (K) = Spec (R). □

9.2.2. Theorem. Let A be a left Noetherian R-algebra and let M be a finitely generated left A-module, then: pd_A (M) = sup {pd_{A_ω} (M_ω), $\omega \in$ Max (R)}.

In particular, l.gldim (A) = sup {l.gldim (A_ω), $\omega \in$ Max (R)}. Furthermore, pd_A (M) < ∞ if and only if pd_{A_ω} (M_ω) < ∞ for all $\omega \in$ Max (R)

Proof. Put n = sup {pd_{A_ω} (M_ω), $\omega \in$ Max (R)}. Proposition 9.1.3. implies pd_R(M) $\geq$ n, so we have equality if n = ∞.

If n < ∞, consider U_n (R). Since A is left Noetherian the finiteness conditions on M necessary in order to apply Proposition 9.2.1. hold, so U_n (R) is an open set such that its complement contains no maximal ideals. Thus U_n (R) = Spec (R) and therefore pd_R (M) $\leq$ n. If pd_{A_ω} (M_ω) < ∞ for all $\omega \in$ Max (R) then

pd_{A_ω} (M_p) < ∞ for all p $\in$ Spec (R), (Proposition 9.1.3.). Therefore we arrive at Spec (R) = $\bigcup_{n\geq 0} U_n$ (R). But Spec (R) is quasi-compact, hence U_n (R) $\subset$

U_{n+1} (R) for some n ∈ IN and then there exists a k ∈ IN such that Spec (R) = U_k (R). By Proposition 9.2.1, pd_R (M) ≤ k follows. □

A similar result holds for the injective dimension of a module.

9.2.3. Theorem. Let A be a left Noetherian R-algebra and M a left A-module. Then inj.dim_A (M) = sup {$\text{inj.dim}_{A_\omega}$ (M_ω), ω ∈ Max (R)}.

Proof. I S is a multiplicative set of R then we have a canonical isomorphism $\text{Ext}^n_{S^{-1}A}$ (S^{-1}N, $S^{-1}M$) ≅ S^{-1} Ext^n_A (X,M), for any finitely generated left A-module N and n ≥ 0. To establish the latter one uses a projective resolution for N containing finitely generated projective left A-modules and we use Proposition 9.1.3. □

§9.3 Rings Integral over their Centres.

Let A be an R-algebra.We say that A is *integral* over R if every element of A satisfies a monic polynomial f ∈ R[X]. First we include some well-known properties of integral extensions: "lying over", "going up", "incomparability".

9.3.1. Proposition (Lying over). If A is integral over R and p is a prime ideal of R then there exists a prime ideal P of A such that P ⋂ R = p.

Proof. First assume tht R is local with maximal ideal p. If pA = A then 1 = $\sum_{i=1}^n \lambda_i a_i$ for some $a_1, ..., a_n$ ∈ A, $\lambda_1, ..., \lambda_n$ ∈ p. Put A' = R[$a_1, a_2, ..., a_n$] ⊂ A, then pA' = A'. But since $a_1, ..., a_n$ are integral over R, A' is a finitely generated R-module. Nakayama's lemma then leads to A' = 0, a contradiction. Hence pA ≠ A and we may select a maximal ideal P of A such that pA ⊂ P and thus we obtain p = P ⋂ R.
In the general case we pass to R_p ⊂ A_p and pR_p and apply Proposition 9.1.2. □

9.3.2. Proposition (Going up). Let A be integral over R. Consider prime ideals $p \subset q$ of R and suppose that P is a prime ideal of A such that $P \bigcap R = p$. Then there exists a prime ideal Q of A such that $P \subset Q$ and $Q \bigcap R = q$.

Proof. Since integrality passes over to factor rings of A we may pass on to the integral extension $R/p \subset A/P$ and then apply Proposition 9.3.1. to find a $\overline{Q} = Q/P$ lying over $\overline{q} = q/p$. □

9.3.3. Proposition (Incomparability). Let A be a ring with Krull dimension which is integral over R. If $P_1 \subsetneq P_2$ are prime ideals of A then
$P_1 \bigcap R \subsetneq P_2 \bigcap R$.

Proof. Evidently we may assume that $P_1 = (0)$, i.e. A is a prime ring. Since P_2 is a nonzero ideal of A it is essential as a left ideal. But A has Krull dimension, hence it is a left Goldie ring, and thus there is a regular element $s \in P_2$.

Integrality of A over R yields the existence of $r_1, ..., r_n \in R$ such that
$s^n + r_1 s^{n-1} + ... + r_n = 0$. We may assume that $r_n \neq 0$ because s is regular. In this case $r_n = -s^n - r_1 s^{n-1} - ... - s r_{n-1}$, i.e. $r_n \in P_2 \bigcap R$ i.e.
$P_2 \bigcap R \neq P_1 \bigcap R$.

Remark. In the situation of the proposition we actually have: if P is a prime ideal $P \subsetneq I$ where I is an ideal, then $P \bigcap R \subsetneq I \bigcap R$.

9.3.4. Proposition. Let A be integral over R having left Krull dimension then A is (left and right) fully bounded.

Proof. Let P be a prime ideal of A. We may assume $P = 0$ i.e. that A is a prime ring. For an essential left ideal I of A there exists a regular element $c \in I$. Since c is integral over R and regular we have $c^n + r_1 c^{n-} + ... + r_n = 0$ with $r_n \neq 0$. Clearly $r_n \in I \bigcap R$ and thus $Ar_n \subset I$ and Ar_n is an ideal.

A similar argument may be used in case Ar_n is an essential right ideal of A. □

9.3.5. Proposition. Let A be left Noetherian and integral over R. A prime ideal P of A is in Spec_α (A) for an ordinal α, if and only if $P \bigcap R \in \text{Spec}_\alpha$ (R). In particular R has cl.Kdim R = cl.Kdim A = Kdim A.

Proof. By the "going up" property it is clear that R satisfies the ascending chain condition for prime ideals and thus R has classical Krull dimension.
We continue by induction on α. The case $\alpha = 0$ is clear. Let $\alpha > 0$ and $P \in \text{Spec}_\alpha$ (A). If $q \in \text{Spec}$ (R) is such that $p \underset{\neq}{\subset} q$ where $p = P \bigcap R$, then the going up property yields the existence of a $Q \in \text{Spec}$ (A) such that $P \underset{\neq}{\subset} Q$ and $Q \bigcap R = q$. Then Q $\in \text{Spec}_\beta$ (A) for some $\beta < \alpha$ and so the induction hypothesis yield $q \in \text{Spec}_\beta$ (R). Therefore $p \in \text{Spec}_\alpha$ (R).
Conversely, let $P \in \text{Spec}$ (A) be such that $p = P \bigcap R \in \text{Spec}_\alpha$ (R). Consider $Q \underset{\neq}{\supset} P$, $Q \in \text{Spec}$ (A). Proposition 9.3.3. entails that $p \underset{\neq}{\subset} q$ where $q = Q \bigcap R$. Hence $q \in \text{Spec}_\beta$ (R) for some $\beta < \alpha$. Using the induction hypothesis we obtain $Q \in \text{Spec}_\beta$ (R) and consequently $P \in \text{Spec}_\alpha$ (R) results. □

9.3.6. Corollary. Assumptions being as in the proposition: for every $P \in \text{Spec}$ (A) we have: Kdim (A/P) = Kdim (A/pA) = cl.Kdim (R/p), where $p = P \bigcap R$.

Proof. Since $Ap \bigcap R = p$ and A/pA being integral over R/p the proposition yields the result. □

9.3.7. Lemma. Let A be a left Noetherian R-algebra integral over R. If p is an ideal of R then I = Ap has the AR-property.

Proof. By Proposition 8.5.3. we know that it will be sufficient to prove that $\mathcal{F}_I$ is stable. Look at a left A-module M which is $\mathcal{F}_I$-torsion and note that it is not restrictive to assume that M is finitely generated. Since E_A (M) is a finite direct sum of injective indecomposable modules we may assume at once that M is uniform Pick $x \neq 0$ in E_A (M) and put L = Ax. Since $\lambda x \neq 0$, $\lambda x \in M$ for some $\lambda \in A$ there also exists an $n \in \mathbb{N}$ such that $I^n (\lambda x) = 0$. For $r \in p$ we consider φ_r: $L \to L$, $y \mapsto ry$. Clearly Ker $(\varphi_r) \neq 0$, hence L is an essential extension of Ker (φ_r). Because L is left Noetherian there must exists a positive integer k such that

Ker $(\varphi_r)^k \bigcap$ Im $(\varphi_r)^k = 0$. Since Ker $(\varphi_r) \subset$ Ker $(\varphi_r)^k$, Im $(\varphi_r)^k = 0$ for some k, i.e. $r^k L = 0$.

On the other hand, $I = r_1 A + ... + r_s A$ for certain $r_1, ..., r_s \in p$. Considering the $k_i \in \mathbb{N}$ for which $r_i^{k_i} L = 0$ we may take $t \in \mathbb{N}$ sufficiently large such that $I^t L = 0$. Proposition 8.5.3. establishes our claims. □

9.3.8. Proposition. Let A be a left Noetherian R-algebra which in integral over R, then the following assertions hold:

1. $J(A) \bigcap R = J(R)$.
2. $\bigcap_{n \geq 0} J(A)^n = 0$.
3. A is semi-local if and only if R is semi-local.
4. If A is semi-local then J(A) has the AR-property.

Proof. 1. Take $\omega \in$ Max (R). The "lying over" property implies that $\omega = M \bigcap R$ for some $M \in$ Max (A), hence $J(R) \supset J(A) \bigcap R$. Consequently, if S is a simple left A-module there exist $x_1, ..., x_t \in S$ such that Ann_A (S) $= l_A(x_1) \bigcap ... \bigcap l_A(x_t)$, (since A is fully left bounded then A has property (H), cf. Section 6.3.). Therefore A/Ann_A (S) is a simple Artinian ring i.e. Ann_A (S) is a maximal ideal. Then Ann_A (S) $\bigcap$ R is a maximal ideal of R, hence $J(A) \bigcap R \supset J(R)$ and thus $J(R) = J(A) \bigcap R$.

2. It suffices to prove that the injective hull, E(S), of a simple left A-module is a semi-Artinian module (in fact it is an Artinian module). Put $M = \mathrm{Ann}_A$ (S); it is a maximal ideal of A, hence $\omega = M \bigcap R$ is a maximal ideal of R. Now R/ω is a field and $A/\omega A$ is integral over R/ω, hence $A/\omega A$ is an Artinian ring.

Envoking Lemma 9.3.7., using the fact that $(\omega A)S = 0$, entails that E(S) is $\mathcal{F}_I$-torsion where $I = \omega A$. Because $A/\omega A$ is Artinian, E(S) is a semi-Artinian A-module.

3. If A is semi-local then part 1. yields that R is semi-local. If the latter property holds then $R/J(R)$ is a semisimple Artinian ring and therefore $A/J(R)A$ is an Artinian ring. From $J(R)A \subset J(A)$ it follows that $A/J(A)$ is an Artinian ring and therefore A is a semi-local ring.

4. By the lemma we know that $J(R)A$ has the AR-property. Since $J(R)A \subset J(A)$ and $A/J(R)A$ is an Artinian ring, it follows that $J(A)^s \subset J(R)A$ for some $s \in \mathbb{N}$ and it is clear from this that J(A) has the AR-property. □

9.3.9. Corollary. Let A be a left Noetherian R-algebra which is integral over R. Then l.gldim (R) = sup {pd_R (S), S a simple left A-module}.

Proof. Put n = sup {pd_R (S), S a simple left A-module}. It is clear that l.gldim R $\geq$ n and we have equality for n = ∞. By Theorem 9.2.2. it will suffice to prove that for any $\omega \in$ Max (R), l.gldim A_w $\leq$ n. But A_w is a semi-local ring in view of Corollary 8.5.7. and we obtain: l.gldim A_w = pd_{A_w} $(A_w/J(A_w))$. Let X be a simple left A_w-module. Since A_w is a left bounded ring, Q = l_{A_w} (X) is a maximal ideal. There exists a unique maximal ideal P of A such that Q = S^{-1}P. Therefore $A_w/Q = S^{-1}$ (A/P). Because A/P is a simple Artinian ring, pd_R (A/P) $\leq$ n. Therefore pd_A $(A_w/Q) \leq$ n and hence pd_{A_w} (X) $\leq$ n.
Consequently, pd_{A_w} $(A_w/J(A_w)) \leq$ n. □

9.3.10. Theorem. Let A be a left Noetherian R-algebra which is integral over R. Suppose that A has finite left global dimension n, then Kdim (R) $\leq$ n.

Proof. It will be sufficient to prove that ht (M) $\leq$ n for a maximal ideal M of A. We continue by induction. The case n = 0 is obvious. Suppose n $\geq$ 1. Put ω = M $\bigcap$ R $\subset$ Max (R). Localizing at R - ω yields ht(M) = ht(MA_w). The assumption l.gldim A_w $\leq$ n allows us to assume that R is local with maximal ideal ω. Then A is semi-local and M a maximal ideal. Consider a prime ideal P with P $\subsetneq$ M and maximal with respect to this property. Put p = P $\bigcap$ R and let P = $P_1, P_2, ..., P_t$ be the prime ideals of R lying over p.
Since p $\subsetneq_\omega$, none of the P_i can be maximal. Moreover, since A/J(A) is a semisimple Artinian ring it follows that J(A) $\not\subset P_i$ for 1 $\leq$ i $\leq$ t. Therefore, $P_i \subsetneq P_i$ + J(A). By the integrality of A/P_i over R/p there exists an s $\in$ R - p such that s $\in$ P_i + J(A). It is easily seen that we may assume s $\in$ J(A). Obviously, $\bar{s}$ is a non-zerodivisor over the ring A/P_i, where $\bar{s}$ is the image of s in A/P_i. We claim that pd_A (A/P_i) $\leq$ n-1. Indeed if pd_A (A/P_i) > n-1 then pd_A (A/P_i) = n. The proof of Theorem 8.3.23. establishes the equivalence: for a finitely generated left A-module M we have pd_R (M) $\leq$ n if and only if Ext^{n+1} (M,N) = 0 for all finitely generated modules N with J(A)N = 0. Now pd_A (A/P_i) = n implies the existence of a finitely generated A-module X with J(A)X = 0 such that Ext^n (A/P_i,X) $\neq$ 0. On the other hand we have the exact sequence:

$$0 \longrightarrow A/P_i \xrightarrow{\varphi_s} A/P_i \longrightarrow \text{coker}(\varphi_s) \longrightarrow 0,$$

where φ_s (x) = sx, x $\in$ A/P_i. This leads to the exact sequence:

$$\text{Ext}^n_A\ (A/P_i,X) \xrightarrow{\varphi_s^*} \text{Ext}^n_A\ (A/P_i,X) \longrightarrow \text{Ext}^{n+1}_A\ (A/P_i,X).$$

Since Ext^{1+n}_A $(A/P_i,X) = 0$, φ_s^* must be epimorphic. But since $s \in J(A)$, $sX = 0$ follows and then $\varphi_s^* = 0$. This entails the contradiction Ext^n_A $(A/P_i,X) = 0$. So we have proved our claim: pd_A $(A/P_i) \leq$ n-1 for all i, $1 \leq i \leq t$.
Localizing A at R - p yields a ring A_p with maximal ideals $P_iA_p = (P_i)_p$, $1 \leq i \leq t$. From $A_p/(P_i)_p \cong (A/P_i)_p$ we may derive that pd_{A_p} $(A_p/P_iA_p) \leq n - 1$. Thus pd_{A_p} $(A_p/J(A_p)) \leq n - 1$, since A_p is semi-local. The Artin-Rees property for $J(A_p)$ then entails the inequality l.gldim $A_p \leq n - 1$.
Finally, the induction hypothesis yields $\text{ht}(P) \leq \text{ht}(PA_p) \leq n - 1$ and so we obtain that $\text{ht}(M) \leq n$. □

Remark In general the inequality established in the theorem is a strict one. Indeed, if $A = \begin{pmatrix} K & K \\ 0 & K \end{pmatrix}$ where K is a field then A is integral over its centre $Z(A) \cong K$ and gl.dim (A) = 1 but Kdim (A) = 0.

Let us point out a result of K. Goodearl, L. Small, [1], stating that for any Noetherian P.I. ring with gldim (R) < ∞ then Kdim (R) $\leq$ gldim (K). Some more recent results will be mentioned in the bibliographical comments of this chapter.

§9.4 Commutative Rings of Finite Global Dimension.

In this section we characterize commutative local regular rings following the presentation of Kaplansky's, [1],[2]. In the sequel of this section R will be a commutative Noetherian ring unless mentioned otherwise. First some preliminary results.

9.4.1. Proposition. Let $p_1, \ldots, p_n$ be a finite number of ideals of R and let S be a subring of R contained in $\bigcup_{i=1}^n p_i$. If at least n - 2 of the ideals $p_1, \ldots, p_n$ are prime ideals then there is a $k \in \{1, \ldots, n\}$ such that $S \subset p_k$.

Proof. By induction on n. For n = 1 the statement is clear.
For n = 2, if $S \not\subset p_2$ and $S \not\subset p_1$ then there exist $x_1, x_2 \in S$ such that $x_i \notin p_i$,

i = 1,2. Then $x = x_1 + x_2 \in S$ and $x \notin p_1 \cup p_2$, a contradiction.

Assume n > 2. For each i we my assume that $S \not\subset p_1 \cup ... \cup \hat{p}_i \cup ... \cup p_n$ (where "^" means "omit"). Of course at most two of the ideals appearing in $p_1 \cup ... \cup \hat{p}_i \cup ... \cup p_n$ may be non-prime ideals. Pick $x_i \in S$, $x_i \notin p_1 \cup ... \cup \hat{p}_i \cup ... \cup p_n$, hence $x_i \in p_i$. Since n > 2, at leats one of the ideals $p_1, ..., p_n$ is a prime ideal, say p_1. Pick $y = x_1 + x_2x_3 ... x_n$. We have $y \in S$ and $y \notin p_k$ for all $k \in \{1, ..., n\}$, contradiction. □

9.4.2. Proposition. Let S be a subring of R, I an ideal of R contained in S, $I \neq S$, and assume that $S - I \subset p_1 \cup ... \cup p_n$ for some prime ideals $p_1, ..., p_n$ of R, then $S \subset p_i$ for some i.

Proof. Obviously $S \subset I \cup p_1 \cup ... \cup p_n$. Since $I \neq S$ we may apply the foregoing proposition and conclude that $S \subset p_k$ for some $k \in \{1, ..., n\}$. □

9.4.3. Proposition (The Generalized Principal Ideal Theorem). Let $I \neq R$ be an ideal generated by n elements in R. Let p be a prime ideal of R minimal over I,then $ht(p) \leq n$.

Proof. By induction on n.

If n = 1 the result is known as the principal ideal theorem (cf. Sections in Chapter 6 and 7). Up to localizing at p we may assume that R is local with unique maximal ideal p. Assume that $ht(p) \geq n$. Then there is a chain $p_0 \subset p_1 \subset ... \subset P_{n+1} = p$ of prime ideals of R. We may assume that this chain of prime ideals cannot be refined.. Cleary $I \not\subset p_n$. If I is generated by $a_1, ..., a_n$ then we may assume that $a_n \notin p_n$ (up to reordering). Therefore, $(a_n, p_n) \supsetneq p_n$ and then $R/(a_n, p_n)$ is a local Artinain ring. The latter implies that $p^m \subset (a_n, p_n)$ for some $m \in \mathbb{N}$. Thus there exists a $t \in \mathbb{N}$, sufficiently large, such that $a_i^t = \lambda_i a_n + b_i$ with $\lambda_i \in R$, $b_i \in p_n$ and i = 2,...,n.

Put $J = (b_2, ..., b_n)$. Then $J \subset p_n$. Since $ht(p_n) \geq n$, the induction hypothesis entails the existence of a prime ideal q such that $J \subset q \subsetneq p_n$. It is evident that (a_n,q) contains some power of each of the $a_1, ..., a_n$, so the minimality assumption on p entails that p is also minimal over (a_n,q).

Consider the ring S = R/q. In this ring we have a chain: $(\overline{a}_n) \subset \bar{p}_n \subsetneq \bar{p}$ (where $^-$ denotes the image in S) but this contradicts the principal ideal theorem. □

9.4.4. Proposition. If a $\in$ J(R) then Kdim (R) $\leq$ Kdim (R/(a)) + 1.

Proof. Since a $\in$ J(R) and (a) has the AR-property, every ideal is closed in the (a)-adic filtration. So by Proposition 4.2.6.3. and Theorem 4.2.6.4. we have Kdim (R) $\leq$ Kdim (G(R)). That Kdim (G(R)) $\leq$ Kdim (R/(a)) + 1 is easily checked. □

9.4.5. Proposition. Let M $\neq$ 0 be a finitely generated R-module, then:

1. Ass(M) is non-empty and finite.
2. $\bigcup$ {p, p $\in$ Ass(M)} is the set of zero-divisors from R on M.

Proof. 1. See Section 5.4.
2. If a $\in$ $\bigcup$ {p, p $\in$ Ass(M)} then say a $\in$ p, p $\in$ Ass(M). But p = l_R(x) for some x $\neq$ 0 in M. From ax = 0 it follows that a is a zero-divisor on M.
Conversely, assume that a $\in$ R is a zero-divisor on M, i.e. there is an x $\in$ M, x $\neq$ 0, such that ax = 0. But then N = Rx $\neq$ 0 and Ass(N) $\neq$ ϕ. Let q $\in$ Ass(N). There exists a λ $\in$ R with λx $\neq$ 0, such that q = $l_R(\lambda$x). However, ax = 0 implies that aλx = 0 and therefore a $\in$ q.
On the other hand Ass(N) $\subset$ Ass(M), so the proof is complete. □

Assume now that R is a local ring with maximal ideals ω. Put v(R) = $\dim_k$ (ω/ω^2), where k = r/ω. Let $\{\bar{a}_1,...,\bar{a}_n\}$ be a k-basis for ω/ω^2 ($\bar{a}_i$ the image of a_i in ω/ω^2). In view of Nakayama's lemma we obtain that $\{a_1,...,a_n\}$ is a minimal system of generators for the ideal w. The ring R is said to be *regular* if n = Kdim (R) (= cl.Kdim (R)).

9.4.6. Theorem. Let R be a local ring with maximal ideal ω. If a $\in$ ω - ω^2 and putting R_1 = R/(a), then v(R_1) = v(R) - 1. In particular, if R is a regular local ring then R_1 is a regular local ring. Conversely if R_1 is a regular local ring and a is not contained in a minimal prime ideal of R then R is regular.

Proof. The ideal ω_1 = ω/(a) is the unique maximal ideal of R_1. Let $x_1^*,...,x_r^*$ be a minimal system of generators for ω and pick any x_i $\in$ ω having x_i^* for its image in w_1, i = 1,...,r. We claim that $\{\bar{a}, \bar{x}_1,...,\bar{x}_r\}$ is a k-basis for ω/ω^2 (where $^-$ denotes taking images in ω/ω^2). Indeed, if there exists a relation: $\bar{\lambda}\bar{a} + \bar{\mu}_1\bar{x}_1 + ... + \bar{\mu}_2\bar{x}_r = 0$ then $\lambda a + \mu_1 x_1 + .. + \mu_2 x_2 \in w^2$ yields $\mu_1^* x_1^* + ... + \mu_2^* x_2^* = 0$. Consequently $\mu_i^* \in \omega^*$ i.e. $\mu_i \in \omega$ for i = 1,...,r. So we obtain $\bar{\lambda}\bar{a} = 0$, hence $\lambda a \in \omega^2$. If $\lambda \notin \omega$ then λ is invertible and therefore a $\in \omega^2$, a contradiction. Therefore $\lambda \in \omega$, and the claim is established.

Now let R be a regular local ring. In view of Proposition 9.4.1. we have:
Kdim (R) $\leq$ Kdim (R_1) + 1, and Proposition 9.4.3. then yields Kdim (R_1) = ht $(w_1) \leq$ v(R_1). Consequently: v(R) = Kdim (R) $\leq$ Kdim (R_1) + 1, and since v(R) = v(R_1) + 1 we obtain Kdim $(R_1) \geq$ v(R_1) and also Kdim (R_1) = v(R_1).
Conversely, suppose that R_1 is a regular local ring. Then Kdim (R_1) = v(R_1) = v(R) - 1. But Kdim(R) $\leq$ Kdim (R_1) + 1 and since a is not contained in a minimal prime ideal of R we have Kdim (R_1) < Kdim (R) and also that Kdim (R_1) = Kdim (R) - 1. Hence, Kdim (R) - 1 = v(R) - 1, or Kdim (R) = v(R). □

In Chapter 8, Section 6 we have called a local ring regular if the maximal ideal ω has a regular set of generators, i.e. $\omega = (a_1, ..., a_n)$ such that a_1 is regular in R, $\bar{a}_i$ is regular in R/$(a_1, ..., a_{i-1})$, i = 2,...,n. The following results provide the justification for the use of terminology that may seem to be ambiguous at first sight.

9.4.7. Theorem. Let R be a local ring with maximal ideal w. Assume that w has a set of regular generators $\{a_1, ..., a_n\}$, then R is a regular local ring.

Proof. Proposition 8.6.1. entails that Kdim (R) = ht(ω) = n. On the other hand we know that v(R) $\leq$ n and ht$(\omega) \leq$ v(R) in view of Proposition 9.4.3. Consequently, Kdim (R) = v(R) = n. □

For the proof of the converse of the foregoing theorem we need the following intermediate rsult:

9.4.8. Theorem. If R is a regular local ring then R is a domain.

Proof. Assume Kdim (R) = n; the proof is by induction on n.
If n = 0 then $\omega = \omega^2$ and Nakayama's lemma yields ω = 0 i.e. R is a field.

Assume that n ≥ 1; then $\omega \neq \omega^2$. Pick an a $\in \omega - \omega^2$. Then R_1 = R/(a) is a regular local ring with Kdim (R_1) = n - 1. The induction hypothesis entails that R_1 is a domain so p = (a) is prime. Suppose that R is not a domain, then there exists a prime ideal q such that q $\subsetneq$ (a) = p. If b $\in$ q then b = λa and $\lambda \in$ q, hence $\lambda = \mu$a with μ $\in$ q and therfore b = μa^2.
In fact, this proves that q $\subset$ p^n for all n $\geq$ 0 but then q $\subset$ $\bigcap_{n\geq 0}$ p^n,because p is contained in J(R) = ω. However q = 0 contradicts the hypothesis (R was not a domain). It follows that (a) is a minimal prime ideal of R, but as a is arbitrary in ωw - ω^2 it follows that ω - ω^2 $\subset$ $p_1 \cup ... \cup p_n$ where the p_i are all the minimal prime ideals of R. Proposition 9.4.2. entails that $\omega \subset p_k$ for some k $\in$ $\{1, ..., n\}$, and thus Kdim (R) = 0, contradiction. □

9.4.9. Theorem. Let R be an n-dimensional regular local ring with maximal ideal ω. Then ω has a regular set of generators of length n.

Proof. Induction on n. If n = 0 then R is a field.
If n $\geq$ 1 then $\omega \neq \omega^2$. Pick a_1 $\in \omega - \omega^2$ and put R_1 = R/(a_1). Since R_1 is an (n-1)-dimensional regular local ring, the induction hypothesis yields that there exists a regular set of generators $a_2^*, ..., a_n^*$ for ω_1 = $\omega/(a_1)$. Since R is a domain, a_1 is regular on R. Clearly ($a_1, ..., a_n$) is a regular set of generators for ω. □

9.4.10. Theorem. A local ring R is regular if and only if R has finite global dimension. If this is the case then gldim (R) = Kdim (R).

Proof. Assume that R is a regular local ring with Kdim (R) = n. Theorem 9.4.9. and Theorem 8.6.3. yield that gldim (R) = n.
Conversely, assume that gldim (R) = n < ∞. If n = 0 then R is a semisimple Artinian ring but if we combine this with the fact that it is a local commutative ring we obtain that it is a field.
If n = 1 and ω is the maximal ideal of R then we obtain that ω is a projective module. But every finitely generated pojective module over a local ring is necessarily free, hence ω is free. Assume that $\{a_1, ..., a_m\}$ is a basis for ω over R. If m $\geq$ 2 then select i,j with i $\neq$ j; since $a_i a_j$ $\in$ Ra_i $\bigcap$ Ra_j = 0 we obtain a_i = 0 or a_j = 0, contradiction. Therefore ω = Ra_1 and thus Kdim (R) = 1. Assume now that n $\geq$ 2. If every element of ω is a zero-divisor then ω $\subset$ $p_1 \cup ... \cup p_2$ where Ass(R) = $\{p_1, ..., p_r\}$. Hence ω $\subset$ p_k for some 1 $\leq$ k $\leq$ n, and therefore ω = p_k. Corollary

9.3.9. entails that n = pd_R (R/ω). Put k = R/ω and look at the exact sequence 0 $\to \omega \to$ R $\to$ k $\to$ 0. Since n > 1 we obtain that gldim (R) $\geq 1 +$ n, contradiction. It follows that ω must contain a nonzero-divisor. In view of Proposition 9.4.2. there exists an a $\in \omega - \omega^2$ which is a nonzero-divisor. Put R_1 = R/(a) and $\omega_1 = \omega/(a)$. Because of Theorem 8.1.17. we have:
pd_{R_1} $(\omega/a\omega_1) \leq pd_R$ (ω) = n - 1. Over R_1, w_1 is isomorphic to a direct summand of ω-aω. Indeed, since a $\notin w^2$ we may select a k-basis $\{\bar{a}, \bar{b}_1, ..., \bar{b}_s\}$ for ω/ω^2. Put I = aω + Rb_1 + ... + Rb_1. Then I + Ra = ω. On the other hand, I $\bigcap$ Ra = aω since for x $\in$ I $\bigcap$ Ra we obtain x = λa = μa + $\mu_1 b_1$ + ... + $\mu_s b_s$, where $\mu_i \in \omega$, i.e. λa - $\mu_1 b_1$ - ... - $\mu_s b_s$ = μa $\in w^2$ and then we obtain a relation for $\bar{a}, \bar{b}_1, ..., \bar{b}_s$ in ω/ω^2, i.e. $\bar{\lambda} = 0$ or $\lambda \in \omega$ and x $\in$ aω follows, as claimed.
Therefore we arrive at: $\omega/a\omega$ = I/a$\omega \bigoplus$ Ra/aω = (I/I $\bigcap$ Ra) $\bigoplus$ ((a)/aω) $\cong$
(I + Ra)/Ra $\bigoplus$ (a)/aω = (ω/Ra) $\bigoplus$ (a)/aω = $\omega_1 \bigoplus$ (a)/aω. Consequently pd_{R_1} $(w_1) \leq$ n - 1. The exact sequence: 0 $\to$ (a) $\to \omega \to \omega_1 \to$ 0, yields that pd_R (w_1) = n - 1, taking into account that pd_R (ω) = n - 1 > 0 and pd_R (a) = 0. By Theorem 9.1.16. we obtain that pd_{R_1} (w_1) = n - r and therefore gldim (R_1) = n - 1. Applying the induction hypothesis we may conclude that R_1 is an (n-1)-dimensional regular local ring and then R is an n-dimensional regular local ring in view of Theorem 9.4.6. □

9.4.11. Corollary. If p is a prime ideal of a regular local ring then R_p is a regular local ring.

Proof. gldim $(R_p) \leq$ gldim (R) < ∞, so the theorem applies. □

§9.5 Exercises.

In this set of exercises we have gathered some of the results of K. Brown, C. Hajarnavis, A. MacEacharn, [1], and of K. Brown, C. Hajarnavis, [1]. The reader might consider this set of exercises more like an addendum (without giving proofs) to the foregoing chapters, in particular 8. and 9.

Throughout this section A is a left Noetherian ring and an R-algebra; M will be a finitely generated left A-module.

(173) If $M \neq 0$, then for a nonzero R-submodule N in M we have that $Ass(N) \neq \phi$ and it is a finite set.
If $Ass_A(M) = \{P_1, ..., P_n\}$ then $Ass_R(M) = \{P_1 \bigcap R, ..., P_n \bigcap R\}$.

(174) If S is a subring of R (probably without unit) consisting of zero-divisors on M then there is a nonzero $x \in M$ such that $Sx = 0$

Let Z_R (M) be the set of all zero-divisors from R on M. The ordered sequence $a_1, ..., a_n$ of R is said to be an *R-sequence* on M if the following properties hold:

1. $(a_1, ..., a_n)M \neq M$.
2. for i = 1,...,n, $a_i \notin Z_R\ (M/(a_1, ..., a_{i-1})M)$.

(175) Suppose M has an R-sequence $a_1, ..., a_n$ which is contained in J(A), then each permutation of it is again an R-sequence.

(176) For an R-sequence $a_1, ..., a_n$ the ideals $Aa_1 \subset Aa_1 + Aa_2 \subset ... \subset ... \subset ... \subset Aa_1 + ... + Aa_n$, is a properly ascending chain.

(177) If I is an ideal of R such that $IM \neq M$, then an R-sequence in I on M has only finitely many terms and it may be extended to a maximal R-sequence.

(178) Maximal R-sequences in I on M have the same length.

If I is an ideal of R then the length of a maximal R-sequence in I on M is called the *R-grade* of I on M and it will be denoted by grade_R (I,M). For an ideal I of A we put $\text{grade}_R\ (I,M) = \text{grade}_R\ (I \bigcap R,M)$

(179) Let S be multiplicatively closed in R with $1 \in S$, $0 \notin S$. If $a_1, ..., a_n$ is an R-sequence on M such that $S^{-1}M \neq (\bar{a}_1, ..., \bar{a}_n)\ S^{-1}M$, then $\bar{a}_1, ..., \bar{a}_n$ is an $S^{-1}R$-sequence on $S^{-1}M$.

(180) If A satisfies the "lying over" property over R and I is an ideal contained in a prime ideal p of R then: $\text{grade}_R\ (I,A) \leq \text{grade}_{R_p}\ (I_p, A_p)$. Moreover there exists a maximal ideal ω of R such that $\text{grade}_R\ (I,A) = \text{grade}_{R_\omega}\ (I_\omega, A_\omega)$.

(181) There exist $P \in$ Spec (A), $p \in$ Spec (R) such that:

1. $I \subset p = P \bigcap R$.
2. $\text{grade}_R\ (I,M) = \text{grade}_R\ (p,M) = \text{grade}_R\ (P,M)$.

(182) Let $a \in J(A) \bigcap R$ and put $J = I + Ra$, then:
1. if $I \in \mathcal{Z}_R(M)$ and $a \notin \mathcal{Z}_R(M)$ then $J \subset \mathcal{Z}_R(M/aM)$.
2. if $I \subset J(A)$ then $\text{grade}_R(J,M) \leq 1 + \text{grade}_R(I,M)$.

(183) Assume that R is a local central subring of A with maximal ideal ω contained in J(A). Let I be an ideal of R such that $\text{grade}_R(I,M) < \text{grade}_R(\omega,M)$. Then there exists a prime ideal P of A and a prime ideal p of R, both containing I, such that:
1. $P \bigcap R = p$
2. $\text{grade}_R(P,M) = \text{grade}_R(p,M) = 1 + \text{grade}_R(I,M)$.

A left Noetherian ring A is said to be *centrally Macaulay* if it contains a central subring R such that for all $M \in \text{Max}(A)$, $\text{ht}(M) = \text{grade}_R(M,A)$, then A is said to be *R-Macaulay*.

(184) Let A be integral over R and assume that A is R-Macaulay. Then the following properties hold:

1. $\text{grade}_R(P,A) = \text{ht}(P)$ for all $P \in \text{Spec}(A)$.
2. $\text{grade}_R(p,A) = \text{ht}(p)$ for all $p \in \text{Spec}(R)$.
3. $\text{ht}(P) = \text{ht}(P \bigcap R)$ for all $P \in \text{Spec}(A)$.
4. if $a_1, \dots, a_t$ is an R-sequence of A and $I = \sum_{i=1}^{t} Aa_i$ then every $P \in \text{Ass}_A(A/I)$ is maximal over I and A/I has a left Artinian left quotient ring.

(185) Let A be a left Noetherian local ring of finite left global dimension n. Suppose that A is integral over its centre $Z(A) = C$. Then the following properties hold:

1. A is a prime C-Macaulay ring.
2. The ideal J(A) contains a maximal C-sequence of length n.
3. $\text{Kdim}(A) = \text{pd}_A(A/J(A)) = \text{grade}_C(J(A)/A) = \text{ht}(J(A)) = n$.
4. $A = \bigcap \{A_p,\ p \text{ a prime ideal of C of rank } 1\}$ and each A_p is hereditary.
5. C is Krull domain, $\text{cl.Kdim}(C) = n$.

If C is also Noetherian then $\text{Kdim}(C) = n$.

(186) Let A be a Noetherian local ring of finite global dimension n and assume that A is a finitely generated Z(A)-module.Then the following statements are equivalent:
1. A is a free C-module.
2. $\text{gldim}(C) < \infty$.
3. $\text{gldim}(C) = n$.

We say that A is *homologically homogeneous* over R if the following conditions hold:

HH.1. A is left Noetherian.

HH.2. A is integral over R.

HH.3. l.gldim (A) $< \infty$.

HH.4. if V,W are irreducible left A-modules having the same annihilator in R then pd_A (V) = pd_A (W).

(187) If A is homologically homogeneous over R and M is a maximal ideal of A then $grade_R$ (M,A) = ht(M) = pd_A (A/M). In particular A is an R-Macaulay ring.

(188) Let $R \subset C = Z(A)$. Then A is homologically homogeneous over R if and only if it is homologically homogeneous over C.

(189) Let A be homologically homogeneous over $R \subset Z(A)$ and let p be a prime ideal of R then A_p is homologically homogeneous over R_p.

(190) If A is homologically homogeneous over R then A is a semiprime ring.

(191) If A is homologically homogeneous then A[X] is homologically homogeneous over Z(A)[X]. If A is homologically homogeneous over R and σ is an automorphism of A of finite order and such that $\sigma |R = 1_R$ then A[X,σ] is homologically homogeneous.

(192) Let G be a finitely generated group with abelian normal subgroup H of finite index, let k be a field such that G contains no element of order p = char k then kG is homologically homogeneous.

Finally let us point out a result of S.P. Smith, concerning the global dimension of the ring of differential operators on a non-singular variety over a field of positive characteristic. Let k be closed of characteristic $p \neq 0$. Let A be the coordinate ring of a non-singular affine algebraic variety X over k, then
gldim D(A) = dim(X), where D(A) is the ring of differential operators over X.

An earlier result of S. Chase established that wdim(D(A)) = dim(X), but since D(A) need not be left Noetherian there is no obvious reason why wdim(D(A)) and gldim (D(A)) should coincide.

For results along this vein we refer to the following additional references:

- I. Bernstein, On the Dimension of Modules and Algebras III, Direct Limits. *Nagoya Math. J.*, **13** (1958), 83-84.
- S. Chase, On the Homological Dimension of Algebras of Differential Operators, *Comm. in Alg.* **5**, 1974, 351–363.

- **S.P. Smith, The Global Homological Dimension of the Ring of Differential Operators on a Non-Singular Variety over a Field of Positive Characteristic, preprint, U. of Warwick.**

Bibliographical Comments to Chapter 9.

In this chapter we present some results connecting Krull dimension and the global homological dimension of a ring.

One of the main results is Theorem 9.3.10. due to Brown, Hajarnavis, MacEacharn, [2]. Other results of a similar nature are also known e.g. in [1] Brown and Warfield prove for a fully bounded Noetherian ring R containing an uncountable set $\mathcal{U}$ of central units such that the differences of different elements of $\mathcal{U}$ are again units, that gldim (R) $< \infty$ entails Kdim (R) $\leq$ gldim (R).

As an application of this result, making use of the Laurent series ring, K. Goodearl, L. Small [1] have proved the following result. If R is any Noetherian P.I.ring with gldim (R) $< \infty$ then Kdim (R) $\leq$ gldim (R).

Other results in this direction have been obtained by: Brown, Hajarnavis, MacEacharn [1],[2]; M. Ramras [1],[2]; R. Resco, L. Small, T. Stafford [1].

For the study of commutative regular local rings in Section 9.4. we followed the presentation given by Kaplansky in [1] and [2].

Most of the exercise are results from K. Brown, C. Hajarnavis, A. MacEacharn [1]; Brown and Hajarnavis [1]. We have listed them in the right order so that it becomes possible to solve them step by step and reconstruct parts of the papers mentioned, although this is not always easy. The reader may consider this set of exercises to be an appendix, bringing the material up-to-date without having to include all details.

Other typical references for the contents of chapter 9 are:

H. Bass [1]; W.D. Blair [1]; M. Chamarie and A. Hudry [1]; S. Chase [1]; K.R. Goodearl [1]; R. Resco, L.W. Small and J.T. Stafford [1]; J.L. Bueso, F. Van Oystaeyen and A. Verschoren [1].

Chapter 10

The Gelfand-Kirillov Dimension.

At the origin of the recent boom in research concerning the so-called Gelfand-Kirillov dimension of algebras we situate two papers published by I.M. Gelfand, A.A. Kirillov, [1],[2]. In these papers the Gelfand-Kirillov conjecture is made: the enveloping algebra of a finite dimensional algebraic Lie algebra has a division algebra of fractions isomorphic to the quotient ring D_n of some Weyl algebra $A_n(k)$.

In this contents it is important to be able to decide whether $D_n \cong D_m$ implies n = m and this problem can be dealt with by making use of a variant of the invariant, introduced in loc. cit., which is now commonly known as the Gelfand-Kirillov dimension (abbreviated: GK-dimension or GKdim).

The general theory of GK-dimension has some similarity with the theory of Krull dimension on some (rather rare) occasions but it is fundamentally different in at least two ways. First, GK-dimension is defined here only for algebras over fields and it is not clear how to extend it to arbitrary rings. Secondly, GK-dimension cannot be interpreted from the lattice point of view.

It makes sense to say that GKdim is a generalisation of the notion of "transcendence degree" for finitely generated commutative domains to the case of non-commutative k-algebras. The recently published monograph by G. Krause, T. Lenagan [1], contains a rather complete introduction to the theory of GK-dimension. This chapter is in large based on Krause-Lenagan book [1], we have added some results of J. Krempa, J. Okninski, cf. [1], concerning GKdim of tensorproducts.

§10.1 Definitions and Basic Properties.

Throughout this section k is a field and A is a k-algebra. If V is a k-vectorspace contained in A we let k[V] stand for the k-algebra generated by V in A. A finite dimenional k-space V contained in A and containing 1 is said to be a *sub-frame of A*. A sub-frame V of A is said to be a *frame* of A if k[V] = A, in this case A is necessarily an affine k-algebra i.e. a finitely generated k-algebra.
Let $V = kv_1 + ... + kv_d$ be a sub-frame of A and let V^i denote the set of monomials of length i in the $v_1, ..., v_d$. We write: $F_n^V(A) = k + V + V^2 + ... + V^n$, then $\{F_n^V(A), n \in \mathbb{Z}\}$ determines an exhaustive filtration on k[V], where we put $F_m^V(A) = 0$ for m $\leq$ -1. The associated graded ring of this filtration is isomorphic to $\bigoplus_{i=-1}^{\infty} V^{i+1}/V^i$. Define: $d_V(n) = \dim_k(F_n^V(A))$. We need a general lemma, the proof of which is just straightforward checking:

10.1.1. Lemma. Let f: $\mathbb{N} \to \mathbb{R}$ be a function which is eventually monotone increasing and positively valued, i.e. there is an $n_0 \in \mathbb{N}$ such that for all $n \geq n_0$, f(n) > 0 and f(n+1) $\geq$ f(n). Then: $\overline{\lim} \frac{log(f(n))}{log(n)} = \inf \{\rho \in \mathbb{R}, f(n) \leq n^\rho$ for almost all n$\}$, where $\overline{\lim}$ stands for limsup (liminf will be denoted by $\underline{\lim}$).

It is obvious from the above equality that the number we have associated to f measures how this function "grows" with n. If V and W are sub-frames of A such that k[V] = k[W] then $W \subset F_m^V(A)$ and $V \subset F_r^W(A)$ for some m,r $\in \mathbb{N}$, so it follows that $d_W(n) \leq d_V(mn)$ and $d_V(n) \leq d_W(rn)$. By the lemma it is clear that $\overline{\lim} \frac{log(d_V(n))}{log(n)} = \overline{\lim} \frac{log\ d_V(mn)}{log(n)}$ (for m fixed). Hence, if we put GKdim k[V] $= \overline{\lim} \frac{log(d_V(n))}{log(n)}$ then this real number depends only on k[V] and not on the choice of the frame V. The *Gelfand-Kirillov dimension of a k-algebra A* is then defined by GKdim (A) = sup {GKdim k[V], V a sub-frame of A}. If A is an affine k-algebra then GKdim (A) = GKdim (k[V]) where V is any frame of A. It is clear from the definition that GKdim (A) = 0 exactly then when every affine subalgebra of A is finite dimensional.

It is possible to replace $\overline{\lim}$ by $\underline{\lim}$ in the foregoing definitions, the dimension defined this way will be denoted by GK$\underline{dim}$ and we call it the *lower GK-dimension.* We shall use GK$\underline{dim}$ when dealing with tensor-products but otherwise it is always the usual GKdim that we use in this chapter.

Let us now consider a left A-module M. A *sub-frame of M* is just a finite dimensional k-subspace; we say that a sub-frame μ of M is a *frame of M* if $A\mu$ =

M; in particular, the existence of a frame entails that M is a finitely generated left A-module. If V is any sub-frame of A then $M(V,\mu) = \bigcup_{m\in\mathbb{N}} V^m\mu$ is a finitely generated k[V]-module. Put $d_V(n,\mu) = \dim_k (V^n\mu)$ and put GKdim $(M(V,\mu)) = \overline{\lim} \frac{\log(d_V(n,\mu))}{\log(n)}$. One easily verifies that this real number only depends on $M(V,\mu)$ and not on the particular choices of V and μ. The *Gelfand-Kirillov dimension of M*, GKdim M, is now defined to be equal to sup {GKdim $(M(V,\mu))$, V a sub-frame of A, μ a sub-frame of M}.

10.1.2. Examples.

1. If the algebra A is algebraic over k then GKdim (A) = 0.

2. Let A be the free k-algebra k<X,Y> and put V = kX + kY. Then $\dim_k(V^i) = 2^i$, hence $d_V(n) = 1 + 2 + ... + 2^n = 2^{n+1} - 1$, therefore GKdim (k<X,Y>) = ∞. Note that V is not a frame because we did not take $1 \in V$ but the calculation is still valid (in fact we may use V = k + kX + kY and a slightly more complicated calculation).

3. Put A = k<X,Y>/(XY) = k<x,y>, where x = Xmod(XY), y = Ymod(XY). Take V = kx + ky. We see that $F_n^V(A)$ is generated by the y^ix^i with $i > 0$, $j > 0$ and $i + j \leq n$, consequently we obtain: $\dim_K (F_n^V(A)) = d_V(n) = (n+1) + n + (n-1) + ... + 1 = \frac{(n+2)(n+1)}{2} = n^2 + 0(n)$.
Hence GKdim (A) = 2 follows.

4. Consider A as a left A-module (write: ${}_AA$) then we have (taking μ = k): GKdim (A) = GKdim (${}_AA$). Moreover, if M is a free left A-module of finite rank then GKdim (A) = GKdim (${}_AM$).

The use of the logarithmic growth mystifies the nature of the real numbers that can actually appear as the GKdim of some k-algebra. In 1976, W. Borho and H. Kraft showed, cf. [1], that for all $r \in \mathbb{R}$ such that $r \geq 2$ there exists a k-algebra A such that GKdim (A) = r. It is very easy to see that except for 0 and 1 there are no GKdim-numbers in the interval [0,1].

G. Bergman proved that numbers in]1,2[cannot appear as the GKdim of some k-algebra, by using properties of words in a free semigroup and order ideals (cf. Remark, p.21, in Krause-Lenagan's book, [1]).

10.1.3. Lemma. Let A be a k-algebra.

1. If B is a k-subalgebra of A then GKdim (B) $\leq$ GKdim (A). If C is an epimorphic image of A then GKdim (C) $\leq$ GKdim (A).
2. Let M be a left A-module, N a submodule of M then:
GKdim (N) $\leq$ GKdim (M) and GKdim (M/N) $\leq$ GKdim (M),
i.e. GKdim (M) $\geq$ max {GKdim (N), GKdim (M/N)}.

Proof. 1. If B $\subset$ A the every sub-frame of B is a sub-frame of A so the result follows. Consider an epimorphism A $\to$ C $\to$ 0. If $\overline{V}$ is a sub-frame of C, say $\overline{V}$ = k$\overline{v}_1$ + ... + k$\overline{v}_m$ then we obtain a sub-frame V of A by choosing arbitrary representatives $v_1, ..., v_m$ for $\overline{v}_1,...,\overline{v}_m$. Since $\dim_k (V^n) \geq \dim_k (\overline{V}^n)$ for all n, the statement of the lemma is clear.
2. Trivial. □

10.1.4. Proposition.

1. If A_1 and A_2 are k-algebras then we have:
GKdim ($A_1 \bigoplus A_2$) = max {GKdim (A_1), GKdim (A_2)}.
2. Let M_i be left A-modules, then:
GKdim ($M_1 \bigoplus ... \bigoplus M_r$) = $\max_i$ {GKdim (M_i)}.

Proof. 1. In view of the lemma, GKdim ($A_1 \bigoplus A_2$) $\geq$ max {GKdim (A_1), GKdim (A_2)}. Put γ = max {GKdim (A_1), GKdim (A_2)}. If $\gamma = \infty$ we have nothing to prove so let us assume that $\gamma < \infty$. If W is a sub-frame of $A_1 \bigoplus A_2$, let U and V be the canonical projections of W on A_1, resp. A_2. By definition of GKdim it is sufficient to establish that $d_W(n) < n^{\gamma+\epsilon}$ for any $\epsilon \neq 0$ in $\mathbb{R}$. For almost all n in $\mathbb{N}$ we have: $d_U(n) < n^{\gamma+\epsilon/2}$, $d_V(n) < n^{\gamma+\epsilon/2}$ hence $d_W(n) \leq d_U(n) + d_V(n) < 2n^{\gamma+\epsilon/2}$. By choosing n large enough such that $n^{\epsilon/2} > 2$ we obtain $2n^{\gamma+\epsilon/2} < n^{\gamma+\epsilon}$, i.e. $d_W(n) < n^{\gamma+\epsilon}$.
2. Easy (in fact one may use an argument very similar to the proof of 1.) □

10.1.5. Corollary. If $I_1, ..., I_d$ are ideals of the k-algebra A then
GKdim (A/$I_1 \bigcap ... \bigcap I_d$) $\leq$ max {GKdim (A/I_j), j = 1,...,d}.

Proof. Embed A/$I_1 \bigcap ... \bigcap I_d$ in $\bigoplus_{j=1}^d$ (A/I_j). Apply the foregoing Proposition, and the lemma. □

Constructing k-algebras of GKdim equal to m ∈ IN is very easy in view of the following lemma.

10.1.6. Lemma.

1. If X is an indeterminate then GKdim (A[X]) = 1 + GKdim (A).
2. Consider a k-derivation δ on A and form the differential polynomial algebra A[X,δ] where multiplication is defined by Xa = aX + δ(a) for all a ∈ A, then we have: GKdim (A[X,δ]) ≥ 1 + GKdim (A).

Proof. Let us first establish 2. Let V be a sub-frame of A and put W = V + kX ⊂ A[X,δ]. Evidently: $V^n + V^nX + \ldots + V^nX^n \subset W^{2n}$.
Thus $d_W(2n) \geq (n+1)\, d_V(n)$ and consequently:
GKdim (A[X,δ]) ≥ $\overline{\lim}\ \frac{log((n+1)d_V(n))}{log(n)} = \lim\limits_{n\to\infty} \frac{log(n+1)}{log(n)} + \overline{\lim}\ \frac{log(d_V(n))}{log(n)}$,
hence GKdim (A[X,δ]) ≥ 1 + GKdim (A).
1. If we show that GKdim (A[X]) ≤ 1 + GKdim (A) then equality follows by taking $\delta = 0$ in 1. If W is a sub-frame in A[X] then W ⊂ V + VX + ... + VX^m for some m ∈ IN and some sub-frame V of A. Therefore $d_W(n) \leq n(m+1)\, d_V(n)$ or GKdim (A[X]) ≤ 1 + GKdim (A).

10.1.7. Theorem. Suppose that δ is a k-derivation of the k-algebra A such that every sub-frame of A is contained in a δ-stable finitely generated subalgebra then GKdim (A[X,δ]) = 1 + GKdim (A).

Proof. Proposition 3.5. p. 28 of Krause-Lenagan's book [1]. □

10.1.8. Corollary.

1. GKdim (A[$X_1, \ldots, X_d$]) = GKdim (A) + d.
2. If A_n(k) is the nth Weyl algebra over the field k (here char k = 0 is assumed) then GKdim (A_n(k)) = 2n.

Proof. 1. Obvious, by induction on d.

2. Recall that $A_d(k) = k[X_1, ..., X_d][Y_1, ..., Y_d, \delta_1, ..., \delta_d]$ where δ_i is the partial derivative $\frac{\partial}{\partial X_i}$, i = 1,...,d. The lemma implies that GKdim $(k[_1, ..., X_d][Y_1, \delta_1] = d + 1$. Adding $Y_2, ..., Y_d$ consecutively and applying the theorem at each step, we arrive at GKdim $(A_n(k)) = 2n$. □

10.1.9. Example. Put A = $\mathbb{R}[[X]]$. Let $r_i \in \mathbb{R}$ be a countable infinite set of $\mathbb{Q}$-independent numbers. Each function $f_i = e^{r_i x}$ may be viewed as an element of A via its Maclaurin series. Therefore A contains subalgebras isomorphic to $\mathbb{R}[X_1, ..., X_n]$ for each n $\in \mathbb{N}$, i.e. GKdim $(\mathbb{R}[[X]]) = \infty$.

We now investigate the GK-dimension of a tensor product and we follow the presentation contained in Krempa-Okninski [1]; in fact these results use the lower GK-dimension and a slight improvement of some results obtained by R. Warfield in [1].

We define GK$\underline{dim}$ (A) by replacing $\overline{\lim}$ in the definition of GKdim (A) by $\underline{\lim}$, and we call this number (in $\mathbb{R}$) the *lower Gelfand-Kirillov dimension.* The equivalent of Lemma 10.1.2. also holds for GK$\underline{dim}$ (A) and it also follows that GK$\underline{dim}$ (A) ≥ 2 when GKdim (A) ≥ 2 (because of . Bergman's result stating that]1,2[contains no numbers of the type GKdim (A) for some A). First we need a lemma:

10.1.10. Lemma. Let A and B be k-algebras, then:
max (GKdim (A), GKdim (B)) $\leq$ GKdim $(A \otimes_k B) \leq$ GKdim A + GKdim B.

Proof. The first inequality is clear (see Lemma 10.1.2.). If C is a finitely generated k-algebra in $A \otimes_k B$ then $C \subset A' \otimes_k B'$ where A',B' are finitely generated k-algebras in A',B' resp.. Hence we may reduce the problem to the finitely generated case.

Suppose A = k[U], B = k[V] where U and V are frames for A,B resp.. Clearly $A \otimes_k B = k[U \otimes 1 + 1 \otimes V]$ and $W = U \otimes 1 + 1 \otimes V$ is finite dimensional over k. Now $U^n \otimes_k V^n \subset W^{2n} \subset U^{2n} \otimes_k V^{2n}$, thus $d_U(n)\ d_V(n) \leq d_W(2n) \leq d_U(2n)\ d_V(2n)$. We now calculate: GKdim $(A \otimes_k B) = \overline{\lim} \frac{\log d_W(2n)}{\log n} =$ $\overline{\lim} \frac{\log(d_U(n) d_V(n))}{\log n} = \overline{\lim} \frac{\log(d_U(n))}{\log n} + \frac{\log d_V(n)}{\log n}$. The latter sum may be smaller than $\overline{\lim} \frac{\log(d_U(n))}{\log n} + \overline{\lim} \frac{\log(d_V(n))}{\log n}$ ($\overline{\lim}$ is a strange thing !) which equals GKdim (A) + GKdim (B). □

Making use of the algebras of type k <X,Y>/{Z} for suitable relations $Z \subset$ <X,Y>, J. Krempa and J. Okninski prove the following theorem: (Theorem 1. in [1]):

10.1.11. Theorem. Consider real numbers $2 \leq \alpha \leq \beta \leq \infty$. There exists a finitely generated k-algebra A such that GK$\underline{dim}$ A $= \alpha$, GKdim (A) $= \beta$.

Proof. Somewhat technical, we refer to Krempa and Okninski, [1]. □

10.1.12. Proposition. Let A and B be k-algebras, then:

1. max (GKdim (A) + GK$\underline{dim}$ (B), GK$\underline{dim}$ (A) + GKdim (B)) $\leq$ GKdim ($A \otimes_k B$) $\leq$ GKdim (A) + GKdim (B).
2. GK$\underline{dim}$ (A) + GK$\underline{dim}$ (B) $\leq$ GK$\underline{dim}$ ($A \otimes_k B$) $\leq$ min (GKdim (A) + GK$\underline{dim}$ (B), GK$\underline{dim}$ (A) + GKdim (B)).

Proof. 1. Put GK$\underline{dim}$ (A) $= \underline{\alpha}$, GK$\underline{dim}$ (B) $= \underline{\beta}$, GKdim (A) $= \overline{\alpha}$, GKdim (B) $= \overline{\beta}$ and let $\underline{\alpha} + \overline{\beta} \geq \underline{\beta} + \overline{\alpha}$ (up to renumbering). When $\overline{\alpha} = \infty$ or $\overline{\beta} = \infty$ then GKdim ($A \otimes_k B$) $= \infty$ and both inequalitis hold. Thus we may assume that $\overline{\alpha}, \overline{\beta} < \infty$. Pick $\varepsilon > 0$ in $\mathbb{R}$. There exist sub-frames $U \subset A$ and $V \subset B$ such that: $\underline{\lim}$ f $\frac{\log d_U(n)}{\log(n)} > \underline{\alpha} - \varepsilon/2$, $\overline{\lim} \frac{\log d_V(n)}{\log(n)} > \overline{\beta} - \varepsilon/2$. We may chose a subsequence n_i such that: $\overline{\lim} \frac{\log(d_V(n_i))}{\log(n_i)} > \overline{\beta} - \varepsilon/2$ while $\lim \frac{\log d_V(n_i)}{\log(n_i)}$ exists. The choice of U is such that $\underline{\lim} \frac{\log d_U(n_i)}{\log(n_i)} > \underline{\alpha} - \varepsilon/2$. If $W = U \otimes 1 + 1 \otimes V$ then the proof of Lemma 10.1.8. learns that: $\frac{\log d_W(2n_i)}{\log(2n_i)} \geq \frac{\log d_U(n_i) + \log d_V(n_i)}{\log(n_i) + \log 2}$. The latter converges to a number larger that $\underline{\alpha} + \overline{\beta} - \varepsilon$. Consequently $\overline{\lim} \frac{\log d_W(n)}{\log(n)} > \underline{\alpha} + \overline{\beta} - \varepsilon$ and as ε is arbitrary it follows that
$\underline{\alpha} + \overline{\beta} = \max(\underline{\alpha} + \overline{\beta}, \underline{\beta} + \overline{\alpha}) \leq$ GKdim ($A \otimes_k B$).
The Lemma 10.1.8. proves the second inequality.
2. Very similar to the proof of 1. □

We say that a k-algebra A has *uniform growth* if GK$\underline{dim}$ (A) = GKdim (A).

10.1.13. Corollary. If one of the algebras A,B has uniform growth then:

1. GKdim ($A \bigotimes_k B$) = GKdim (A) + GKdim (B).
2. GK$\underline{dim}$ ($A \bigotimes_k B$) = GK$\underline{dim}$ (A) + GK$\underline{dim}$ (B).

If the k-algebra A is such that there is a k-subalgebra A_0 of the form k[V] for some finite dimensional k-space V, such that GKdim (A) = GKdim (A_0) = $\lim_{n\to\infty} \frac{\log d_V(n)}{\log(n)}$ then A has uniform growth. From this observation it follows that the Corollary above covers Proposition 3.11. of Krause-Lenagan, [1], originally due to R. Warfield. In particular, k-algebras of GK-dimension less that or equal to 2 have uniform growth. On the other hand there exist algbras of uniform growth such that their GK-dimension cannot be realized on a finitely generated subalgebra.

In Krempa-Okninski [1] there is also an affirmative answer to a question of R. Warfield, [1], i.e. if $2 + \beta \leq \gamma \leq \alpha + \beta$ for some $\alpha, \beta, \gamma \in \mathbb{R}$ then γ = GKdim ($A \bigotimes_k B$) for some K-algebras A and B having GKdimension α and β resp..

10.1.14. Proposition. Let I be an ideal of a k-algebra A and suppose that I contains a left (or right) regular element of A then GKdim (A/I) + 1 $\leq$ GKdim (A).

Proof. Put $\overline{A}$ = A/I and let $\overline{V}$ be a sub-frame of $\overline{A}$ such that $1_{\overline{A}} \in \overline{V}$. Put V = k + kc + k$a_1$ + ... + ka_n. Let $C(I)_n$ be the complement of $I \bigcap V^n$ in the k-space V^n, i.e. $\overline{V}^n = V^n + I/I \simeq V^n/V^n \bigcap I \equiv C(I)_n$. Since $C(I)_n \bigcap cA = 0$, $C(I)_n + C(I)_n c + ... + C(I)_n c^n$ is a direct sum of k-spaces in A and it is contained in V^{2n}; therefore: $\dim_k (V^{2n}) \geq n\dim_k (C(I)_n) = n\dim_k (\overline{V}^n)$.
It follows that $d_V(2n) \geq n d_{\overline{V}}(n)$ and we calculate: GKdim (A/I) + 1 =
$sup_{\overline{V}} \overline{\lim} \frac{\log(d_{\overline{V}}(n))}{\log(n)} + 1 = sup_{\overline{V}}(\overline{\lim} \frac{\log d_{\overline{V}}(n)}{\log(n)} + \lim_{n\to\infty} \frac{\log(n)}{\log n}) =$
$sup_{\overline{V}}(\overline{\lim} \frac{\log(d_{\overline{V}}(n).n)}{\log(n)} \leq sup_V(\overline{\lim} \frac{\log d_V(2n)}{\log(n)}) \leq$ GKdim (A). □

10.1.15. Corollary. Let A be a k-algebra such that for every P $\in$ Spec (A), A/P is a left Goldie ring, then GKdim (A) $\geq$ GKdim (A/P) + ht(P).

Proof. Consider a strictly decending chain of prime ideals of A, P = $P_0 \supset P_1 \supset ... \supset P_m$. For each i, P_i/P_{i+1} is an essential left ideal in the left Goldie ring A/P_{i+1} hence it contains a left regular element and GKdim (A/P_i) + 1 $\leq$ GKdim (A/P_{i+1}) follows from the lemma. By induction we then arrive at GKdim

(A/P) + m $\leq$ GKdim (A/P_m) and if we consider a maximal chain the result follows.

That GKdimension is well-behaved with respect to central localization is the essence of the following proposition:

10.1.16. Proposition. Let S be a multiplicative set of central regular elements in a k-algebra, then GKdim $(S^{-1}A)$ = GKdim (A).

Proof. That GKdim (A) $\leq$ GKdim $(S^1 A)$ is a consequence of $A \hookrightarrow S^{-1}A$. Conversely, let W be a sub-frame in $S^{-1}A$. One may select a common denominator, c say, for a chosen k-basis of W. Then cW $\subset$ A so we may consider
V = cW + kc + k in A. Obviously $W^n \subset c^{-n}V^n$ for all n $\in \mathbb{N}$, thus
$\dim_k (W^n) \leq \dim_k (V^n)$ and hence GKdim $(S^{-1}A) \leq$ GKdim (A). □

This result may succesfully be applied to derive properties of GK-dimension for commutative k-algebras.

10.1.17. Lemma. Let B be a k-subalgebra of a commutative k-algebra A and suppose that A is finitely generated as a B-module, then GKdim (A) = GKdim (B).

Proof. We only have to show that GKdim (B) $\geq \overline{\lim} \frac{\log d_V(n)}{\log(n)}$ for every finite dimensional k-subspace V of A.
Clearly we may restrict attention to those V in A containing a set of B-generators for A, say $x_1, ..., x_m$. Let $v_1, ..., v_t$ be a k-basis for V. Write $v_i = \sum_{j=1}^m v_{ij}x_j$, i = 1,...,t and $v_i v_k = \sum_{l=1}^m v_{ik}^l x_l$ for all v,j in {1,...,t}, where v_{ij} and v_{ik}^l are i B. Take for W the k-space generated by 1,v_{ij}, v_{ik}; i,j,k $\in$ {1,...,t}, l $\in$ {1,...,m}. It is easy to verify that: $V + V^2 \subset x_1 W + ... + x_m W$, hence $V^n \subset x_1 W^{2n-1} + ... + x_m W^{2n-1}$ for all n $\in \mathbb{N}$. From $d_V(n) \leq m d_W(2n-1)$ the inequality GKdim (A) $\leq$ GKdim (B) results. □

10.1.18. Corollary. Let k $\subset$ K be fields and let A be a commutative k-algebra which is algebraic over K, then (we assume K $\subset$ A):

1. GKdim (A) = GKdim (K).
2. GKdim (K) = tr.deg$_k$ (K).

Proof. 1. Easy verification by reduction to finitely generated algebras.
2. Use 1. to reduce the problem to a purely transcendental extension K over k. Up to applying Proposition 10.1.14. it suffices to envoke Corollary 10.1.7.(1). □

10.1.19. Theorem. For a commutative k-algebra A, the GKdim (A) is either infinite or a natural number. When A is fiitely generated then GKdim (A) = clKdim (A).

Proof. In view of Noether's Normalization lemma we may view A as a finitely generated module over a polynomial ring B = k$[X_1, ..., X_d]$.
We know that cl.Kdim (A) is the equal to d and this proves the second statement. The first statement follows easily from this by taking the sup over the finitely generated subalgebras □

One of the main differences between the behaviour of Krull dimension and GK-dimension is that GKdim is not necessarily exact on modules i.e. there exist exact sequences 0 → M' → M → M" → 0 such that GKdim (M) $\neq$ max {GKdim (M'), GKdim (M")}.

On the other hand, the GK-dimension will be left and right symmetric for finitely generated (left and right) bimodules; moreover, for left Noetherian k-algebras A the GK-dimension is ideal invariant in the sense that GKdim (I $\otimes_A$ M) is smaller than or equal to GKdim (M) for any finitely generated left A-module M and any ideal I of A. More detail in these properties may be found in I. Bernstein [1], A. Joseph, L.Small [1] as well as in our basic reference (Krause-Lenagan's book [1]).

10.1.20. Proposition. Let A be a k-algebra, M a left A-module.

1. If IM = 0 for an ideal I of A then GKdim ($_AM$) = GKdim ($_{A/I}M$)
2. If M is finitely generated and $\alpha \in$ End$_A$ (M) is injective then GKdim (M/α(M)) $\leq$ GKdim (M) − 1.
3. GKdim (M) $\leq$ GKdim (A).

4. If M_i, i = 1,...,r, are submodules of M then: GKdim (M) = max {GKdim(M_i), i = 1,...,r}.

Proof. 1. Obvious.

2. Let $\overline{\mu}$ be a frame of $\overline{M}$ = M/α(M). Since α(M) is finitely generated too, we may lift $\overline{\mu}$ to a frame of M such that $\overline{\mu} = \mu$ mod α(M). Let $C_n(\alpha)$ be a k-complement for $V^n\mu \bigcap \alpha(M)$ in $V^n\mu$, where V is a sub-frame of A, for every n ∈ ℕ . It is clear that $C_n(\alpha) \equiv V^n\overline{\mu}$ (compare to the proof of Proposition 10.1.14.). Since $C_n(\alpha) \bigcap \alpha(M)$ = 0 it is clear that $C_n(\alpha) + \alpha(C_n(\alpha)) + ... + \alpha^i(C_n(\alpha))$ is a direct sum. Since $\alpha(\mu)$ is finite dimensional and μ being a frame there must exist a sub-frame W of A, W ⊃ V, such that $\alpha(\mu) \subset W\mu$. Consequently: $\bigoplus_{j=0}^n \alpha^j\ (C_n(\alpha)) \subset \bigoplus_{j0}^n \alpha^j\ (V^n\mu) = \bigoplus_{j=0}^n V^n\alpha^j(\mu) \subset \bigoplus_{j=0}^n W^jW^n\mu \subset W^{2n}\mu$. This leads to $\dim_k\ (W^{2n}\mu) \geq$ (n+1) $\dim_k\ (C_n(\alpha))$ = (n+1) $\dim_k\ (V^n\mu)$ and the result is then evident from the definition of GKdim.

3. If $\mu \subset$ M, V ⊂ A are sub-frames then $\dim_k\ (V_n\mu) \leq \dim_k(\mu).\dim_k(V^n)$ and the statement follows.

4. Follows from Lemma 10.1.3. □

10.1.21. Example (G. Bergman). Let us reconsider the k-algebra A introduced in 10.1.2.(3), A = k<X,Y>/(XY), with GKdim (A) = 2. Let M be the left A-module Au + Av where uy^nx^{n+1} = 0 for all n ≥ 0 and vy^nx is zero unless n is a square, then vy^nx = u$y^{\sqrt{n}}$x.

Take V = k ⊗ kx ⊗ ky, x = Xmod (XY), y = Ymod (XY) and μ = ku ⊕ kv. It is clear that the monomials u$y^jx^i$with i ≥ j, i+j ≤ n together with the vy^i, i ≤ n, form a k-basis for $V^n\mu$. There are exactly n+1 monomials of the second kind whereas the number of monomials of the first kind depends on the parity of n, i.e. it is $(n+2)^2/4$ when n is even and it is (n+1)(n+3)/4 if n is odd. Since these are quadratic functions of n it follows that GKdim (M) = 2.

Now we consider the left A-submodule N of M generated by kv. A k-basis for V^nv consists of all y^iv, i ≤ n, together with the y^jx^iu, 1 ≤ i ≤ j and i + j^2 ≤ n i.e. i ≤ $(\sqrt{n}-j)(\sqrt{n}+j)$. It is clear that there are at most n monomials of the second kind and thus GKdim (N) = 1. Considering M/N, we see that k$\overline{u}$, where $\overline{u}$ = u + N, is a frame for M/N.

Since y^ix^ju = 0 for o ≤ i < j and y^ix^ju = $y^{i^2}\ x^j$v ∈ N for all i ≥ 0, j ≥ 1 we see that {$y^i\overline{u}$, 0 ≤ i ≤ n} is a k-basis for $V^n\overline{u}$. It follows from the latter fact that GKdim (M/N) = 1. This provides an example of an exact sequence of left A-modules: 0 → N → M → M/N → 0, such that GKdim (M) = 2 ≠ max {GKdim (N), GKdim (M/N)} = 1.

In some sense the failure of exactness of GKdim on modules is made up for by its good properties with respect to bimodules. If A and B are rings then we write ${}_AM_B$ to indicate that M is a left A-module and a right B-module i.e. an A-B-bimodule, and we simply refer to M as a bimodule. If necessary we distiguish between frames for ${}_AM$ and M_B by writing ${}_A\mu$ or μ_B according to the case considered.

10.1.22. Proposition (W. Borho [1], T. Lenagan [8]). Let A and B be k-algebras and let ${}_AM_B$ be a bimodule. Suppose that ${}_AM$ is finitely generated, then we obtain:

1. GKdim (M_B) = GKdim $(B/\mathrm{Ann}_B(M))$.
2. GKdim $(M_B) \leq$ GKdim $({}_AM)$.

Proof. 1. Put $M = Am_1 + ... + Am_r$. Then $\mathrm{Ann}_B(M) = \bigcap_{i=1}^r \mathrm{Ann}_B(m_i)$.
Hence, $B/\mathrm{Ann}_B(M)$ may be embedded in $\bigoplus_{i=1}^r M_B$ and therefore
GKdim $(B/\mathrm{Ann}_B(M))$ = GKdim $((B/\mathrm{Ann}_B(M))_B) \leq$ GKdim (M_B).
The converse inequality follows from Proposition 10.1.20., (1) and (3).
2. Let μ' be a sub-frame of M, V_B a sub-frame of B. Since ${}_AM$ is finitely generated we may enlarge $\mu'$to a frame ${}_A\mu$ of M, i.e. $\mu' \subset {}_A\mu$. Obviously $\mu' V_B^n \subset {}_A\mu V_B^n$ for all $n \in \mathbb{N}$. Since $A({}_A\mu) \supset ({}_A\mu)V_B$ (note: $A({}_A\mu) = M$) there exists a sub-frame V_A of A such that $V_A({}_A\mu) \supset ({}_A\mu)V_B$. For $n \in \mathbb{N}$ we then obtain: $\mu' V_B^n \subset ({}_A\mu)V_B^n \subset V_A^n({}_A\mu)$, so the result follows. □

10.1.23. Corollary. If both ${}_AM$ and M_B are finitely generated then
GKdim $({}_AM)$ = GKdim (M_B).

10.1.24. Proposition. If B is a k-subalgebra of A such that A_B is finitely generated then GKdim (A) = GKdim (B).

Proof. The foregoing corollary implies that GKdim (A) = GKdim $({}_AA)$ = GKdim (A_B). Proposition 10.1.20.(3) yields GKdim $(A_B) \leq$ GKdim (B) and then GKdim (A) $\leq$ GKdim (B) entails the desired equality. □

10.1.25. Proposition. Let M be a left A-module and let ${}_AN_A$ be a bimodule which is finitely generated as a right A-module, then we have: GKdim $(N \otimes_A M) \leq$ GKdim (M).

Proof. Consider sub-frames $\mathcal{E}$ of $N \otimes_A M$ and V of A. It is clear that there exist a sub-frame μ of M and a frame η_A of N such that $\mathcal{E} \subset \eta_A \otimes \mu = \{\sum' f \otimes g, f \in \eta_A, g \in \mu\}$. Since $V\eta_A$ is finite dimensional there exists a larger sub-frame $W \supset V$ such that $V\eta_A \subset \eta_A W$, hence $V^n\eta_A \subset \eta_A W^n$ for all $n \in \mathbb{N}$. Consequently $V^n\mathcal{E} \subset V^n(\eta_A \otimes \mu) \subset \eta_A \otimes W^n\mu$ and the latter is a k-linear image of $\eta_A \otimes_k W^n\mu$. It follows that $\dim_k(V^n\mathcal{E}) = \dim_k(\eta_A)\dim_k(W^n\mu)$. This yields: $\overline{\lim} \frac{\log(\dim_k(V^n\mathcal{E}))}{\log(n)} \leq \overline{\lim} \frac{\log(\dim_k(W^n\mu))}{\log(n)} \leq$ GKdim (M).
Passing to the supremum for all $\mathcal{E}$ and V, the assertion follows. □

§10.2 GKdimension of Filtered and Graded Algebras.

Let A be a $\mathbb{Z}$ -graded k-algebra and consider a graded A-module $M = \bigoplus_{n\in\mathbb{Z}} M_n$. Throughout this section we assume that *all gradations have finite type* i.e. for all $n \in \mathbb{N}$, A_n is a finite dimensional k-space and M_n is a finite dimensional k-space.

We put: $A(n) = \bigoplus_{i=-n}^{n} A_i$, $M(n) = \bigoplus_{i=-n}^{n} M_i$, $d_A(n) = \dim_k(A(n))$, $d_M(n) = \dim_k(M(n))$.

10.2.1. Lemma. Let A be a graded k-algebra, M a graded left A-module.

1. If $V \ni 1$ is a subframe in A and μ is a subframe in M, then GKdim (M) $\leq \overline{\lim} \frac{\log(d_M(n))}{\log(n)}$
2. If A is a finitely generated k-algebra and M is a finitely generated left A-module then GKdim (M) $= \overline{\lim} \frac{\log(d_M(n))}{\log(n)}$

Proof. 1. There exists an $m \in \mathbb{N}$ such that $V \subset A(n)$, $\mu \subset M(n)$, hence $V^n\mu \subset A(mn)\mu \subset M(mn+m) \subset M(2mn)$ for every nonzero $n \in \mathbb{N}$. Therefore $d_V(n,\mu) \leq d_M(2mn)$ and this establishes the assertion.
2. For m large enough $V = A(m)$ generates A (while $1 \in V$ is obvious) and $\mu = M(m)$ generates M as a left A-module. We will show that $M(n) \subset V^n\mu$ for all nonzero $n \in \mathbb{N}$

by showing that $M_{-n} + M_n \subset V^n\mu$. In fact it suffices to prove that $M_n \subset V^n\mu$ because $M_{-n} \subset V^n\mu$ will follow by a completely similar argumentation. Now $\{V^n\mu,\ n \in \mathbb{N}\}$ is an exhaustive filtration on M, hence there is an $r \in \mathbb{N}$ such that $M_n \subset V^r\mu$ i.e. each $x \neq 0$ in M_n may be written as a sum of nonzero monomials $v_s.....v_1.v_0$ with homogeneous $v_0 \in \mu$, $v_i \in V$, $i = 1,...,s$, where $s \leq r$. We may assume that each such monomial has minimal length i.e. $v_1v_0 \notin \mu = M(m)$ and $v_{i+1}v_i \notin V = A(m)$ for $i \geq 1$, or equivalently
$|\deg(v_{i+1}v_i)| > m$ for $i \geq 0$. Since deg $(v_s.....v_1.v_0)$ equals n at least one of the $\deg(v_i)$ must be positive, say $\deg(v_i) > 0$. Assume that $\deg(v_{i+1}) > 0$ while $\deg(v_i) \leq 0$, for some i. Then we calculate: $|\deg(v_{i+1}v_i)| \leq \max\{|\deg(v_i)|, |\deg(v_{i+1})|\} \leq m$ (if $i = s$ consider v_sv_{s-1}). It follows that all v_i must have positive degree and consequently $n \geq s+1$ and $v_s.....v_0 \in V^n\mu$ proving that $M_n \subset V^n\mu$. □

A filtration on A (on M) is said to be a *finite filtration* if each of the filtering additive subgroups F_nA (F_nM) is a finite dimensional k-space. It is obvious that a finite filtration is automatically discrete (i.e. $F_kM = 0$ for all $k \leq n_0$ for some $n_0 \in \mathbb{N}$). For a discrete filtration on M it is well known that M will be a Noetherian (filtered) left A-module when the associated graded G(A)-module G(M) is Noetherian (cf. Section 4.2.6.)

Recall that a filtration $\{F_nM,\ n \in \mathbb{N}\}$ is said to be a *good* filtration if G(M) is a finitely generated G(A)-module (these will also be called "standard" filtrations in Krause-Lenagan,[1]).

10.2.2. Lemma. Let A be a filtered k-algebra with filtration $\{F_nA,\ n \in \mathbb{Z}\}$, M a filtered left A-module with filtration $\{F_nM,\ n \in \mathbb{Z}\}$. Then GKdim (G(M)) $\leq$ GKdim (M).

Proof. Let $V_{G(A)}$ be a sub-frame of G(A) containing 1. and let $\overline{\mu}$ be a sub-frame in G(M). There is a sub-frame V of A and a sub-frame μ in M such that $\overline{\mu} \subset G(\mu)$, $V_{G(A)} \subset G(V)$. Then $V^n_{G(A)}\ \overline{\mu} \subset G(V)^nG(\mu) \subset G(V^n)G(\mu) \subset G(V^n\mu)$. But $G(V^n\mu) = \bigoplus_{i\in\mathbb{Z}} (V^n\mu \bigcap M_i) + M_{i-1}/M_{i-1} \cong \bigoplus_{i\in\mathbb{Z}} (V^n\mu \bigcap M_i)/(V^n\mu \bigcap M_{i-1})$. Since $V^n\mu$ is finite dimensional over k it follows that $\dim_k (V^n_{G(A)}\ \overline{\mu}) \leq \dim_k (V^n\mu)$ for all $n \in \mathbb{N}$. □

Note that for a finitely generated left A-module M such that $M = A\mu$ for some finite dimensional k-space μ we necessarily have that the filtration $F_nM = (F_nA)\mu$ is a good filtration; we will refer to this particular filtration as a *standard fitration.*

10.2.3. Proposition. Let A be a finitely filtered k-algebra such that G(A) is finitely generated and let M be a finitely generated left A-module filtered by $\{F_nM, n \in \mathbb{Z}\}$ such that G(M) is a finitely generated G(A)-module. Put $d_{FM}(n) = \dim_k (F_nM)$ then GKdim G(M) = GKdim (M) = $\overline{\lim} \frac{log(d_{FM}(n))}{log(n)}$.

Proof. Since M is also finitely filtered $F_iM = 0$ for all $i < -q$ for some $q \in \mathbb{N}$. Thus for all $n \geq q$ we obtain $G(M)(n) \cong F_n(M)$ (as k-spaces) and thus $d_{FM}(n) = d_{G(M)}(n)$ for all $n \geq q$.
The hypotheses imply that M is finitely generated as a left A-module. Let V be a frame for A and let μ be a frame of M. There is a $p \in \mathbb{N}$ such that $V \subset F_pA$, $\mu \subset F_pM$. Hence, for all $n \geq 1$ we obtain: $d_V(n,\mu) \leq d_{FM}(2pn)$.
Combining both inequalities established here with the foregoing lemma, the proof is complete if we calculate the $\overline{\lim} \frac{log(--)}{log(n)}$ of the appropriate functions. □

10.2.4. Corollary. The proposition applies in particular to standard filtrations on finitely generated left A-modules.

It is well-known (and easy to proof) that good filtrations on a finitely generated (filtered) A-module are equivalent, so on the level of filtered properties the choice of a good filtration is free so we may work with a standard filtration when we prefer to do so.

Finally we include a theorem due to P. Tauvel [2] which states that GK-dimension is exact on modules in a particular (but useful) situation.

10.2.5. Theorem (P. Tauvel). Let A be a finitely filtered k-algebra such that the associated graded ring G(A) is a finitely generated k-algebra which is left Noetherian. If $0 \to M' \to M \to M'' \to 0$ is an exact sequence of finitely generated left A-modules then GKdim (M) = max {GKdim (M'), GKdim (M")}.

Proof. Let μ be a frame for M and let $\{F_iM = A_i\mu, i \in \mathbb{Z}\}$ be the corresponding

standard filtration. Thus the associated graded module G(M) is finitely generated over G(A) hence it is also left Noetherian. The induced filtrations F_iM' = F_iM $\bigcap$ N and F_iM" = F_iM + M'/M' yield an exact sequence of graded left G(A)-modules: 0 → G(M') → G(M) → G(M") → 0.
Since G(A) is left Noetherian, all appearing modules are finitely generated; all filtrations are finite filtrations and, for all n ∈ $\mathbb{N}$: d_{FM}(n) = $\dim_k$ (F_nM) = $\dim_k$ (F_nM $\bigcap$ M') + $\dim_k$ (F_nM + M'/M') = $d_{FM'}$ (n)+$d_{FM''}$(n).
So we arrive at: $\overline{\lim}\ \frac{log(d_{FM}(n))}{log\ n} = \max\{\overline{\lim}\ \frac{log(d_{FM'}(n))}{log(n)}, \overline{\lim}\ \frac{log(d_{FM''}(n))}{log(n)}\}$ We apply Proposition 10.2.3. and obtain: GKdim (M) = GKdim (G(M)) = $\overline{\lim}\ \frac{log\ d_{FM}(n)}{log(n)}$ = $\max\{\overline{\lim}\ \frac{log(d_{FM'}(n))}{log(n)}, \overline{\lim}\ \frac{log(d_{FM''}(n))}{log(n)}\}$ = max {GKdim (G(M')),GKdim (G(M"))} = max {GKdim (M'),GKdim(M")}. □

Obvious applications of these filtered techniques will be found in the study of enveloping algebras of (solvable) Lie algebras, Weyl algebras an other rings with commutative associated graded ring. In the next section we give an idea about these applications.

§10.3 Applications to Special Classes of Rings.

In this section we will hint at some applications of the theory of GK-dimension but we do not intend to provide full detail on the classes of rings we treat here. A more complete treatment may be found in Krause-Lenagan [1] or the references given there.

10.3.1 Rings of Differential Operators and Weyl Algebras.

A differential polynomial ring A[,δ] where δ is a derivation of A may be viewed as a ring of differential operators on A. Also the Weyl algebra A_n(k) (where k is a field of characteristic zero) may be viewed as a ring of differential operators on the polynomial ring k[$X_1, ..., X_n$]. For a detailed account of the theory of rings of differential operators the reader may consult J-E. Björk's book [1].

All these rings have the property that their associated graded ring is commutative. Even more is true, they are *almost commutative* rings in the sense of M. Duflo i.e. {F_nA, n ∈ $\mathbb{Z}$ } is a filtration such that: F_0A = k; F_1A is finite dimensional over k and A is generated as a k-algebra by F_1A; G(A) is commutative. If A is almost commutative then G(A) is affine hence Noetherian, then A is also (left and right) Noetherian. The class of almost commutative algebras consists exactly of the epimorphic images of universal enveloping algebras of finite dimensional Lie algebras over k. By P. Tauvel's theorem (10.2.5.), GKdim is exact for modules over an almost commutative ring.

Let A be a filtered k-algebra which is almost commutative and let M be a left A-module which is finitely filtered and such that G(M) is a finitely generated G(A)-module,then for large n the function d_{FM}(n) = $\dim_k$ (F_nM) = $d_{G(M)}$(n) = $\dim_k$ (F_0M $\oplus$ F_1M/F_0M $\oplus$ ... $\oplus$ F_nM/F_{n-1}M) is a polynomial in n with rational coefficients, called the ***Hilbert-Samuel polynomial.*** In view of Proposition 10.2.3., the degree of the Hilbert-Samuel polynomial equals GKdim ($_AM$).

10.3.1.1. Observation. Let M be a finitely generated module over an affine commutative k-algebra, then GKdim (M) = Kdim (M).

Indeed, since both dimensions are exact on modules in this particular case it suffices to establish the claim for M = A. In view of Theorem 10.1.19. we have
GKdim (A) = clKdim (A) but for a commutative Noetherian ring we have established earlier that Kdim (A) = clKdim (A). □

We relate GKdim and Kdim for almost commutative algebras.

10.3.1.2. Lemma. Let A be almost commutative. Assume that every finitely generated M ∈ A-mod with Kdim (M) = m has GKdim (M) ≥ m+r for some r ∈ $\mathbb{N}$
For every finitely generated N ∈ A-mod we then have
GKdim (N) ≥ Kdim (N) + r.

Proof. By induction on Kdim N = α.
If α = m we have nothing to prove.
Consider an N with Kdim N = α > m and assume that the lemma holds for A-modules of Kdim β with m ≤ β < α. Consider an infinite descending chain of submodules of N, N = $N_0 \supset N_1 \supset N_2 \supset \ldots$, such that Kdim N_i/N_{i+1} = α - 1.
By the induction hypothesis: GKdim (N_i/N_{i+1}) ≥ α - 1 + r and
GKdim (N) > α - 1 + r i.e. GKdim (N) ≥ α + r. □

10.3.1.3. Corollary. If A is almost commutative and M is a finitely generated nonzero left A-module, then GKdim (M) $\geq$ Kdim (M).

Proof. Put m = r = 0 in the lemma. □

10.3.1.4. Corollary. If A is almost commutative then GKdim (A) $\geq$ Kdim (A) + s(A), where s(A) = min {GKdim (S), S a simple left A-module}.

Proof Put m = 0, r = s(A) in the lemma. □

10.3.1.5. Corollary. If A is almost commutative and simple then for every finitely generated A-module M, GKdim (M) $\geq$ Kdim (M) + 1 (we restrict to the case where $\dim_k$ (A) is not finite).

Proof. Ann_A (M) = 0 hence A acts faithfully on M by k-linear transformations Ψ_a: m $\to$ am, hence $\dim_k$ (M) = ∞ and therefore GKdim (M) $\geq$ 1. Apply the lemma with m = 0 and r = 1. □

10.3.1.6. Theorem (I.N. Bernstein). If M is a nonzero left module over the Weyl algebra A_n(k) then GKdim (M) $\geq$ n.

We do not go deeper into the theory of the Bernstein number, holonomic modules and other very interesting topics in the theory of rings of differential operators but we refer to J. Björk [1] for a detailed account of the algebraic theory of these rings (as well as for some more analytical items).

We conclude this section by a remark concerning the Weyl algebras, refering to J. Björk [1], Krause-Lenagan [1], for the details.

If D_n(k) is the quotient division algebra of the Weyl algebra A_n(k) then GKdim (D_n(k)) = ∞. In fact consider A_1(k) = k[x,y] with xy-yx = 1, then a result of Makar-Limanov yields that $(xy)^{-1}$ and $(xy)^{-1}(1-x)^{-1}$ generate a free subalgebra of D_1(k), hence GKdim (D_1(k)) = ∞; the result for D_n(k) follows from the embedding D_1(k) $\hookrightarrow$ D_n(k). For this reason it is necessary to introduce another invariant, the *Gelfand-Kirillov transcendence degree* of a k-algebra A defined by GKtdeg (A) = $\sup_V \inf_b$

GKdim (k[bV]) where V ranges over the sub-frames of A and b ranges over the regular elements of A.

I.M. Gelfand and A.A. Kirillov established that GKtdeg (D_n) = 2n. If A is an almost commutative k-algebra then for a $\in F_pA$, b $\in F_qA$ we define

$(ab)_{p+q} = a_p b_q$. If a,b,c,d $\in$ $F_{p_1}A$, $F_{p_2}A$, $F_{p_3}A$, $F_{p_4}A$ are such that $ab^{-1} = cd^{-1}$ in the quotient division algebra D of A then $a_{p_1} b_{p_2}^{-1} = c_{p_3} d_{p_4}^{-1}$ in the quotient division algebra Q(G(A)). It is equally clear that the filtration of A extends to a filtration of D by putting $F_nD = \{ab^{-1}$, fdeg(ab^{-1}) "=" fdeg(a) - fdeg(b) $\leq$ n$\}$ where fdeg is the filtration degree i.e. fdeg(x) is the integer d such that x $\in$ F_dA - $F_{d-1}(A)$ and this may be unambiguously(!) extended to D in the way indicated by the equality fdeg(ab^{-1}) = fdeg(a) - fdeg(b).

10.3.1.7. Observation. If W is a sub-frame of D, let σ(W) be the k-vector-space generated by the principal parts of the w $\in$ W (i.e. the w_p where p is such that w $\in$ F_pD - $F_{p-1}D$). The multiplicativity of the operation "taking principal parts" yields that $(\sigma(W))^n \subset \sigma(W^n)$. Moreover if x $\in$ D then $x_p\ \sigma$(W) $\subset$ σ(xW), where p is such that x $\in$ F_pD - $F_{p-1}D$.

We use these observations to prove:

10.3.1.8. Theorem (I.M Gelfand, A.A.Kirillov). Let D_n(k) be the n-th Weyl field, then GKtdeg(D_n(k)) = 2n.

Proof. Let V be a sub-frame of D_n(k). Then xV $\subset$ A_n(k) for some common denominator, x $\neq$ 0, say, for a set of basis elements of V. Thus GKdim (k[GV]) $\leq$ GKdim (A_n(k)) = 2n.

Conversely, put V = k.1 + kx_1 + ... + kx_n + ky_1 + ... + ky_n (where the x_i, y_i now denote the standard generators for A_n(k)). By the commutativity of G(A_n(k)) it follows that (using the observation) $(b^m)_p\ \sigma(V)^m \subset \sigma((bV)^m)$, for every b $\neq$ 0 in D_n(k) and for every m $\in \mathbb{N}$. Consequently we obtain:

$\dim_k ((k+bV)^m) \geq \dim_k ((bV)^m) = \dim_k (\sigma((bV)^m)) \geq \dim_k ((b^m)_p\ \sigma(V)^m) = \dim_k ((\sigma(V))^m)$, (where p = fdeg($b^m$)). But the reader will easily convince himself that $\sigma(V)$ is a frame for G(A_n(k)) and we know that GKdim (G(A_n(k))) = GKdim (k[$X_1, ..., X_{2n}$]) = 2n. It follows that GKdim (k[bV]) $\geq$ 2n, and the result follows. □

Since GKtdeg (A) = GKtdeg (B) when A ≅ B it follows that $D_n(k)$ cannot be isomorphic to $D_m(k)$ when n ≠ m and this is one of the main applications of the invariant GKtdeg.

10.3.2 Some Remarks on Enveloping Algebras of Lie Algebras. (Addendum)

Another class of rings where GK-dimension may have some applications consists of the enveloping algebras of Lie algebras. These algebras have been featured in ring theoretic considerations on many occasions, in particular they consitute a very nice class of almost commutative algebras (considering finite dimensional Lie algebras only). However, it is perhaps not very logical to separate a treatment of universal enveloping algebras from the theory of Lie algebras where it should be naturally embedded. Of course the latter can only be studied thoroughly from a non-associative point of view and therefore we cannot even try to give a self-contained account of some results and problems concerning enveloping algebras and GK-dimension, e.g.: how can GK-dimension be used in the description of the primitive ideals of solvable or semisimple Lie algebras?

For the theory of enveloping Lie algebras we refer to J. Dixmier [1].

Throughout we write $\mathcal{G}$ for a Lie algebra over the field k, $U(\mathcal{G})$ will be its (universal) enveloping algebra. We define a positive $\mathbb{Z}$ -filtration on $U(\mathcal{G})$ by putting $F_0(U(\mathcal{G})) = k$, $F_n(U(\mathcal{G})) = \mathcal{G}^n$, $n \in \mathbb{N}$. The following theorem (cf. J. Dixmier, [1] ch.2) is fundamental in the study of enveloping algebras:

10.3.2.1. Theorem (Poincaré, Birkhoff, Witt).

1. $G(U(\mathcal{G}))$ is the symmetric algebra on $\mathcal{G}$, denoted by $S(\mathcal{G})$.
2. If $\{x_1, ..., x_n, ...\}$ is an ordered k-basis of $\mathcal{G}$ then $\{x_{i(1)}.....x_{i(m)}$, with $m \in \mathbb{N}$, $i(1) \leq ... \leq i(m)\} \bigcup \{1\}$ is a k-basis for $U(\mathcal{G})$.

If $\mathcal{G}$ is finite dimensional then the filtration defined abov is a finite filtration and $G(U(\mathcal{G}))$ is finitely generated. By the theorem $G(U(\mathcal{G}))$ is isomorphic to $S(\mathcal{G})$ which is now nothing but the polynomial ring in $\dim_k (\mathcal{G})$ variables. Consequently if $\mathcal{G}$ is finite dimensional we may apply Proposition 10.2.3. and conclude that GKdim $(U(\mathcal{G})) = \dim_k(\mathcal{G})$.

10.3.2.2. Theorem. Let I be an ideal of $U(\mathcal{G})$ where $\mathcal{G}$ is a finite dimensional Lie algebra over k. Consider a left $U(\mathcal{G})/I$-module M, then GKdim (M) $\in \mathbb{N}$ and GKdim (M) $\leq \dim_k(\mathcal{G})$.

Proof. The filtration introduced on $U(\mathcal{G})$ induces a finite filtration on $U(\mathcal{G})/I$ and $G(U(\mathcal{G})/I) \cong S(\mathcal{G})/G(I)$ is a commutative affine, hence Noetherian, k-algebra.
Therefore GKdim $(S(\mathcal{G})/G(I)) \in \mathbb{N}$. If N is a finitely generated $U(\mathcal{G})/I$-module then we may select a frame η for N and define the standard filtration $V^n\eta = F_nN$ with respect to some frame V for $U(\mathcal{G})/I$.
Again from Proposition 10.2.3. we have that GKdim (G(M)) = GKdim (M) and G(M) is finitely generated over the affine commutative k-algebra $S(\mathcal{G})/G(I)$,thus GKdim (G(M)) $\leq \dim_k(\mathcal{G})$ follows. □

The relation between almost commutative k-algebras and algebras of the form $U(\mathcal{G})$ is fully expressed in the following result:

10.3.2.3. Theorem. The class of almost commutative k-algebras consist exactly of the epimorphic imges of (universal) enveloping algebras of finite dimensional Lie algebras over k.

Proof. Assume that A is almost commutative. If a,b $\in A_1$ then ab-ba $\in A_1$ because of the commutativity of G(A), hence we may use [a,b] = ab-ba to define a Lie algebra structure on A_1 and it is a finite dimensional one over k. Write $\mathcal{A}_1$ for this Lie algebra. By the universal property of the enveloping algebra, the embedding $\mathcal{A}_1 \to$ A extends to a unique k-algebra morphism $U(\mathcal{A}_1) \to$ A which must also be surjective because $\mathcal{A}_1$ generates A as a k-algebra.
Conversely, assume that $\mathcal{G}$ is a finite dimensional Lie algebra over k and let A be an epimorphic image of $U(\mathcal{G})$ given by π: $U(\mathcal{G}) \to$ A, a k-algebra epimorphism. Putting $F_nA = \pi(\mathcal{G}_n)$ yields a discrete finite filtration of A such that k+F_1A is a frame for A (since k+$\mathcal{G}$ is a frame for $U(\mathcal{G})$). The morphism π induces a surjective morphism $G(\pi)$: $G(U(\mathcal{G})) \to$ G(A) (this is easy verified). By the Poincaré, Birkhoff, Witt theorem $G(U(\mathcal{G}))$ is a commutative polynomial ring, so we have obtained all the properties necessary to include that A is almost commutative. □

Using that Kdim $(U(\mathcal{G})) \geq$ clKdim $(U(\mathcal{G}))$ together with Corollary 10.3.1.3.

(plus Lie's theorem by passing to $\mathcal{G} \otimes_k \bar{k}$, for an algebraic closure $\bar{k}$ of k, where g is a solvable Lie algebra) one easily derives the following result:

10.3.2.4. Proposition. If $\mathcal{G}$ is a solvable Lie algebra of $\dim_k(\mathcal{G}) = n$ then GKdim $(U(\mathcal{G}))$ = Kdim $(U(\mathcal{G}))$ = clKdim $(U(\mathcal{G})) = n$.

The forgoing proposition may also be derived in a rather staightforward way from the fact that $U(\mathcal{G})$, for a solvable $\mathcal{G}$, is an iterated (n-times if $\dim_k(\mathcal{G}) = n$) Ore extension, here one may derive that clKdim $(U(\mathcal{G})) \geq n$ without making use of Lie's theorem, the other equalities follow from the quoted facts plus GKdim $(U(\mathcal{G}))$ = n, what has been observed after Theorem 10.3.2.1.

In order to give an idea of how the GK-dimension may be used further in the theory of enveloping algebras we list a few important results without proofs. The interested reader may consult Krause-Lenagan [1] Ch.9.

10.3.2.5. Theorem. Let $\mathcal{G}$ be a finite dimensional solvable Lie algebra over an algebraically closed field k with char(k) = 0 and let P be a prime ideal of $U(\mathcal{G})$, then: ht(P) + GKdim $(U(\mathcal{G})/P$ = GKdim $(U(\mathcal{G})) = \dim_k(\mathcal{G})$.

For a $\mathcal{G}$ as in the theorem every nonzero ideal of $U(\mathcal{G})/I$ for any ideal I contains a normalizing element (a consequence of Lie's theorem). This plays an important part in the proof of:

10.3.2.6. Theorem. Let $\mathcal{G}$ be as above and let $P \subsetneq Q$ be prime ideals of $U(\mathcal{G})$ such that $ht(Q/P) = 1$ then: 1 + GKdim $(U(\mathcal{G})/Q)$ = GKdim $(U(\mathcal{G})/P)$.

Recall that a ring R is catenary if for any pair of prime ideals $P \subsetneq Q$ of R all maximal strictly ascending chains of prime ideals between P and Q have the same length.

10.3.2.7. Corollary. For a Lie algebra $\mathcal{G}$ as in the foregoing theorem, $U(\mathcal{G})$ is a catenary ring.

10.3.2.8. Theorem (O. Gabber). Let $\mathcal{G}$ be a finite dimensional algebraic Lie algebra over an algebraically closed field k with char(k) = 0, let $M \in U(\mathcal{G})$-mod be finitely generated, then GKdim $(U(\mathcal{G})/\mathrm{Ann}_{U(\mathcal{G})}(M)) \leq 2$ GKdim (M).

Finally let us mention a result of K. Brown, S.P. Smith, [1] concerned with the symmetry of Kdim, but the proof relies heavily on the symmetry of GK-dimension and suitable relations between Kdim and GKdim.

10.3.2.9. Theorem (K. Brown, S.P. Smith). Let $\mathcal{G}$ be a finite dimensional algebraic solvable Lie algebra over an algebraically closed field of characteristic zero. Suppose that M is a $U(\mathcal{G})$-bimodule which is finitely generated both as a left and as a right $U(\mathcal{G})$-module, then Kdim $({}_{U(\mathcal{G})}M)$ = Kdim $(M_{U(\mathcal{G})})$.

10.3.3 P.I. Algebras (Addendum).

Just like in Section 4.2.8. we refer to C. Procesi [1] for an extensive treatment of the theory of rings satisfying polynomial identities.

For a (semi)prime P.I. ring R the total ring of fractions, Q say, may be obtained by a central localization. If the P.I. ring R is prime then Q is a central simple algebra over Z(Q) = K which is the field of fractions of Z(R) = C. If A is a prime P.I. algebra over k we define: tdeg_k (A) = tdeg_k (K).

10.3.3.1. Theorem. If A is a prime P.I. algebra over k then
GKdim (A) = tdeg_k (A).

Proof. Since Q is obtained by localizing A at Z(A)-{0} it follows from 10.1.16. that GKdim (A) = GKdim (Q). Since Q is finite dimensional over K, GKdim (Q) = GKdim (K) = tdeg_k (K), all of these equalities have been established in Section 10.1. □

An *affine P.I. algebra* will be a k-algebra which is finitely generated as a k-algebra and such that it is a P.I. ring. The theorem above implies that

GKdim (A) $\in \mathbb{N}$ when A is a prime affine P.I. algebra. (indeed, by Amitsur's theorem A may be embedded in M_n (C) for some commutative ring C which may be taken to be an affine k-algebra, hence GKdim (A) $< \infty$). In fact a theorem of Berele, [1], learns that the mentioned result remains valid if we drop the "prime" condition in the statement (cf. Corollary 10.7. in Krause-Lenagan [1] for a short proof using A. Braun's theorem on the nilpotency of the Jacobson radical in an affine P.I. algebra). Since prime homomorphic images of any P.I. ring are left and right Goldie rings, Corollary 10.1.15. entails that clKdim (A) $\leq$ GKdim (A) for any P.I. algebra A. In order to establish the converse inequality we first observe:

10.3.3.2. Lemma. If A is a simple affine P.I. algebra over k then GKdim (A) = 0.

Proof. Z(A) is a field, K say; and A is a finite K-module since it is a central simple algebra. Therefore K is an affine k-algebra (look at structure constants for some K-basis of A and at the coefficients in K appearing in the expression of a finite set of k-algebra generators for A, add all these to k and observe that A is finitely generated over the Noetherian(!) ring one obtains). Since K is a field and an affine k-algebra, a classical result of Zariski's yields that K is algebraic (and even finite) over k hence the statement: GKdim (A) = 0. □

10.3.3.3. Theorem (M-P. Malliavin [1]). If A is an affine prime P.I. algebra over k then GKdim (A) = clKdim (A) = tdeg_k (A).

Proof. We only have to establish that clKdim (A) $\geq$ GKdim (A), the other inequalities have been derived before. We may use induction on GKdim (A), the case GKdim (A) = 0 being clear (if the domain Z(A) is algebraic over k then Z(A) is a field and A is simple).
Assume GKdim (A) = d > 0. There is a t $\in$ K = Z(Q(A)) which is transcendental over k and ct $\in$ Z(A) for some c $\neq$ 0 in Z(A). Passing from A to B = A[c^{-1}] preserves all the imposed conditions and as A $\rightarrow$ B is in fact a central localization we have: tdeg_k B = GKdim_k (B) = GKdim_k (A) = tdeg_k (A), and also clKdim (B) $\leq$ clKdim (A) (cf. Chapter 5.3. a.o.).
We have to show that clKdim (B) $\geq$ GKdim (B). Put L = k(t), (t $\in$ B(!)), S = k[t] - {0} and A' = S^{-1}B. Obviously A' is a prime P.I. ring and it is a finitely generated L-algebra. Now, GKdim_L (A') = tdeg_L (A') = d - 1 and thus

clKdim (A') $\geq$ d - 1 because of the induction hypothesis. Any strictly descending chain of n prime ideals of A' intersects to a strictly descending chain of n prime ideals of B, say A' $\underset{\neq}{\supset} P_0 \underset{\neq}{\supset} \cdots \underset{\neq}{\supset} P_{n-1} \supset 0$, B $\underset{\neq}{\supset} P_0 \cap B \underset{\neq}{\supset} \cdots \underset{\neq}{\supset} P_{n-1} \cap B \supset 0$.
If M is a maximal ideal of B then the image of t in B/M is algebraic over k in view of the lemma, thus k[t] $\cap$ M $\neq 0$ and A'M = A'.
But A'($P_0 \cap$ B) = $P_0 \neq$ A', thus $P_0 \cap$ B cannot be maximal in B and therefore clKdim (B) $\geq$ d (because clKdim (A') $\geq$ d - 1 and we have to go up at least 1 in view of the preceding observation). □

10.3.3.4. Theorem (M. Lorenz, L.Small [1]). Let A be a Noetherian P.I. algebra with nilpotent radical N, then GKdim (A) = GKdim (A/N) = max {GKdim (A/P), P a minimal prime of A}.

10.3.3.5. Corollary. Let A be a Noetherian P.I.-algebra such that GKdim (A) $< \infty$ then GKdim (A) $\in \mathbb{N}$. If A is moreover affine then GKdim (A) = clKdim (A).

Proof. By the theorem, GKdim (A) = GKdim (A/P) for some minimal prime ideal P of A; so the first statement follows from Theorem 10.3.3.1. If A is affine then so is A/P and hence we may apply Theorem 10.3.3.3. to obtain:
GKdim (A/P) = clKdim (A/P). Now clKdim (A/P) $\leq$ clKdim (A) yields GKdim (A) $\leq$ clKdim (A) while on the other hand clKdim (A) $\leq$ GKdim (A) folows from Corollary 10.1.15. □

Finally we show how to derive from Theorem 10.3.3.4. that GKdim is exact for finitely generated modules over a Noetherian P.I.-algebra.

10.3.3.6. Lemma. Let A be a Noetherian P.I.-algebra over k and let M be a finitely generated left A-module, then GKdim (M) = GKdim (A/Ann(M)).

Proof. Since M is an A, Ann(M)-module, GKdim (M) $\leq$ GKdim (A/Ann(M)). Since a Noetherian P.I. ring is fully (left and right) bounded it satisfies condition H (cf. Section 6.3., 6.4.), hence Ann(M) is an intersection

$\mathrm{Ann}(m_1) \bigcap ... \bigcap \mathrm{Ann}\,(m_q)$ for some $m_i \in M$, $i = 1,...,q$.
We obtain an embedding $A/\mathrm{Ann}(M) \hookrightarrow Am_1 \bigoplus ... \bigoplus Am_q \hookrightarrow M \bigoplus ... \bigoplus M$ and hence GKdim (A/Ann(M)) $\leq$ GKdim (M). □

10.3.3.7. Lemma. If I and J are ideals of a Noetherian P.I.-algebra over k such that IJ = 0 then we have: GKdim (A) = max{GKdim (A/I),GKdim (A/J)}.

Proof. By Theorem 10.3.3.4., GKdim (A) = GKdim (A/P) for some minimal prime ideal P of A. But either I $\subset$ P or J $\subset$ P, say I $\subset$ P. In this case: max {GKdim (A/I),GKdim(A/J)} $\geq$ GKdim (A/I) $\geq$ GKdim (A/P) = GKdim (A).
The reverse inequality is evident. □

10.3.3.8. Proposition (T. Lenagan). Let A be a Noetherian P.I.-algebra over k and consider an exact sequence, $0 \to M' \to M \to M'' \to 0$, of finitely generated left A-modules, then GKdim (M) = max {GKdim (M'),GKdim(M")}.

Proof. Pass to A/Ann(M) and apply Lemma 10.3.3.6., i.e. we may assume that M is a faithful A-module and that GKdim (M) = GKdim (A). Now (Ann(M').Ann(M"))M = 0, thus Ann(M')Ann(M") = 0. Applying both lemmas above we obtain: GKdim (M) = GKdim (A) = max {GKdim (A/Ann(M")), GKdim (A/Ann(M'))} = max {GKdim (M'), GKdim (M")}.

The effect of the foregoing properties is that GKdim for Noetherian P.I.-algebras shares several formal properties with Kdim. Note that Noetherian affine P.I.-algebras are very close to being finite modules over the centre (it depends only on the invertibility of the pi-degree in k) so in this case the relation with clKdim is a very expected one.
In conclusion we mention:

10.3.3.9. Proposition. Let A be a Noetherian P.I.-algebra. Write s(A) for the smallest GKdim (S) where S is any simple left A-module, then: GKdim (M) $\geq$ Kdim (M) + s(A), for any finitely generated M $\in$ A-mod. □

10.3.3.10. Proposition (L. Small,T. Stafford, R. Warfield [1]). Affine algebras of Gelfand-Kirillov dimension one are P.I.-algebras.

§10.4 Exercises.

(193) Let R = k[X_1, X_2] and let σ be the automorphism of R determined by $\sigma(X_1)$ = $X_2 + X_1^2$, $\sigma(X_2) = X_1$. Put A = R[X,σ]. Prove that GKdim (A) = ∞.
Similar for A = k[X][Y,σ] where σ: X $\rightarrow$ X^2 is an injective endomorphism of k[X].

(194) Put R = k[$X_1, ..., X_n$] and consider the automorphism σ of R determined by $\sigma(X_1) = \alpha_1 X_1 + p_1(X_2, ..., X_n)$, $\sigma(X_2) = \alpha_2 X_2 + p_2(X_3, ..., X_n)$, ... , $\sigma(X_n)$ = $\alpha(X_n) + \lambda$ where $\alpha_1,...,\alpha_n, \lambda \in$ k. Proof that GKdim (A) = 1 + GKdim (R), where A = R[T,σ].

(195) Let R be a k-algebra and σ a k-algebra automorphism of R which is locally algebraic (i.e. every a $\in$ A is contained in an affine k-subalgebra of A which is σ-invariant). Prove that GKdim (A[X,σ]) $\leq$ 1 + GKdim (A).

(196) Let A $\subset$ B be the k-algebras and I an ideal of A which is invertible in B, say $A_1 = I^{-1} \supset$ A. Consider the subring of $A_1[X, X^1]$ defined as follows: $\sum_{n\in\mathbb{Z}} I^n X^n$ = $\check{A}$(I) (generalized Rees ring). Prove that GKdim ($\check{A}$(I)) = 1 + GKdim (A).

(197) Let A = $\bigoplus_{n\in\mathbb{Z}} A_n$ be a strongly graded ring (i.e. $A_n A_m = A_{n+m}$ for all n,m $\in \mathbb{Z}$. Then ϕ: $\mathbb{Z} \rightarrow$ Pic (A_0), n $\mapsto$ [A_n] is a group morphism, so Im(Φ) is either finite or infinite cyclic. If Im Φ is finite prove then that GKdim (A) = GKdim (A_0) + 1.
If Im Φ is infinite investigate whether GKdim (A) = ∞ or else GKdim (A) = GKdim (A_0) + 1.

(198) Let A and B be k-algebras such that GKdim (A) $\geq$ GKdim (B) > 2. Prove that: GKdim (A) + 2 $\leq$ GKdim (A $\otimes_k$ B) $\leq$ GKdim (A) + GKdim (B). Hint: use the following result (in itself not so easy to prove): let W be a set of words in a free semigroup X = < $x_1, ..., x_m$ > such that each subword of a word in W is also a word of W. If for some d $\in \mathbb{N}$, W contains at most d words of length d then W contains at most d^3 of length h for any h $\geq$ d. (G. Bergman).
Use this to show that for a sub-frame V of B, d_n(V) $\geq$ $\frac{1}{2}$ (n+1)(n+2) and calculate GKdim (A $\otimes_k$ B).

(199) In the Weyl algebra $A_1(k)$, $\delta = \{1,x,x^2,...\}$ is a non-central Ore set $(A_1(k) = k[x,y]/(xy-yx = 1))$. Show that: GKdim $(S^{-1}A_1(k)) =$ (GKdim $A_1(k)) = 2$.

(200) Let A and B be k-algebras and ${}_AM_B$ a bimodule then GKdim $\begin{pmatrix} A & M \\ 0 & B \end{pmatrix} \leq$ GKdim (A) + GKdim (B).
Hint: Enlarg a sub-frame V of $\begin{pmatrix} A & M \\ 0 & B \end{pmatrix}$ to one, V^* say, generated by $\begin{bmatrix} U^* & 0 \\ 0 & 0 \end{bmatrix}$, $\begin{bmatrix} 0 & X^* \\ 0 & 0 \end{bmatrix}$, $\begin{bmatrix} 0 & 0 \\ 0 & W^* \end{bmatrix}$ where U^*,W^* are sub-frames of A,B resp. and X^* is a sub-frame of M. Calculate that: $V^n \subset < U^n,\ W^n,\ U^n \times W^n >$ and $\dim_k (U^n \times W^n) = \dim_k (X(U^n \otimes_k (W^{0p})^n) \leq \dim_k (X) \dim_k (U^n) \dim_k (W^n)$, then continue the proof.

(201) If I,J are ideals of the k-algebra A such that IJ = 0 then GKdim (A) $\leq$ GKdim (A/I) + GKdim (A/J).
Hint: A result of J. Lewin yields that A may be embedded in $\begin{pmatrix} A/I & M \\ 0 & A/J \end{pmatrix}$ for some bimodule ${}_{A/I}M_{A/J}$, ...

(202) Let N be a nilpotent ideal, of nilpotency index m, in the k-algebra A then GKdim (A) $\leq$ m. GKdim (A/N). (use 201.,200.)

(203) Put A_m = k<X,Y>$/(y)^m$. Prove that GKdim (A_m) = m. Hint: use 202. to obtain GKdim $(A_m) \leq$ m. For the reverse inequality put V = k + kx + ky, x = X mod $(Y)^m$, y = Y mod $(Y)^m$. There are $\binom{n}{m-1}$ monomials of degree n in x, degree m - 1 in y; these monomials are all in V^n and they are linearly independent modulo V^{n-1}. Since $\binom{n}{m-1}$ is a polynomial of degree m - 1 in n we can derive from this that GKdim $(A_m) \geq$ m. Check all these claims and write a detailed proof.

(204) A module M over a k-algebra is *homogeneous* if for all nonzero submodules N of M we have that GKdim M = GKdim N. Prove that a prime Goldie k-algebra A is homogeneous as a left A-module.
Hint: show that GKdim (U) = GKdim (A) for a uniform left ideal U.

(205) Let M be a homogeneous A-module such that GKdim (M/N) < GKdim (M) for all submodules N, then N is essential in M.
Hint: If X $\neq$ 0 is such that N $\bigcap$ X = 0 then X embeds in M/N.

(205) (W. Borho, H. Kraft) Consider an infinite ascending chain of finite field extensions $k = k_0 \subsetneq k_1 \ldots \subsetneq k_n \subsetneq \ldots$, $K = \bigcup_i k_i$. Let A be the subring $\bigoplus_{i=0}^{\infty} k_i x^i$ of K[x]. Prove that A is a homogeneous subalgebra of K[x] with GKdim (A) = 1. Prove also that $\overline{\lim} \frac{\log(d_A(n))}{\log(n)} = \infty$ (here A is not finitely generated and that causes all these troubles).

Bibliographical Comments to Chapter 10.

In this chapter we aimed to present a coherent survey of the general theory of GK-dimension. The basic reference is G. Krause, T Lenagan's book [1] enriched with some results of J. Krempa. J. Okninski [1]. The application of GKtdeg (-) to the Weyl fields stems from the papers [1],[2] by I.M. Gelfand, A.A. Kirillov and this is also the origine of the whole theory. For some work on the GKdimension of differential polynomial rings we refer to M. Lorenz [1]. Another general reference is the paper by W. Borho, H. Kraft [1]. For the properties of GK-dimension of modules the reader may also consult: I. Bernstein [1]; A. Joseph, L. Small [1]. Some examples and counter-examples are due to G. Bergman cf. Example 10.1.21. The connections of GK-dimensions of algebras A and B via some bimodule M depends on results of W. Borho [1], T. Lenagan [8]. Exactness of GK-dimension for finitely generated modules over finitely filtered algebras A with affine Noetherian associated graded ring stems from P. Tauvel [1].

The role of the GK-dimension in the study of rings of differential operators is important but this is somewhat ambiguous because one is really dealing with the Bernstein dimension and the Bernstein number. The reader may read J.-E. Björk's book [1] and convince himself that the GK-dimension is only present in the form of the other invariants mentioned above.

We have presented the material of Section 10.3.1. for almost commutative algebras, following the approach of G. Krause, T. Lenagan [1].

Section 10.3.2. must be considered as an addendum. Here we point out some applications of GK-dimension in the theory of enveloping algebras of (solvable) Lie algebras. This is a very extensive subject where there is now a growing interest in applying purely ring theoretical methods. A self-contained treatment would lead us much to far but we thought it might be useful for the reader to see some of the applications of GK-dimension in this field, even if full detail cannot be made available in this book.

The proofs we did include require only knowledge of the definition of an enveloping algebra and two very basic theorems concerning those i.e. the Poincaré-Birkhoff-Witt theorem and Lie's theorem. We included some results due to O. Gabber [1]; K. Brown, S.P. Smith [1].

Section 10.3.3. may also be considered as an addendum, although P.I. rings have been considered before in Section 4.2.8. One of the main results here is M.-P. Malliavin's theorem (10.3.3.3.) and further we included some results of M. Lorenz,

L. Small [1] and T. Lenagan [8] concerning Noetherian P.I. algebras. The latter class of rings is particularly well-behaved where GK-dimension is concerned and there exist here several similarities between the theory of GK-dimension and the theory of Kdim.

REFERENCES.

Albu T., [1] Sur la dimension de Gabriel des modules, *Seminar F. Kasch*, Bericht nr. 21 (1974).

Albu T., C. Năstăsescu, [1] Relative finiteness conditions in module theory, *Monographs on Pure and Appl. Math.*, vol. **84** (1984) Marcel Dekker, New-York.

[2] Décompositions primaires dans les catégoris de Grothendieck commutatives I, *J. Reine Angew. Math*, **280** (1976), 172–190; and II, *J. Reine Angew. Math.*, **282**, (1976), 172–185.

Auslander M., [1] On the dimension of modules and algebras III, *Nagaya Math. J.*, **9** (1955), 67–77.

Anderson F, K. Fuller, [1] Rings and Categories of Modules (1974), Springer-Verlag, New-York–Heidelberg–Berlin.

Bass H., [1] Algebraic K-theory, W.A. Benjamin Inc. (1968).

[2] Descending chains and the Krull ordinal of commutative Noetherian rings, *J. Pure Appl. Algebra*, **1** (1971), 347–366.

Berele A., [1] Homogeneous Polynomial Identities, *Israël J. of Math.*, **42** (1982), 258–272.

Bernstein I.N., [1] Modules over a ring of differential operators, Study of the fundamental solutions of equations with constant coefficients, *Fund. Anal. Appl.*, **5** (1971), 89–101.

[2] On the dimension of modules and algebras III, Direct limits, *Nagoya Math. J.*, **13** (1958), 83–84.

Björk J.E., [1] Rings of Differential Operators, vol **21** (1979), *Math Library*, North-Holland.

Blair W.D., [1] Right Noetherian rings integral over their centres, *J. Algebra*, **27** (1973), 187–198.

Bergman G., [1] On Jacobson Radicals of Graded Rings (preprint)
[2] A note on growth functions of algebras and semigroups, *mimeographed notes*, University of California, Berkeley (1978).

Bergman G., I.M. Isaac, [1] Rings with fixed-point free group actions, *Proc London Math. Soc.*, **27** (1973), 69–87.

Birkoff G., [1] Lattice Theory, *Amer. Math Soc.*.

Bitt D.J., [1] Normalizing Extensions II, Ring theory Antwerp 1980, *Lecture Notes in Math.*, Springer-Verlag, **825**.

Bitt D.J., C. Robson, [1] Normalizing Extensions I, Ring Theory Antwerp 1980, *Lecture Notes in Math.*, Springer-Verlag, **825**.

Boratynsky M., [1] A change of rings theorem and the Artin Rees property, *Proc Amer. Math. Soc.*, **53** (1975), 307–320.

Borho W., [1] Invariant dimension and restricted extension of Noetherain rings, *Lecture Notes in Math.*, vol **924**, New-York, Springer-Verlag, (1982)

Borho W., H. Kraft, [1] Uber die Gelfand-Kirillov Dimension, *Math. Ann.*, **220**, (1976), 1–24.

Bourbaki N., [1] Théorie des ensembles 2-e ed., Hermann, Paris, (1983)
[2] Algèbre chap. 2–8, Hermann, Paris, (1982)
[3] Algèbre commutative, chap 1–7, Hermann, Paris, (1964).

Brown K.A., C.R. Hajarnavis, [1] Homologically homogeneous rings, *Trans Amer. Math. Soc.*, vol **281** (1984), 197–208.

Brown K.A., R.B. Warfield Jr., [1] Krull and global dimension of fully bounded Noetherian rings, *Proc. Amer. Math. Soc.*, **92** (1984), 169–174.

Brown K.A., C.R. Hajarnavis, A.B. MacEacharn, [1] Noetherian rings with finite global dimension, *Proc. London Math. Soc.*, vol **XLN** (1982), 349–371.
[2] Rings of finite global dimension integral over their centres, *Comm. in Algebra*, **11 (1)** (1983), 67–93.

Brown K.A., S.P. Smith, [1] Bimodules over Solvable Lie Algebra, preprint Warwick.

Bueso J.L., F. Van Oystaeyen, A. Verschoren, [1] Relating Homological Dimensions to Relative Homological Dimension, *Comm. in Algebra* (to appear).

Cartan H., S. Eilenberg, [1] Homological Dimension, Princeton Univ. Press, Princeton, N.J. (1956).

Chamarie M., A. Hudry, [1] Anneaux noetheriens a droite entiers sur un sous anneau de leur centre, *Comm. in Algebra*, **6** (1978), 203–222.

Chase S., [1] On the Homological Dimension of Algebras of Differential Operators, *Comm. in Algebra*, **5** (1974), 351–363.

Cauchon G., [1] Les T-anneaux, la condition (H) de Gabriel et ses consequences, *Comm. in Algebra*, **4** (1976), 11–50.

Chatters A.W., [1] Localization on P.I. Rings, *J. London Math. Soc.*, **2** (1970), 763–768.

Cohn P.M., [1] Algebra II.

Chatters A.W., A.W. Goldie, C.R. Hajarnavis, T.H. Lenagan, [1] Reduced Rank in Noetherian Rings *J. Algebra*, **61** (1979), 582–589.

Cohen M., S. Montgomery, [1] Group graded rings, smash products and group actions, *Trans.Amer. Math. Soc.*, **282** (1985), 237–258.

Dade E., [1] Group-Graded Rings and Modules, *Math.Z.*, **174** (1980), 241–262.
[2] Compounding Clifford's Theory, *Ann. of Math.*, **2** (1976), 236–290.

Dixmier J., [1] Enveloping Algebras, *North-Holand Math. Library*, vol **14** (1977).

Eakin P.M.Jr., [1] The converse to a well-known theorem on Noetherian rings, *Math. Ann*, **177** (1968), 278–282.

Faith C., [1] Algebra, Rings, Modules and Categories I, Springer-Verlag, New-York (1973)
[2] Algebra II – Ring Theory, Springer-Verlag, Bew York (1976)
[3] Injective modules and injective quotient rings, *Lecture Notes in pure and appl. Math.*, Marcel Dekker, New York, vol **72** (1982).

Fisher J., [1] Chain Conditions for Modular Lattices with Finite Group Actions, *Canad. J. math.*, **31** (1979), 558–564.

Fisher J., C. Lanski, J.K. Park [1] Gabriel Dimension of finte normalizing extension, *Comm. in Algebra*, **8,16** (1980), 1493–1504.

Fisher J., S. Montgomery, [1] Semiprime skew group rings, *J. Algebra*, **52** (1978), 241–247.

Fisher J., J. Osterburg, [1] Semiprime ideals in rings with finite group actions, *J. Algebra*, **50** (1978), 488–502.

Fields K.L., [1] On the global dimension of skew polynomial rings, *J. Algebra*, **13** (1969), 1–4.

[2] On the global dimension of residual rings, *Pacific J. Math.*, **32,2** (1970), 345–349

Formanek E., A. Jategaonkar, [1] Subrings of Noetherian rings, *Proc. Amer. Math. Soc.*, **46** (1974), 181–186.

Formanek E., [1] Noetherian P.I.rings, *Comm. Algebra*, **1**, (1974), 79–86.

Gabber O., [1] Equidimensionalité de la varieté characteristique Exposé rédigé par T. Levasseur, Paris VI, (1982).

Gabriel P., [1] Des Catégories abéliens, *Bull. de Math. France*, **90** (1962), 323–448.

Gelfand J.M., A.A. Kirillov, [1] Sur les corps liés aux algèbres enveloppantes des algèbres de Lie, *Publ. Math. I.H.E.S.*, **31** (1966), 5–19.

[2] Fields associated with enveloping algebras of Lie algebras, *Doklady*, **167**, 407–409.

Gilmer R., [1] Commutative Semigroup rings, *Chicago Lect. in Math.*, Chicago (1984).

Golan J.S., [1] Localization of noncommutative rings, Marcel Dekker, New York (1975).

[2] A Krull-like dimension for noncommutative rings, *Israel Journal*, **19** (1974), 297–304.

[3] Decomposition and dimension in module categories, *Lecture Notes in Math.*, vol **33**, Marcel Dekker (1977).

Goldie A.W., [1] The structure of prime rings under ascending chain conditions, *Proc. Lond. Math. Soc.*, **8** (1958), 589–608.

[2] Semi-prime rings with maximum condition, *Proc. London Math. Soc.*, **3** (1960), 201–220.

[3] Localization on non-commutative noetherian rings, *J. Algebra*, **5** (1967), 89–105.

[4] A survey of progress in non-commutative noetherian rings, *ring theory 1978*, Antwerp, *Lecture Notes in Math.*, **51**, Marcel Dekker.

Goldie A.W., L. Small, [1] A study in Krull Dimension, *J. Algebra*, **25** (1973), 152–157.

Gomez Pardo L., C. Nâstâsescu, Spliting property for graded rings, to appear in *Comm. in Algebra.*

Goodearl K.R., L.W. Small, [1] Krull versus global dimension in noetherian P.I. rings, *Proc Amer. Math. Soc.*, **92,2** (1984), 175–178.

Goodearl K.R., R. Warfield, [1] Krull Dimension of Differential Operator Rings, *Proc. London Math. Soc.*, **3,48** (1982), 49–70.

Goodearl K.R., [1] Global Dimension of Differential Operator Rings II, *Trans Amer. Math. Soc.*, **209** (1975), 65–85.

Gordon R., J.C. Robson [1] Krull Dimension, *Amer. Math. Soc. Memoires*, **133** (1973).

[2] The Gabriel Dimension of a Module, *Journal of Algebra*, **29** (1974), 459–473.

[3] Semiprime Rings with Krull Dimension are Goldie, *J. Algebra*, **25** (1973), 519–521.

Gordon R., [1] Gabriel and Krull Dimension, *Proc. of Univ. Oklahoma, Ring Theory Conf.*

[2] Artinian Quotient Rings of F.B.N. Rings, *J. Algebra*, **35** (1975), 304–307.

[3] Primary Decomposition on Right Noetherian Rings, *Comm. in Algebra*, **39** (1976), 100–130.

[4] Some Aspects of Non-commutative Noetherian Rings, Noncommutative Ring Theory, Kent-State 1975, *Lecture Notes in Math.*, **545**, Springer-Verlag, New-York.

Gordon R., L. Small, [1] Affine P.I. Rings have Gabriel Dimension, *Comm. in Algebra*, **12(11)** (1984), 1291–1300.

Greszczuk P., [1] On G-systems and G-graded rings, *Proc. Amer. Math. Soc.*, **95,3** (1985), 348–352.

Greszczuk P., E. Puczylowski, [1] On Goldie and dual Goldie dimension, *J. Pure and Appl. Algebra*, **31**, nr 1-3 (1984), 47–54.

[2] On infinite Goldie dimension of modular lattices and modules, *J. Pure Appl. Algebra.*

[3] Goldie dimension and chain condition for modular lattices with finite group actions, to appear in *Canadian Math. Bull.*.

Gulliksen T., [1] The Krull ordinal and Noetherian localization of large Polynomial Rings.

Hart R., [1] Krull dimension and Global dimension of simple Ore Extension, *Math.Z.*, **121** (1971) 341–345.

Herstein I.N., [1] A counter example on noetherian Rings, *Proc. Nat. Act.Sci.*, **54** (1965), 1036–1037.

Hodges T.J., [1] The Krull Dimension of Skew Laurent Extensions of Commutative Noetherian Rings, *Comm. in Algebra*, **12,11** (1984), 1301–1311.

Hodges T.J., J.C. Connell, [1] On Ore and skew Laurent extension of Noetherian rings, *J. Algebra*, **73** (1981), 56–64.

Ion D.I., C. Nâstâsescu, [1] Anneaux gradués semi-simples, *Revue Roum. Math. Pure et Appl.*, **4** (1978), 573–588.

Ion D.I., N. Radu, [1] Algebr a, *Edit. Didactica in Pedagogic* a, Bucharest (1980).

Jategaonkar A.V., [1] Principal Ideal Theorem for Noetherian P.I.Rings, *J. of Algebra*, **35, 1-3** (1975), 17–22.

[2] Relative Krull Dimension and Prime Ideals in Right Noetherian Rings, *Comm. in Algebra*, **4** (1974), 429–468.

[3] Relative Krull Dimension and Prime Ideals in Right Noetherian Rings: an Addendum, *Comm. in Algebra*, **10(4)** (1982), 361–366.

[4] Jacobson's Conjecture and Modules over Fully Bounded Noetherian Rings, *J. Algebra*, **30** (1974), 103–121.

[5] Left Principal Ideal Rings, *Lect. Notes in Math.*, **123** (1970), Springer-Verlag, New-York.

Jacobson N., [1] Structure of Rings, *Colloquium Publications*, **37**, Am. Math. Soc. Providence (1974).

Joseph A., [1] Dimension en algèbre non-commutative, *Mimeographed notes*, Univ. Paris VI (1980).

Joseph A., L. Small, [1] An additivity principle for Goldie rank, *Israel J. of Math.*, **31** (1978), 89–101.

Kaplansky I., [1] Commutative Rings, Allyn and Bacon, Boston (1970).

[2] Fields and Rings, Chicago and London (1969).

Kharchenko V., [1] Galois extensions and quotient rings, *Alg. and Logic*, **13** (1975), 265–281.

Krempa J., J. Okninschi [1] Gelfand-Kirillov dimension of a tensor product (preprint).

Krause G., [1] Descending chains of submodules and the Krull dimension of noetherian modules, *J. Pure and Appl. Algebra*, **3** (1973), 385–397.

[2] On fully left bounded left noetherian rings, *J. Algebra*, **23** (1972), 88–99.

[3] On the Krull-dimension of left noetherian left Matlis-rings, *Math. Z.*, **118** (1970), 207–214.

[4] Krull dimension and Gabriel dimension of idealizers of semimaximal left ideals, *J. London Math. Soc.*, **12** (1976), 137–140.

Krause G., T.H. Lenagan, [1] Growth of algebras and Gelfand-Kirillov dimension, *Research Notes in Mathematics*, Pitman Adv. Publ. Program, vol **116** (1985).

Krause G., T.H. Lenagan, S.T. Stafford, [1] Ideal Invariance and artinian quotient rings, *J. Alg*, **55** (1978), 145–154.

Lambek J., [1] Lectures on rings and modules, Blaisdell Waltham (1966).

Lambek J., G. Michler, [1] The torsion theory at a prime ideal of a right noetherian ring, *J. Algebra*.

Lanski C., [1] Gabriel dimension and rings with involution, *Huston J. Math.*, **4** (1978), 397–415.

Lemonnier B., [1] Dimension de Krull et codeviation. Application au theorem d'Eakin, *Comm. in Algebra*, **6(16)** (1978).

[2] Deviation des ensembles et groupes abeliens totalement ordonné, *Bull Sc. Math.*, **96,4** (1972), 288–303.

[3] Sur une classe d'anneaux definie a partir de la deviation, *C.R. Acad. Sci. Paris*, **274** (1972), 297–299.

[4] Dimension de Krull et codéviation. Quelques applications en théorie des modules, Thèse.

Lenagan T.H., [1] Modules with Krull Dimension, *Bull. London Math. Soc.*, **12, part 1, 34** (9176), 39–40.

[2] Krull Dimension and Invertible Ideals in Noetherian Rings, *Proc Edinburgh Math. Soc.*, **20** (1976), 81–86.

[3] The Nil-radical of a Ring with Krull Dimension, *Bull. London Math. Soc.*, **5** (1973), 307–311.

[4] Reduced Rank on Rings with Krull Dimension, Ring Theory Proc. Antwerp Conference 1978, *Lecture Notes in Pure and Appl. Math.*, Marcel Dekker.

[5] Noetherian Rings with Krull Dimension One, *J. London Math. Soc.*, **15** (1977), 41–47.

[6] Artinian Quotient Rings of Macaulay Rings, Noncomm. Ring Theory, Kent State 1975, *Lecture Notes in Math.*, **545**, Springer-Verlag, New-York.

[7] Artinian ideals in noetherian rings, *Proc. Amer. Soc*, **51(2)** (1975).

[8] Gelfand-Kirillov dimension and affine P.I.-rings, *Comm. in Alg.*, **10** (1981), 87–92.

[9] Gelfand-Kirillov dimension in enveloping algebras, *Quarterly J. Math. Oxford*, **32** (1981), 69–80.

[10] Gelfand-Kirillov dimension is exact for Noetherian P.I. algebras, *Canadian Math. Bull.*, **27** (1984), 247–250.

Lesieur L., R. Croisot, [1] Algèbre noetherienne non-commutatif, *Mém. Sci. Math.*, **154** (1963).

Lorenz M., [1] On the Gelfand-Kirillov dimension of skew polynomial rings, *J. Algebra*, **77** (1982), 186–188.

Lorez M., L. Small, On the Gelfand-Kirillov dimension of noetherian P.I. algebras. *Contemp. Math.*, **13**, 199–205, Am. Math. Soc (1982).

Lorenz M., D.S. Passman, [1] Observations on Crossed Products and Fixed Rings, *Comm. in Algebra*, **8**, nr **8** (1980), 743–780.

[2] Two Applications of Maschke's Theorem, *Comm. in Algebra*, **8(19)** (1980), 1853–1866.

Makar-Limanov L., [1] The skew field of fractions of the Weyl algebra contains a free subalgebra, *Comm. in Alg.*, **11** (1983), 2003–2006.

Malliavin M.P., [1] Dimension de Gelfand-Kirillov des algebres à identité polynomiales, *C.R. Acad. Sci Paris*, **282** (1976), 679–681.

Miller R.W., M. Teply [1] The descending chain condition relative to a torsion theory, *Pacific J. Math.*, **83** (1979), 207–220.

Michler G., [1] Radikale und Sockel, *Math. Ann*, **167** (1966), 1–48.

[2] Primringe mit Krull-dimension eins, *J. Reine Angew. Math.*, **239–240** (1970), 366–384.

Montgomery S., [1] Fixed Rings of Finite Automorphism Groups of Associative Rings, *Lect. Notes in Math.*, vol **818**, Springer-Verlag, New-York.

Năstăsescu C., [1] Quelques observations sur la dimension de Krull, *Bull. Math. de la Soc. Math. Roum.*, **20(68) 3-4** (1976), 291–293.

[2] Strongly graded ring of finite groups, *Comm. in Algebra*, **11(10)** (1980), 1033–1071.

[3] La filtration de Gabriel I, *Annali della Scuola Normale Sup. Pisa*, **27** (1973), 457–470.

[4] La filtrazione di Gabriel II, *Rend. Sem. Mat. Univ. Padova*, **50** (1973), 189–195.

[5] Group Rings of Graded Rings. Applications, *J. Pure and Appl. Algebra*, **33** (1984), 313–335.

[6] Anneaux et Modules gradés, *Revue Roum. Math. Pure et Appl.*, **7** (1977), 911–931.

[7] Conditions de finitude pour les modules I, *Rev. Roum. Math. Pure et Appl.*, **24** (1979), 745-758 and II, *Rev. Roum. Math. Pure et Appl.*, **25** (1980), 615–630.

[8] Théorème de Hopkins pour les catégories de Grothendieck, Ring Theory Antwerpen 1980, *Lect. Notes in Math.*, **825**, Springer-Verlag.

[9] Quelques remarques sur la dimension homologique des anneaux éléments reguliers, *J. of Algebra*, **19** (1971), 470–485.

[10] Ternia dimensium in algebra necomutativa, Edit. Acad. R.S.R. (1983).

Năstăsescu C., N. Popescu [1] Anneaux semi-artiniens, *Bull.Soc. Math. France*, **96** (1968), 357–368.

Năstăsescu C., F. Van Oystaeyen [1] Graded and Filtered Rings and Modules, *Lecture Notes in Math.*, **758** (1979), Springer-Verlag.

[2] On strongly Graded Rings and Crossed Product, *Comm. in Algebra*, **10** (1982), 2095–2106.

[3] Graded Ring Theory, *Math. Library*, **28** (1982), North-Holland.

Năstăsescu C., S. Raianu [1] Gabriel dimension of Graded Rings, *Rend. Sem. Math. Padova*, **71** (1984), 195–208.

[2] Gabriel Dimension of Graded Rings II (to appear)

[3] Stability conditions for commutative rings with Krull dimension, *Methods in Ring Theory*, **129** (1983), Reidel Publ. Comp.

[4] Finiteness conditions for graded modules (gr-$\Sigma(\Delta)$-injective modules), *J. Algebra* (1986).

Okninski J., [1] Commutative monoid rings with Krull dimension (preprint).

Park J.K., Skew group rings with Krull dimension, *Math. J. Okayama Univ.*, **25** (1983), 75–80.

Passman D.S., [1] The Algebraic Structure of group rings, A. Wiley Int. Publ. (1977).

[2] It's essentially Maschke's theorem, *Rocky Mountain J. Math.*, **13** (1983), 37–54.

Popescu N., [1] Abelian categories with application to rings and modules, London-New-York Academic Press (1973).

[2] Théorie de la decomposition primaire dans les anneaux semi-noetheriens, *J. Algebra*, **23** (1972), 482-492.

Procesi C., [1] Rings with polynomial identities, Monograph (1973), Marcel-Dekker,

New-York.

Quinn D., [1] Group-graded rings and duality, *Trans Am. Math. Soc.*, vol. **292** (1985), 155-168.

[2] Thesis, University of Wisconsin (1985).

Rentschler R., P. Gabriel, [1] Sur la dimension des anneaux et ensembles ordonnées, *C.R. Acad. Sci. Paris*, **265** (1967), 712-715.

Resco R., L.W. Small, J.T. Stafford, [1] Krull and global dimension of semiprime noetherian P.I. rings, *Trans Am. Math. Soc.*, **274** (1985), 285-295.

Ramras M., [1] Center of an order with finite global dimension, *Trans Am. Math. Soc.*, **210** (1975), 249-257.

[2] Orders with finite global dimension, *Pacific J. Math.*, **50** (1974), 583-587.

Rowen L., [1] Polynomial identities in ring theory, Academic Press, New-York (1980).

Sarah B., K. Varadarajan [1] Dual Goldie dimension II, *Comm. in Algebra*, **7**, nr **17** (1979), 1885-1900.

Segal D., [1] On the Residual Simplicity of certain Modules, *Proc. London Math. Soc.*, **34** (1977), 327-353.

Shamsuddin A., [1] Ph.D. Thesis, University of Leeds (1976).

Small L.W., [1] Orders in Artinian Rings, *J. Algebra*, **4** (1966) 13-4.

[2] Change of Rings Theorem, *Proc. Amer. Math. Soc.*, **19** (1968), 662-666.

[3] An example in Noetherian rings, *Proc. Nat. Sci. U.S.A.*, **54** (1965), 1035-1036.

Small L.W., J.T. Stafford, R.B. Warfield [1] Affine Algebras of Gelfand-Kirillov dimension one are P.I, preprint, Leeds (1984).

Smith P.F. [1] Localization and the A.R. property, *Proc. London Math. Soc*, **22 22** (1971), 39-68.

[2] On the dimension of group rings, *Proc. London Math. Soc*, **25** (1972), 288-302.

Stafford J.T. [1] Stable Structure of noncommutative Noetherian Rings, *J. Alg.*, **47** (1977), 244-265 and *J. Alg.*, **52** (1978), 218-235.

[2] Module Structure of Weyl algebras, *J. London Math. Soc.*, **44** (1982), 385-404.

Stenström Bo., [1] Radicals and socles of lattices, *Archiv der Math.*, **XX 3** (1969).

[2] Rings of quotients, *Grundlehren math. Wissenschaft*, **217** (1975), Springer-Verlag, Berlin-New-York

Tauvel P., [1] Sur les quotients premiers de l'algèbre envelopante d'une algèbre de Lie résoluble, *Bull. Soc. Math. France*, **106** (1978), 177-205.

[2] Sur la dimension de Gelfand-Kirillov, *Comm. in Alg.*, **10**, 939-963.

Ulbrich K.H., [1] Vollgraduierte Algebren (1978), Dissertation.

Van den Bergh M., [1] On a Theorem of S. Montgomery and M. Cohen, *Proc. Amer. Math. Soc.*, **94** (1985), 562-564.

Van Oystaeyen F., [1] Prime Spectra in non-commutative Algebra, *Lecture Notes in Math.*, **444**, Springer-Verlag, 1975

[2] Localization of Fully Left Bounded Rings. *Comm. in Alg.*, **4** (1976), 271-284.

[3] On Graded Rings and Modules of Quotients, *Comm. in Alg.*, **6** (1978), 1923-1959.

[4] Graded Prime Ideals and Left Ore Condition, *Comm in Alg.*, **8** (1980), 861-868.

[5] On Clifford systems and generalized crossed products, *J. Alg.*, **87** (1984), 396-415.

[6] Crossed Products over Arithmetically Graded Rings, *J. Alg.*, **80**, nr. **2** (1083), 537-551.

Van Oystaeyen F., A. Verschoren [1] Relative invariants of rings, the commutative theory, *Monographs in Pure and Appl. Math.*, Marcel Dekker (1983).

[2] Non-commutative Algebraic Geometry, *Lecture Notes in Math.*, vol **887**, Springer-Verlag (1981).

Varadarajan K., [1] Dual Goldie Dimension, *Comm. in Alg.*, **7** nr. **6** (1979), 565-610.

Walker R., [1] Local Rings and Normalzing Sets of Elements, *Proc. London Math. Soc.*, **(3)24** (1972), 27-45.

Warfield R.B., [1] The Gelfand-Kirillov dimension of a tensor product, *Math. Z.*, **185** (1984), 441-447.

Woods S.M., [1] Existence of Krull dimension in group rings, *J. London Math.Soc*, **9** (1975), 406-410.

INDEX

H

I

J

T

U

V

W

Z